T0155858

Communications
in Computer and Information Science   2036

Rationale

The CCIS series is devoted to the publication of proceedings of computer science conferences. Its aim is to efficiently disseminate original research results in informatics in printed and electronic form. While the focus is on publication of peer-reviewed full papers presenting mature work, inclusion of reviewed short papers reporting on work in progress is welcome, too. Besides globally relevant meetings with internationally representative program committees guaranteeing a strict peer-reviewing and paper selection process, conferences run by societies or of high regional or national relevance are also considered for publication.

Topics

The topical scope of CCIS spans the entire spectrum of informatics ranging from foundational topics in the theory of computing to information and communications science and technology and a broad variety of interdisciplinary application fields.

Information for Volume Editors and Authors

Publication in CCIS is free of charge. No royalties are paid, however, we offer registered conference participants temporary free access to the online version of the conference proceedings on SpringerLink (http://link.springer.com) by means of an http referrer from the conference website and/or a number of complimentary printed copies, as specified in the official acceptance email of the event.

CCIS proceedings can be published in time for distribution at conferences or as post-proceedings, and delivered in the form of printed books and/or electronically as USBs and/or e-content licenses for accessing proceedings at SpringerLink. Furthermore, CCIS proceedings are included in the CCIS electronic book series hosted in the SpringerLink digital library at http://link.springer.com/bookseries/7899. Conferences publishing in CCIS are allowed to use Online Conference Service (OCS) for managing the whole proceedings lifecycle (from submission and reviewing to preparing for publication) free of charge.

Publication process

The language of publication is exclusively English. Authors publishing in CCIS have to sign the Springer CCIS copyright transfer form, however, they are free to use their material published in CCIS for substantially changed, more elaborate subsequent publications elsewhere. For the preparation of the camera-ready papers/files, authors have to strictly adhere to the Springer CCIS Authors' Instructions and are strongly encouraged to use the CCIS LaTeX style files or templates.

Abstracting/Indexing

CCIS is abstracted/indexed in DBLP, Google Scholar, EI-Compendex, Mathematical Reviews, SCImago, Scopus. CCIS volumes are also submitted for the inclusion in ISI Proceedings.

How to start

To start the evaluation of your proposal for inclusion in the CCIS series, please send an e-mail to ccis@springer.com.

Christophe Cruz · Yanchun Zhang · Wanling Gao
Editors

Intelligent Computers, Algorithms, and Applications

Third BenchCouncil International Symposium, IC 2023
Sanya, China, December 3–6, 2023
Revised Selected Papers

 Springer

Editors
Christophe Cruz
Université de Bourgogne
Dijon, France

Yanchun Zhang ⓘ
Victoria University
Melbourne, VIC, Australia

Wanling Gao ⓘ
Chinese Academy of Sciences
Beijing, China

ISSN 1865-0929   ISSN 1865-0937 (electronic)
Communications in Computer and Information Science
ISBN 978-981-97-0064-6   ISBN 978-981-97-0065-3 (eBook)
https://doi.org/10.1007/978-981-97-0065-3

This Springer imprint is published by the registered company Springer Nature Singapore Pte Ltd.
The registered company address is: 152 Beach Road, #21-01/04 Gateway East, Singapore 189721, Singapore

Paper in this product is recyclable.

IC 2023: Provide a Pioneering AI Technology Roadmap

This volume contains the papers presented at IC 2023: the 2023 BenchCouncil International Symposium on Intelligent Computers, Algorithms, and Applications, held in December 2023 in conjunction with the 2023 BenchCouncil International Federated Intelligent Computing and Chip Conference (FICC 2023). The mission of IC 2023 was to provide a pioneering technology roadmap by exploring and advancing the state of the art and state of the practice in processors, systems, algorithms, and applications for machine learning, deep learning, spiking neural networks, and other AI techniques across multidisciplinary and interdisciplinary areas.

The IC conference covers a wide range of topics in intelligent computers, algorithms, and applications in various fields such as computer science, civil aviation, medicine, finance, education, and management. Its multidisciplinary and interdisciplinary emphasis provides an ideal environment for developers and researchers from different areas and communities to discuss practical and theoretical work. The IC 2023 call for papers received 50 papers of high-quality submissions. Through a thorough review process, where each paper underwent double-blind review by a minimum of three experts, the program committee selected 26 papers for the IC 2023 conference. The papers featured in this compilation have been revised based on the suggestions of the program committee members. Additionally, this volume includes five excellent abstracts.

During the conference, the International Open Benchmark Council (BenchCouncil) unveiled two editions of the top 100 AI achievements: AI100 (1943–2021, RFC) and AI100 (2022–2023, RFC). The AI100 (1943–2021, RFC) aims to recognize a centennial edition of the top 100 AI achievements that have made significant impacts and played a crucial role in the advancement of AI and related fields over the past century. On the other hand, the AI100 (2022-2023, RFC) edition focuses on highlighting the latest AI achievements from 2022 to 2023, which have already started to make an impact or are expected to do so in the near future. For more information, you can visit the following link: https://www.benchcouncil.org/evaluation/ai/.

The IC 2023 program featured a plenary session that was shared with Chips 2023, OpenCS 2023, and Bench 2023. This session included twelve keynote lectures delivered by esteemed speakers such as Lieven Eeckhout from Ghent University, Bruce Perens from the Open Source Initiative, D. K. Panda from Ohio State University, Yungang Bao from the Institute of Computing Technology, Chinese Academy of Sciences, Steve Furber from the University of Manchester, Jürgen Schmidhuber from the King Abdullah University of Science and Technology (KAUST), Xianyi Zhang from PerfXLab, Fangcheng Fu from Peking University, Yaodong Cheng from the Institute of High Energy Physics, Chinese Academy of Sciences, Jianhui Tao from TAOS Data, and Hajdi Cenan and Davor Runje from Airt.ai. Additionally, the program included presentations on the Call for 2022–2023 achievements (CFA).

We express our deep gratitude to all the authors who dedicated their efforts to writing, revising, and presenting their papers at the IC 2023 conference. We highly value the invaluable support provided by the IC 2023 Organization Committee and extend our heartfelt thanks for their tireless efforts and significant contributions in upholding the exceptional standards of the IC 2023 Symposium.

December 2023

Christophe Cruz
Yanchun Zhang
Wanling Gao

Organization

General Chairs

Weiping Li   Civil Aviation Flight University of China, China
Tao Tang   BNU-HKBU United International College, China
Frank Werner   Otto-von-Guericke-University, Germany

Program Chairs

Christophe Cruz   Université de Bourgogne, France
Yanchun Zhang   Victoria University, Australia
Wanling Gao   ICT, Chinese Academy of Sciences, China

Program Vice-chairs

Jungang Xu   University of Chinese Academy of Sciences, China
Yucong Duan   Hainan University, China

Area Chairs

AI Algorithms

Hideyuki Takahashi   Tohoku Gakuin University, Japan
Faraz Hussain   Clarkson University, USA
Chunjie Luo   University of Chinese Academy of Sciences, China

AI Systems

Pengfei Chen   Sun Yat-sen University, China
Jason Jia   Amazon, USA
Xiaoguang Wang   University of Illinois Chicago, USA

AI for Ocean Science and Engineering

Guoqiang Zhong   Ocean University of China, China
Hui Yu   University of Portsmouth, UK

AI in Finance

Co-chairs

Changyun Wang   Renmin University of China, China
Michael Guo   Durham University, UK

Program Co-chairs

Zhigang Qiu   Renmin University of China, China
Shinan Cao   University of International Business and
                                     Economics, China

AI for Education

John Impagliazzo   Hofstra University, USA
Xuesong Lu   East China Normal University, China
Stéphane Bressan   National University of Singapore, Singapore

AI for Law

Minghui Xiong   Zhejiang University, China
Bart Verheij   University of Groningen, The Netherlands

AI for Materials Science and Engineering

Siqi Shi   Shanghai University, China
Turab Lookman   AiMaterials Research LLC, USA
Yue Liu   Shanghai University, China

AI for Science

Tao Zhou	Institute of Computational Mathematics and Scientific/Engineering Computing, Chinese Academy of Sciences, China
Weile Jia	Institute of Computing Technology, Chinese Academy of Sciences, China

AI for Civil Aviation

Weiping Li	Civil Aviation Flight University of China, China
Lin Zou	Civil Aviation Flight University of China, China

AI for Medicine

Co-chairs

Zhenchang Wang	Beijing Friendship Hospital, Capital Medical University, China
Jie Lu	Xuanwu Hospital, Capital Medical University, China
Jinlyu Sun	Peking Union Medical College Hospital, China

Vice-chair

Zhifei Zhang	Capital Medical University, China

AI for Space Science and Engineering

Ziming Zou	National Space Science Center, Chinese Academy of Sciences, China
Liming Song	Institute of High Energy Physics, Chinese Academy of Sciences, China

AI for High Energy Physics

Co-chairs

Yaodong Cheng   Institute of High Energy Physics, Chinese Academy of Sciences, China

Yaquan Fang   Institute of High Energy Physics, Chinese Academy of Sciences, China

Program Chair

Xinchou Lou   University of Texas at Dallas & Institute of High Energy Physics, Chinese Academy of Sciences, China

AI and Security

Bo Luo   University of Kansas, USA

Yu Wen   Institute of Information Engineering, Chinese Academy of Sciences, China

Publicity Chairs

Fei Teng   Southwest Jiaotong University, China

Zheng Yuan   King's College London, UK

Roy Lee   Singapore University of Technology and Design, Singapore

Ming Gao   East China Normal University, China

Yuan Cheng   Fudan University, China

Juan Li   Central South University, China

Tianwen Xu   Zhejiang University, China

Yicheng Liao   Zhejiang University, China

Xiao Chi   Zhejiang University, China

Zhengqun Yang   Shanghai University, China

Han Lv   Beijing Friendship Hospital, Capital Medical University, China

Xiaoyan Hu   National Space Science Center, Chinese Academy of Sciences, China

Yanjie Fu   University of Central Florida, USA

Weiwei Tang	National Space Science Center, Chinese Academy of Sciences, China
Pengyang Wang	University of Macau, China
Haijun Yang	Shanghai Jiao Tong University, China
Xingtao Huang	Shandong University, China
Huilin Qu	European Organization for Nuclear Research (CERN), Switzerland

Program Committee

Diego Oliva	University of Guadalajara, Mexico
Yogendra Arya	J.C. Bose University of Science and Technology, India
Nazar Khan	Punjab University, Pakistan
Yingjie Shi	Beijing Institute of Fashion Technology, China
Sansanee Auephanwiriyakul	Chiang Mai University, Thailand
Xiexuan Zhou	Max Planck Institute of Biochemistry, Germany
Zihan Jiang	Huawei, China
Xiaoguang Wang	University of Illinois Chicago, USA
Pengfei Zheng	Huawei Ltd., China
Yushan Su	Princeton University, USA
Runan Wang	Imperial College London, UK
Jindal Anshul	Technical University of Munich, Germany
Hui Dou	Anhui University, China
Saiyu Qi	Xi'an Jiaotong University, China
Wuxia Jin	Xi'an Jiaotong University, China
Chuan Chen	Sun Yat-sen University, China
Shajulin Benedict	Indian Institute of Information Technology Kottayam, India
Vishvak Murahari	Princeton University, USA
Partha Pratim Roy	Indian Institute of Technology Roorkee, India
Rachid Hedjam	Bishop's University, Canada
Xin Li	China University of Petroleum (East China), China
Zhimin Wang	Ocean University of China, China
Chi Zhang	Ocean University of China, China
George Alexandridis	Reading University, UK
Haoyu Gao	Renmin University of China, China
Yi Huang	Fudan University, China
Fuwei Jiang	Central University of Finance and Economics, China

Dimitris Petmezas	Durham University, UK
Georgios Sermpinis	University of Glasgow, UK
Yanmei Sun	University of International Business and Economics, China
Evangelos Vagenas-Nanos	University of Glasgow, UK
Quan Wen	Georgetown University, USA
Ke Wu	Renmin University of China, China
Teng Zhong	University of International Business and Economics, China
Dexin Zhou	CUNY Baruch College, USA
Xiaoneng Zhu	Shanghai University of Finance and Economics, China
Yifeng Zhu	Central University of Finance and Economics, China
Yunshi Lan	East China Normal University, China
Shenggen Ju	Sichuan University, China
Zhenya Huang	University of Science and Technology of China, China
Tiancheng Zhang	Northeastern University, China
Zheng Yuan	King's College London, UK
Thomas Heinis	Imperial College London, UK
Roy Lee	Singapore University of Technology and Design, Singapore
Sadegh Nobari	Startbahn, Japan
Alison Clear	Eastern Institute of Technology, New Zealand
Tony Clear	Auckland University of Technology, New Zealand
Judith Gal-Ezer	Open University of Israel, Israel
Natalie Kiesler	DIPF—Leibniz-Institute, Germany
Dezhen Xue	Xi'an Jiaotong University, China
Jinjin Li	Shanghai Jiao Tong University, China
Lei Li	Southern University of Science and Technology, China
Maxim Avdeev	Australian Nuclear Science and Technology Organization & University of Sydney, Australia
Yanjing Su	University of Science and Technology Beijing, China
Zhi Wei Seh	Institute of Materials Research and Engineering, A*STAR, Singapore
Zhijun Fang	Donghua University, China
Zijian Hong	Zhejiang University, China
Guihua Shan	Computer Network Information Center, Chinese Academy of Sciences, China

Zhiqin Xu	Shanghai Jiao Tong University, China
Chi Zhou	Shenzhen University, China
Lijun Liu	Osaka University, Japan
Di Fang	University of California, Berkeley, USA
Xiaojie Wu	Bytedance Inc., USA
Tong Zhao	Institute of Computing Technology, Chinese Academy of Sciences, China
Michael Schultz	Bundeswehr University Munich, Germany
Paolo Tortora	Università di Bologna, Italy
Carlos E. S. Cesnik	Duke University, USA
Michael I. Friswell	Swansea University, UK
Song Fu	Tsinghua University, China
Jae-Hung Han	KAIST, South Korea
Jacques Periaux	Universitat Politècnica de Catalunya, Spain
Domenico Accardo	University of Naples Federico II, Italy
Rafic M. Ajaj	Khalifa University, United Arab Emirates
Gang Chen	Xi'an Jiaotong University, China
Mou Chen	Nanjing University of Aeronautics and Astronautics, China
Wing Chiu	Monash University, Australia
Han Lv	Beijing Friendship Hospital, Capital Medical University, China
Peng Wang	Beijing Ditan Hospital, Capital Medical University, China
Chaodong Wang	Xuanwu Hospital, Capital Medical University, China
Longxin Xiong	Nanchang Ninth Hospital, China
Mingzhu Zhang	Beijing Tongren Hospital, Capital Medical University, China
Yi Li	Peking Union Medical College Hospital, China
Shenhai Wei	First Hospital of Tsinghua University, China
Hongxu Yang	GE Healthcare, The Netherlands
Xiaohong Liu	Shanghai Jiao Tong University, China
Bingbin Yu	German Research Center for Artificial Intelligence-Robotic Innovation Center, Germany
Menghan Hu	East China Normal University, China
Shuo Li	Case Western Reserve University, USA
Tao Tan	Macao Polytechnic University, China
Yue Wu	Shanghai Ninth People's Hospital, Shanghai Jiao Tong University, China
Siuly Siuly	Victoria University, Australia
Enamul Kabir	University of Southern Queensland, Australia

Muhammad Tariq Sadiq	University of Brighton, UK
Smith K. Khare	Aarhus University, Denmark
Mohammed Diykh	University of Thi-Qar, Iraq
Supriya Angra	Torrens University, Australia
Abdulkadir Şengür	Firat University, Turkey
Varun Bajaj	PDPM-Indian Institute of Technology, Design and Manufacturing, India
Ömer Faruk Alçin	Malatya Turgut Ozal University, Turkey
K. Venkatachalam	University of Hradec Králové, Czech Republic
Ivan Lee	University of South Australia, Australia
Feng Xia	RMIT University, Australia
Zhiguo Gong	University of Macau, China
Hong Yang	Guangzhou University, China
Qian Zhou	Nanjing University of Posts and Telecommunications, China
Wenjun Tan	Northeastern University, China
Zongcheng Ling	Shandong University, China
Yanjie Fu	University of Central Florida, USA
Jiajia Liu	University of Science and Technology of China, China
Xiaoxi He	University of Macau, China
Xinchou Lou	University of Texas at Dallas, USA & Institute of High Energy Physics, China
Haijun Yang	Shanghai Jiao Tong University, China
Xingtao Huang	Shandong University, China
Huilin Qu	European Organization for Nuclear Research (CERN), Geneva
Bruce Mellado	University of the Witwatersrand, South Africa
Fabio Hernandez	Institute of Nuclear Physics of Lyon, France
Yanwei Liu	Institute of Information Engineering, Chinese Academy of Sciences
Hongjia Li	Institute of Information Engineering, Chinese Academy of Sciences
Zhiqiang Xu	Jiangxi University of Science and Technology, China
Liwei Chen	Institute of Information Engineering, Chinese Academy of Sciences, China
Yanni Han	Institute of Information Engineering, Chinese Academy of Sciences, China
Duohe Ma	Institute of Information Engineering, Chinese Academy of Sciences, China
Bin Wang	Nankai University, China
Bin Wei	Zhejiang University, China

Haixiao Chi	Zhejiang University, China
Heng Zheng	University of Illinois Urbana-Champaign, USA
Huimin Dong	University of Luxembourg, Luxembourg
Juan Li	Central South University, China
Lili Lu	Gansu University of Political Science and Law, China
Liuwen Yu	University of Luxembourg, Luxembourg
Ni Zhang	Sichuan University, China
Peicheng Wu	Zhejiang University, China
Shiyang Yu	Nankai University, China
Tianwen Xu	Zhejiang University, China
Xiang Zhou	Zhejiang University, China
Xiao Chi	Zhejiang University, China
Xiaobiao Xiong	Sun Yat-sen University, China
Xin Sun	Zhejiang Laboratory, China
Yang Feng	Zhejiang University, China
Yun Liu	Tsinghua University, China

Contents

AI for High Energy Physics

AI for Law

Abstracts

AI Algorithms and Systems

Efficient and Scalable Kernel Matrix Approximations Using Hierarchical Decomposition

Keerthi Gaddameedi⬝[iD], Severin Reiz[(✉)][iD], Tobias Neckel,
and Hans-Joachim Bungartz[iD]

Technical University of Munich, School of Computation, Information and Technology,
Munich, Germany
{keerthi.gaddameedi,s.reiz}@tum.de
https://www.cs.cit.tum.de/sccs/

Abstract. With the emergence of Artificial Intelligence, numerical algorithms are moving towards more approximate approaches. For methods such as PCA or diffusion maps, it is necessary to compute *eigenvalues* of a large matrix, which may also be *dense* depending on the kernel. A global method, i.e. a method that requires all data points simultaneously, scales with the data dimension N and not with the intrinsic dimension d; the complexity for an exact dense eigendecomposition leads to $\mathcal{O}(N^3)$. We have combined the two frameworks, datafold and GOFMM. The first framework computes diffusion maps, where the *computational bottleneck* is the eigendecomposition while with the second framework we compute the eigendecomposition *approximately* within the iterative Lanczos method. A hierarchical approximation approach scales roughly with a runtime complexity of $\mathcal{O}(Nlog(N))$ vs. $\mathcal{O}(N^3)$ for a classic approach. We evaluate the approach on two benchmark datasets – scurve and MNIST – with strong and weak scaling using OpenMP and MPI on *dense* matrices with maximum size of $100k \times 100k$.

Keywords: Numerical algorithms · Manifold learning · Diffusion maps · Hierarchical matrix · Strong Scaling

1 Introduction

1.1 Motivation

Data-driven approaches to solve real-world problems have led to a rapid increase in data sizes. The potential of such approaches is limited by the current state of computational power. The memory requirements of dense matrices (i.e. matrices with mostly non-zero entries) is $\mathcal{O}(N^2)$. Similarly, the time complexity for operations such as *mat-vec* is $\mathcal{O}(N^2)$. Therefore, these operations become computationally infeasible when the size of the matrices is large. As a solution to this, we aim to find low-rank approximations of these matrices using

---

K. Gaddameedi and S. Reiz—Equal contribution, joint first author.

© The Author(s), under exclusive license to Springer Nature Singapore Pte Ltd. 2024
C. Cruz et al. (Eds.): IC 2023, CCIS 2036, pp. 3–16, 2024.
https://doi.org/10.1007/978-981-97-0065-3_1

hierarchical algorithms. The fast multipole method (GOFMM) is a novel algorithm for approximating dense symmetric positive definite (SPD) matrices so that the quadratic space and time complexity reduces to $\mathcal{O}(Nlog(N))$ with a small relative error. Dense SPD matrices appear in areas such as scientific computing, data analytics and statistical inference. GOFMM is *geometry-oblivious*, meaning that it does not require the geometry information or the knowledge of how the data has been generated [23]. It just requires the distribution of data as input.

Real-world problems require solving high dimensional data. Data-driven models assume an intrinsic geometry in the data, referred to as a manifold which can be used to extract essential information of lower dimension. datafold is a Python package that provides these data-driven models to find an explicit manifold parametrization for point cloud data [13] by using kernel matrices. Kernels correspond to dot products in a high dimensional feature space, and one uses them to efficiently solve non-linear cases in machine learning [12].

In this paper we enable datafold functionalities to be used in conjunction with GOFMM to scale execution.

1.2 Proposed Approach

Manifold learning approaches learn the intrinsic geometry of high-dimensional data without the use of predetermined classifications (unsupervised learning). There are several manifold learning algorithms such as isomap [21], locally linear embedding [19], Hessian embedding [7] etc., but we focus on *diffusion maps*. Like PCA, diffusion maps also consists of a kernel matrix computation that describes the relation of data points in the space. A Markov chain is defined using the kernel matrix which is then decomposed to compute the eigenvalues and eigenvectors. These eigenvalues and eigenvectors are used to find a lower dimension than the dimension of the ambient space.

datafold provides data-driven models based also on diffusion maps for finding a parametrization of manifolds in point cloud data and to identify non-linear dynamical systems from time series data [13]. Since the eigendecomposition of the kernel matrix is very expensive, especially for huge matrices, hierarchical approaches are applied to be able to reduce the quadratic complexity to $\mathcal{O}(Nlog(N))$.

The framework GOFMM provides hierarchical algorithms for large, dense, symmetric and positive-definite matrices. Let $K \in \mathbb{R}^{N \times N}$ be a dense kernel matrix for manifold data that is to be approximated. Let it also be symmetric and positive-definite. The goal is to find an approximation $\widetilde{K}$ such that the construction and any matrix-vector multiplications take only $\mathcal{O}(Nlog(N))$ work. The approximation must also satisfy the condition that the relative error between the approximated and exact matrix remains small,

$$\frac{||\widetilde{K} - K||}{||K||} \leq \epsilon, \quad 0 < \epsilon < 1, \tag{1}$$

where ϵ is a user-defined tolerance. We use then the *implicitly restarted Arnoldi* iteration to perform an iterative eigendecomposition. The matrix-vector multiplications in every iteration is performed using hierarchical methods from GOFMM, where the dense matrix is compressed once at the beginning and then evaluated in each iteration. The relative error of the resulting eigenvalues with a reference solution is recorded. The scalability of the combined integrated software is tested on multiple cores.

1.3   Related Work and Contributions

There has been a growing interest on randomized computation of matrix decompositions [11,16]. They also occur in theoretical deep learning, for example, with shallow Gaussian processes [8] or for finding weights in deep neural networks by solving a system of linear equations [3] or for Hessian approximations in second-order optimization [18]. Naturally, approximate matrix calculations are suitable for data applications, especially when the *modelling* error (e.g., of neural networks) are bigger than the *numerical* error. However, often matrices like kernels from radial basis functions, may not be global low-rank and only allow for low-rank treatment for off-diagonal matrices with the so-called $\mathcal{H}$-arithmetic [9,10]. Hence, our target here is matrices that have globally significant rank, but allow for approximations on the off-diagonals. To our knowledge, the most prominent framework for hierarchical structured matrices is STRUMPACK [15]; GOFMM [24] shows some superiority for kernel matrices against STRUMPACK, underlining that GOFMM a good candidate for diffusion maps kernels. In addition, eigendecompositions of dense kernel matrices are the computational bottleneck of diffusion maps, limiting the global size. Existing work from our group integrated the GOFMM and the datafold frameworks.

   Contributions of this paper include

1. To our knowledge, first $\mathcal{H}$-arithmetic in iterative eigendecompositions
2. Analysis of dense kernel matrices from diffusion maps *enabling* bigger sizes
3. Versatile approach in software engineering to allow for better reproducibility and portability

Our approach is *using* the framework GOFMM, and *extends* datafold by offering a hierarchical variant for the eigendecomposition.

2   Methods

2.1   Diffusion Maps

Diffusion Maps is a non-linear technique of dimensionality reduction. It tries to obtain information about the manifold encoded in the data without any assumptions on the underlying geometry. As opposed to using the Euclidean distance or the geodesic distance in isomaps, diffusion maps use an affinity or similarity matrix obtained by using a kernel function that produces positive and symmetric values. Given a dataset $X = \{x_1, x_2, x_3, \ldots, x_n\}$ and a Gaussian kernel function, a similarity matrix can be computed as

$$W_{ij} \;=\; w(i,j) \;=\; e^{\frac{-||x_i - x_j||_2^2}{\sigma^2}}, \tag{2}$$

where x_i and x_j are a pair of data points and σ is the radius of the neighborhood around the point x_i. As outlined in Algorithm 1, the similarity matrix is then normalized with the density Q (degree of vertex) and the density parameter α to capture the influence of the data distribution on our approximations. For $\alpha = 0$, the density has maximal influence on how the underlying geometry is captured and vice versa for $\alpha = 1$. Therefore, normalization is done with $\alpha = 1$, and then a Markov chain is defined to obtain the probabilities of transitioning from one point to another. Then, the transition matrix P^t is obtained by performing random walks for t time steps. Afterwards, an eigendecomposition is performed on the transition matrix to compute the eigenpairs which are further used to obtain the underlying lower dimension of the dataset. The computational complexity of diffusion algorithms in standard form is $O(N^3)$, and the eigendecomposition is the most expensive part of the algorithm. Hence, we tackle this by using hierarchical matrix approximations.

Algorithm 1. Diffusion Maps [5]

1: Compute W_{ij} ▷ Similarity matrix

2: Compute normalized weights $W_{ij}^{\alpha} = \frac{W_{ij}}{Q_i^{\alpha} \cdot Q_j^{\alpha}}$ ▷ Q^{α}: Influence of density

3: Define Markov chain $P_{ij} = \frac{W_{ij}^{\alpha}}{Q_i^{\alpha}}$ ▷ P: Transition matrix

4: Perform t random walks to obtain P^t

5: Perform eigendecomposition on P^t ▷ λ_r: eigenvalues, ψ_r: eigenvectors

6: Lower dimension $d(t) = \max\{\, l : \lambda_l^t < \delta \lambda_1^t \,\}$ ▷ δ: Predetermined precision factor

---

Hierarchical Partitioning. If K is a kernel matrix, the hierarchically low-rank approximation $\widetilde{K}$ of K is given as [2,10]

$$\widetilde{K} = D + S + UV, \tag{3}$$

where D is a block-diagonal matrix with every block being an hierarchical matrix (short: $\mathcal{H}$-matrix), S is a sparse matrix and U, V are low rank matrices. The $\mathcal{H}$-matrix $\widetilde{K}$ is to be computed such that the error from Eq. 1 ranges in the order of the user defined tolerance $0 < \epsilon < 1$. The construction of $\widetilde{K}$ and matrix-vector product both take $\mathcal{O}(N \log N)$ operations. We then incorporate these hierarchical approximations into the diffusion maps algorithm to improve the computational costs of the eigendecompositions.

2.2 Implicitly Restarted Arnoldi Method

Implicit restarted Arnoldi method is a variation of Arnoldi process which builds on the *power iteration* method which computes $Ax, Ax^2, Ax^3 ...$ for an arbitrary vector x, until it converges to the eigenvector of the largest eigenvalue of matrix A. To overcome the drawbacks of so many unnecessary computations for a single eigenvalue and its corresponding eigenvector, the Arnoldi method aims to save the successive vectors as they contain considerable information that can be further exploited to find new eigenvectors. The saved vectors form a *Krylov* matrix which is given as [14]

$$\mathcal{K}_n = \text{Span}[x, Ax, A^2x...A^{n-1}x]. \tag{4}$$

Orthonormal vectors $x_1, x_2, x_3...$ that span a *Krylov* subspace are extracted using *Gram-Schmidt* orthogonalization from each column of *Krylov* matrix. The k-step Arnoldi iteration is given in Algorithm 2 [20]. H is the orthogonal projection of A in the *Krylov* subspace. It is observed that eigenvalues of the upper Hessenberg matrix H (the so-called *Ritz* values) converge to the eigenvalues of A. When the current iterate $r_j = 0$, the corresponding Ritz pair becomes the eigenpair of A.

Algorithm 2. k-step ArnoldiFactorization(A,x)

1: $x_1 \leftarrow \frac{x}{\|x\|}$ ▷ Computes first Krylov vector x_1
2: $w \leftarrow Ax_1$ ▷ Computes new candidate vector
3: $\alpha_1 \leftarrow x_1^H w$
4: $r_1 \leftarrow w - \alpha_1 x_1$
5: $X_1 \leftarrow [x_1]$ ▷ Orthonormal basis of Krylov subspace
6: $H_1 \leftarrow [\alpha_1]$ ▷ Upper Hessenberg matrix
7: **for all** $j = 1...k-1$ **do**   ▷ For k steps, compute orthonormal basis X
8:                             ▷ and the projection of matrix A on the new basis
9:     $\beta_j \leftarrow \|r_j\|$; $x_{j+1} \leftarrow \frac{r_j}{\beta_j}$
10:     $X_{j+1} \leftarrow [X_j, x_{j+1}]$; $\hat{H}_j \leftarrow \left[H_j, \ \beta_j e_j^T \right]^T$
11:                        ▷ e_j is the standard basis of coordinate vector space
12:     $z \leftarrow Ax_j$
13:     $h \leftarrow X_{j+1}^H z$; $r_{j+1} \leftarrow z - X_{j+1}h$ ▷ Gram-Schmidt Orthogonalization
14:     $H_{j+1} \leftarrow [\hat{H}_j, h]$
15: **end for**

One of the drawbacks of Arnoldi process is that the number of iterations taken for convergence is not known prior to the computation of well-approximated Ritz values [20]. This causes the computation of the Hessenberg matrix to be of complexity $\mathcal{O}(k^3)$ at the k-th step. A more efficient approach is *implicitly restarted Arnoldi method* which uses an implicitly shifted *QR-iteration*. It avoids storage and numerical instabilities associated with the standard approach by compressing the necessary information from very large *Krylov* subspace into a fixed size k-dimensional subspace.

The Arnoldi factorization of length $m = k + p$ has the form

$$AX_m = X_m H_m + r_m e_m^T . \tag{5}$$

The implicit restarting method aims to compress this to length k by using QR steps to apply p shifts resulting in [20]

$$AX_m^+ = X_m^+ H_m^+ + r_m e_m^T Q , \tag{6}$$

where $V_m^+ = V_m Q$, $H_m^+ = Q^T H_m Q$ and $Q = Q_1 Q_2 ... Q_p$. Q_j is the orthogonal matrix associated with the corresponding shift μ_j. The first $k - 1$ values of $e_m^T Q$ are zero and thus the factorization becomes

$$AX_k^+ = X_k^+ H_k^+ + r_k^+ e_k^T . \tag{7}$$

The residual r_m^+ can be used to apply p steps to obtain back the m-step form. A polynomial of degree p of the form $\prod_1^p (\lambda - \mu_j)$ is obtained from these shifts. The roots of this polynomial are used in the QR process to filter components from the starting vector.

Implicitly Restarted Lanczos Method. Consider the Equation (5) for Arnoldi factorization. X_m are orthonormal columns and H_m is the upper Hessenberg matrix. If A is a Hermitian matrix, it becomes Lanczos factorization. So Arnoldi is basically a generalization to non-hermitian matrices. For Lanczos method, H_m is a real, symmetric and tridiagonal matrix and the X_m are called Lanczos vectors. The algorithms hence remain the same as the ones described for Arnoldi. The method scipy.sparse.linalg.eigs uses Arnoldi iteration since it deals with real and symmetric matrices while scipy.sparse.linalg.eigsh invokes implementation of Lanczos methods.

3   Implementation

Manifold learning data is generated in Python using datafold and then the diffusion maps algorithm is invoked. The eigendecompositions contained in diffusion maps are performed using a subclass of LinearOperator. LinearOperator is instantiated with a matvec implementation from GOFMM. This is done by writing an interface using the Simplified Wrapper Interface Generator (SWIG[1]) to access the GOFMM methods written in C++ from a Python script. In this section, we further delve into the details of how each part has been implemented.

3.1   Integration of datafold and GOFMM

The software architecture of datafold contains integrated models that have been implemented in a modularized fashion and an API that has been templated from scikit − learn library. The architecture as shown in Fig. 1 consists of three layers and describes the hierarchy of the workflow.

datafold.appfold is the highest level in the workflow hierarchy and contains meta-models that provide access to multiple sub-models captured in the class. The second layer datafold.dynfold provides models that deal with point cloud manifold or the dynamics of time series data. Finally, the last layer datafold.pcfold consists of fundamental algorithms such as eigensolvers, distance matrix computations etc. along with objects and data structures associated with them. The

---

[1] swig.org.

| layer 1
datafold.appfold | layer 2
datafold.dynfold | layer 3
datafold.pcfold |

Fig. 1. Workflow hierarchy of datafold

software maintains a high degree of modularity with this workflow and there-fore allows usage of each layer's methods to be used on their own. We have a docker file with commands to install all the run-time dependencies followed by installation of GOFMM and datafold. The docker image containing GOFMM and datafold has been converted to Charliecloud in order to be viable with the linux cluster at LRZ. The docker image is then converted to a charliecloud image using the command `ch-builder2tar <docker-image> /dir/to/save`. Then, the charliecloud image is exported to the linux cluster and unpacked with the command `ch-tar2dir <charliecloud-image> /dir/to/unpack`. Once the compressed image is unpacked, the environment variables are set and GOFMM is compiled. Finally, the SWIG interface file is compiled to generate Python ver-sions of GOFMM's C++ methods.

3.2 LinearOperator

SciPy [22] is an open-source free software with modules for common tasks of scien-tific computing such as linear algebra, solvers, interpolation etc. It contains seven matrix and array classes for different types of representations such as sparse row matrix, column matrix, coordinate format etc. It also accommodates methods to build various kinds of sparse matrices and two submodules csgraph and linalg. The submodule linalg provides an abstract interface named LinearOperator that uses iterative solvers to perform matrix vector products. This interface consists of methods such as matmat(x), matvec(x), transpose(x) for matrix-matrix multi-plication, matrix-vector multiplication and transposition of a matrix. A concrete subclass of LinearOperator can be built by implementing either one of _matvec or _matmat methods and the properties shape and dtype. Depending on the type of matrices at hand, corresponding matvec methods may also be implemented.

scipy.sparse.linalg also provides methods for computing matrix inverses, norms, decompositions and linear system solvers. The functionality we are inter-ested in are the matrix decompositions. In Table 1, we can take a look at various decomposition methods that are present in the module. The method we use to decompose data obtained from datafold is scipy.sparse.linalg.eigsh [22]. This method requires either an ndarray, a sparse matrix or LinearOperator as parame-ters. It optionally takes k, which is the number of desired eigenvalues and eigen-vectors. It solves $Ax[i] = \lambda_i x[i]$ and returns two arrays - λ_i for eigenvalues and k vectors $X[:, i]$, where i is the column index corresponding to the eigenvalue.

Table 1. Matrix Factorizations in scipy.sparse.linalg.

scipy.sparse.linalg.**eigs**	Computes eigenvalues and vectors of square matrix
scipy.sparse.linalg.**eigsh**	Computes eigenvalues and vectors of real symmetric or complex Hermitian matrix
scipy.sparse.linalg.**lobpcg**	Locally Optimal Block Preconditioned Conjugate Gradient Method
scipy.sparse.linalg.**svds**	Partial Singular Value Decompositions
scipy.sparse.linalg.**splu**	LU decomposition of sparse square matrix
scipy.sparse.linalg.**spilu**	Incomplete LU decomposition of sparse square matrix
scipy.sparse.linalg.**SuperLU**	LU decomposition of a sparse matrix

scipy.sparse.linalg.eigsh is a wrapper for the ARPACK functions SSEUPD and DSEUPD which use the implicitly restarted Lanczos method to solve the system for eigenvalues and vectors [1].

4   Results

Several experiments have been performed using datasets such as uniform distribution[2], s-curve[3], swiss-roll[4] and MNIST [6]. Accuracy measurements for the datasets s-curve and MNIST have been presented in Subsect. 4.1. Accuracy has been measured by computing Frobenius norm between eigenvalue computations of scipy solver and GOFMM and additionally, resultant eigenvectors have been plotted to provide a qualitative analysis. Furthermore, experiments were conducted to analyze performance through both weak and strong scaling in Subsect. 4.2. Due to varying computational requirements, weak scaling experiments have been conducted on CoolMUC-2 linux cluster of LRZ[5] and strong scaling on the supercomputer SuperMUC-NG[6]. Efficiency and scalability of our approach were analyzed by examining results obtained with large problem sizes.

4.1   Eigenvalue and Eigenvector Computations

The experiments were performed on the CoolMUC-2 cluster of the Leibniz Supercomputing Centre[5] . It has 812 28-way Intel Xeon E5-2690 v3 ("Haswell") based nodes with 64GB memory per node and FDR14 Infiniband interconnect.

---

[2] https://numpy.org/doc/stable/reference/random/generated/numpy.random.uniform.html.

[3] https://scikit-learn.org/stable/modules/generated/sklearn.datasets.make_s_curve.html.

[4] https://scikit-learn.org/stable/modules/generated/sklearn.datasets.make_swiss_roll.html.

[5] https://doku.lrz.de/coolmuc-2-11484376.html.

[6] https://doku.lrz.de/hardware-of-supermuc-ng-11482553.html.

S-Curve. A 3D S-curve dataset[3] is generated using scikit − learn [17] with 16384 points in the dataset. A 3D S-curve has an underlying intrinsic dimension of 2 and we apply diffusion maps algorithm to compute this. Since our focus lies in the eigendecompositions of the kernel matrix, eigenpairs are computed using two solvers. The first set of values are computed using the scipy solver and these are taken as reference values. The approximations of our GOFMM matvec implementation are computed, and the error values in the Frobenius norm are observed to be in the range of $9e − 4$. We can compare the embeddings obtained from both solvers by fixing the first non-trivial eigenvector and comparing it to the other eigenvectors. Eigenvector comparison for both scipy solver and GOFMM can be observed to be very similar in Fig. 2.

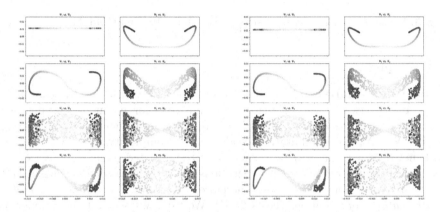

Fig. 2. Eigenvector comparison for scipy solver on the left and GOFMM on the right for **scurve**

MNIST. The MNIST database (Modified National Institute of Standards and Technology database) [6] is a large database of handwritten digits that is commonly used for training various image processing systems. MNIST has a testing sample size of 10,000 and a training size of 60,000 where each sample has 784 dimensions.

Due to a large dataset with 784 dimensions for each sample, MNIST makes a fitting application for hierarchical algorithms. Sample sizes of up to 16384 are loaded from MNIST followed by diffusion maps algorithm applied to the dataset resulting in a kernel matrix of size $16k \times 16k$. As previously mentioned, the goal is to perform efficient eigendecompositions using hierarchical algorithms. Therefore, eigenpairs are computed using scipy and GOFMM and we observe that for a matrix size of 8192, eigenvector comparison for both solvers look qualitatively similar as can be seen in Fig. 3. The Frobenius norm of the difference of the first five eigenvalues is also in the range of $1e − 4$. The parameters required to obtain the results show that the approach is very problem-dependent. As already mentioned in [23], problems with dense matrices are better suited to hierarchical approaches.

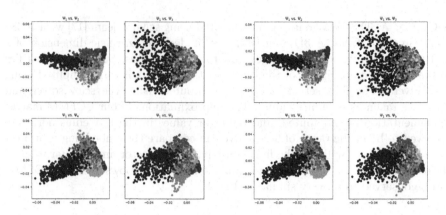

Fig. 3. Eigenvector comparison for scipy solver on the left and GOFMM on the right

4.2 Scaling

Complexity Analysis. As we have established previously in 2.1, computational bottleneck of diffusion maps algorithm (e.g. for manifold learning, see Algorithm 1) is the *eigenvector* (EV) computations. In general, for a matrix of size $N \times N$, EV computations scale with a complexity of $\mathcal{O}(N^3)$. In the past, matrices with large sizes in datafold were restricted to sparse matrices. A sparse matrix only requires $\mathcal{O}(N)$ operations per iteration as one assumes a constant number of non-zero entries. Therefore with N rows, *sparse matrix-vector multiplication* operation only costs $\mathcal{O}(N)$. We usually also limit the number of iterations necessary for the Arnoldi method to a factor of ~100, resulting in an overall computational complexity of $\mathcal{O}(N)$.

However, there exist numerous kernels that do not result in sparse matrices and hence dense matrices are necessary. For a hierarchical approximate *dense matrix-vector multiplication*, we need around $\mathcal{O}(N \log N)$ operations.

Including FLOP counts in [23,25], we have looked at performance measurements for problem sizes up to 200k[7]. Owing to the need for a high number of nodes, scaling experiments were performed on the Intel Xeon Platinum 8174 ("Skylake") partition of SuperMUC-NG[6] which has 6,336 thin nodes and 144 fat nodes with 48 cores per node.

Weak Scaling. In weak scaling, the computational effort per resource (core/node) stays constant. We scale the problem size with respect to nodes and hence, for algorithms with linear complexity, the problem size per node stays constant. But since matrix size scales quadratically, doubling the problem size would require that we scale the number of nodes quadratically in order to maintain a constant computational load per node. This quickly becomes infeasible

---

[7] GOFMM can work with dense matrices of 200k starting with at least 2 nodes, prohibiting the use of a bigger matrix size for strong scaling analysis.

due to limited computational resources. Therefore we scale the nodes linearly instead and provide corresponding ideal runtimes through the dotted lines in Fig. 4.[8]

Weak scaling for GOFMM compression in Fig. 4a results in a runtime complexity between $\mathcal{O}(N)$ (linear) and $\mathcal{O}(N^2)$ (quadratic). Although with this inaccurate complexity estimate we cannot measure the parallel efficiency and communication overhead of GOFMM, it still shows us that it scales really well with increasing problem size and thereby proving that $\mathcal{H}$-matrix approximation is very *beneficial* for large matrices compared to an exact dense multiplication. Figure 4b shows runtimes for matrix-multiplication (also referred to as evaluation) with GOFMM for increasing problem size and nodes. We observe that the runtime for a problem size of $6.25k$ with 1 node is $0.10s$ and for a problem size of $6.25k * 16 \approx 100k$ with 16 nodes is about $0.26s$. Assuming $\mathcal{O}(N \log N)$ computational complexity, ideal scaling would result in a runtime of $0.10s * \log(16) \approx 0.12s$. Instead, the runtime of $0.26s$ we obtained, results in a parallel efficiency of $\frac{0.12}{0.26} \approx 50\%$.

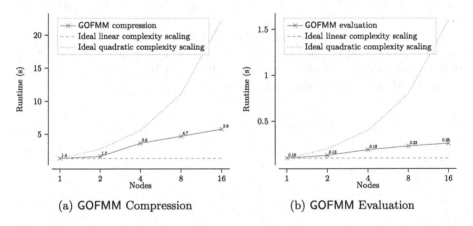

(a) GOFMM Compression   (b) GOFMM Evaluation

Fig. 4. Weak scaling measurements of a Gaussian kernel matrices generated synthetically with 6-D point clouds with roughly 6.25k×6.25k with 1 node, 12.5k×12.5k with 2 nodes, up to ∼100k×100k with 16 nodes. Memory and runtime for exact multiplication of a dense matrix scales quadratically, hence the 2-node problem would correspond to 4-times the memory/computational cost (dotted) in total. Note that the above figures have log scale on the x-axis and linear scale on the y-axis.

To summarize, we see a difference between GOFMM's $\mathcal{O}(N \log N)$ runtime complexity and a quadratic complexity for large matrices with sizes above $25k \times 25k$ (see behavior in Fig. 4).

---

[8] In theory, for a matrix with an off-diagonal rank of r_O, GOFMM has a computational complexity of $\mathcal{O}(N \cdot r_O)$. But with certain adaptive rank selection and a certain accuracy, it potentially increases with problem size and thus for simplicity, we refrain to $\mathcal{O}(N \log N)$.

Strong Scaling. In strong scaling, the problem size stays constant while increasing the computational resources and this can be challenging due to diminishing computational work per node and increasing communication overhead.

Figure 5 shows strong scaling measurements for GOFMM compression and evaluation for a $100k \times 100k$ synthetic kernel matrix. In Fig. 5a on the **left** we see the one-time compression time (For parameter see[9]). Compression algorithm for a 6D random Gaussian kernel matrix of size $100k \times 100k$ takes *13 s* on one node while multiplication with a vector of size $100k \times 512$ has a runtime of *1.35 s*. We can also observe that the parallel efficiency for both algorithms ranges down to 4% and 11% with 128 nodes and that there is no performance gain when nodes higher than 16 are used. As mentioned previously, it is not unusual for efficiency to have tendencies of stagnation or deterioration with strong scaling due to problems such as increasing communication overhead and load imbalance.

Having a limit on maximum acceptable efficiency is not unusual for parallel code either; also to reiterate, growth in runtimes are possible as communication times are increasing. For this reason we highlight similar runtime scaling for matrix evaluation and the one-time matrix compression cost also mentioned in [24].

We see a similar tendency in Fig. 5b starting with 52% efficiency with 16 nodes, implying that the runtime is 8-times slower than on a single node. Note that we also run a problem size of $100k \times 100k$ with 16 nodes for weak scaling in Fig. 4 and get similar results as expected.

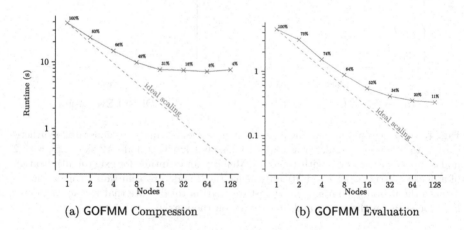

(a) GOFMM Compression   (b) GOFMM Evaluation

Fig. 5. Strong scaling measurements of a Gaussian kernel matrices generated synthetically with 6-D point clouds, all roughly of size 100k-by-100k. Next to the data cross is the parallel efficiency in percent. Results run on Skylake partition of SuperMuc-NG. Each node has 48 cores, 128 nodes hence corresponds to 6144 cores.

---

[9] GOFMM parameters: `max_leaf_node_size` = 768, `max_off_diagonal_ranks` = 768, `user_tolerance` = $1E-3$, `num_neighbors` = 64.

5   Conclusion

With ever-growing applications with non-linear high-dimensional data in Machine learning and AI, it becomes more and more difficult to process this data efficiently. We utilize a manifold learning algorithm (of datafold) to compute the underlying lower dimension of such data and propose an approach to reduce the computational complexity of certain operations contained in such algorithms. We present a proof-of-concept that hierarchical methods can be applied to large matrices in aforementioned algorithms. Since datafold is written in Python and GOFMM is written in C++, the overhead caused by the SWIG interface are unknown. This also causes more limitations on the ability to fully utilize GOFMM's MPI functionality. In ongoing work we integrate other kernels that require dense matrices, and thus are more suitable to the approach and make use of GOFMM to its full potential.

Acknowledgements. This material is based upon work supported by the Technical University of Munich, by the International Graduate School of Science and Engineering (IGSSE) with funds from Deutsche Forschungsgemeinschaft (DFG) through SPPEXA [4] and by the Competence Network for Scientific High Performance Computing in Bavaria (KONWIHR) with funds from Bayerisches Staatsministerium für Wissenschaft und Kunst (STMWK).

References

1. ARPACK Software. http://www.caam.rice.edu/software/ARPACK/
2. Bebendorf, M.: Hierarchical Matrices, 1st edn. Springer Publishing Company, Incorporated (2008). https://doi.org/10.1007/978-3-540-77147-0_3
3. Bolager, E.L., Burak, I., Datar, C., Sun, Q., Dietrich, F.: Sampling weights of deep neural networks. arXiv preprint arXiv:2306.16830 (2023)
4. Bungartz, H.-J., Nagel, W.E., Neumann, P., Reiz, S., Uekermann, B.: Software for Exascale Computing: some remarks on the priority program SPPEXA. In: Bungartz, H.-J., Reiz, S., Uekermann, B., Neumann, P., Nagel, W.E. (eds.) Software for Exascale Computing - SPPEXA 2016-2019. LNCSE, vol. 136, pp. 3–18. Springer, Cham (2020). https://doi.org/10.1007/978-3-030-47956-5_1
5. Coifman, R.R., Lafon, S.: Diffusion maps. Appl. Comput. Harmonic Anal. **21**, 5–30 (2006). https://doi.org/10.1016/j.acha.2006.04.006, https://www.sciencedirect.com/science/article/pii/S1063520306000546, special Issue: Diffusion Maps and Wavelets
6. Deng, L.: The MNIST database of handwritten digit images for machine learning research. IEEE Signal Process. Mag. **29**(6), 141–142 (2012)
7. Donoho, D.L., Grimes, C.: Hessian eigenmaps: Locally linear embedding techniques for high-dimensional data. Proc. Natl. Acad. Sci. **100**(10), 5591–5596 (2003). https://doi.org/10.1073/pnas.1031596100, https://www.pnas.org/doi/abs/10.1073/pnas.1031596100
8. Garriga-Alonso, A., Rasmussen, C.E., Aitchison, L.: Deep convolutional networks as shallow gaussian processes. arXiv preprint arXiv:1808.05587 (2018)
9. Grasedyck, L., Kressner, D., Tobler, C.: A literature survey of low-rank tensor approximation techniques. GAMM-Mitteilungen **36**, 53–78 (2013)

10. Hackbusch, W.: Hierarchical Matrices: Algorithms and Analysis. Springer-Verlag, Berlin Heidelberg (2015). https://doi.org/10.1007/978-3-662-47324-5
11. Halko, N., Martinsson, P.G., Shkolnisky, Y., Tygert, M.: An algorithm for the principal component analysis of large data sets. SIAM J. Sci. Comput. **33**(5), 2580–2594 (2011). https://doi.org/10.1137/100804139, https://doi.org/10.1137/100804139
12. Hofmann, T., Schölkopf, B., Smola, A.J.: Kernel methods in machine learning. Ann. Stat. **2008**, 1171–1220 (2008)
13. Lehmberg, D., Dietrich, F., Köster, G., Bungartz, H.J.: datafold: data-driven models for point clouds and time series on manifolds. J. Open Source Softw. **5**(51), 2283 (2020)
14. Lehoucq, R.B., Sorensen, D.C., Yang, C.: ARPACK USERS GUIDE: solution of large scale eigenvalue problems by implicitly restarted arnoldi methods. SIAM, Philadelphia, PA (1998)
15. Liu, Y., Ghysels, P., Claus, L., Li, X.S.: Sparse approximate multifrontal factorization with butterfly compression for high-frequency wave equations. SIAM J. Sci. Comput. **43**(5), S367–S391 (2021). https://doi.org/10.1137/20M1349667
16. Martinsson, P.G., Rokhlin, V., Tygert, M.: A randomized algorithm for the decomposition of matrices. Appl. Comput. Harmonic Anal. **30**(1), 47–68 (2011). https://doi.org/10.1016/j.acha.2010.02.003, https://www.sciencedirect.com/science/article/pii/S1063520310000242
17. Pedregosa, F., et al.: Scikit-Learn: machine learning in Python. J. Mach. Learn. Res. **12**, 2825–2830 (2011)
18. Reiz, S., Neckel, T., Bungartz, H.J.: Neural nets with a newton conjugate gradient method on multiple GPUs. In: International Conference on Parallel Processing and Applied Mathematics, pp. 139–152. Springer (2022). https://doi.org/10.1007/978-3-031-30442-2_11
19. Roweis, S.T., Saul, L.K.: Nonlinear dimensionality reduction by locally linear embedding. Science **290**(5500), 2323–2326 (2000). https://doi.org/10.1126/science.290.5500.2323, https://www.science.org/doi/abs/10.1126/science.290.5500.2323
20. Sorensen, D.C.: Implicitly restarted Arnoldi/Lanczos methods for large scale eigenvalue calculations. SIAM J. Matrix Anal. Appl. **13**, 357–385 (1992)
21. Tenenbaum, J., Silva, V., Langford, J.: A global geometric framework for nonlinear dimensionality reduction. Science **290**, 2319–2323 (2000)
22. Virtanen, P., Gommers, R., Oliphant, T.E., Haberland, M., Reddy, T., et al.: SciPy 1.0: fundamental algorithms for scientific computing in Python. Nature Methods (2020)
23. Yu, C.D., Levitt, J., Reiz, S., Biros, G.: Geometry- oblivious FMM for compressing dense SPD matrices. In: Proceedings of SC17, Denver, CO, USA (2017)
24. Yu, C.D., Reiz, S., Biros, G.: Distributed-memory hierarchical compression of dense SPD matrices. In: SC18: International Conference for High Performance Computing, Networking, Storage and Analysis, pp. 183–197 (2018). https://doi.org/10.1109/SC.2018.00018
25. Yu, C.D., Reiz, S., Biros, G.: Distributed O(N) linear solver for dense symmetric hierarchical semi-separable matrices. In: IEEE 13th International Symposium on Embedded Multicore/Many-core Systems-on-Chip (MCSoC) (2019)

Second-Order Gradient Loss Guided Single-Image Super-Resolution

Shuran Lin[1,2], Chunjie Zhang[1,2(✉)], and Yanwu Yang[3]

[1] Beijing Key Laboratory of Advanced Information Science and Network Technology,
Beijing Jiaotong University, Beijing 100044, China
`cjzhang@bjtu.edu.cn`
[2] Institute of Information Science, Beijing Jiaotong University, Beijing 100044, China
[3] School of Management, Huazhong University of Science and Technology, Wuhan
430074, China

Abstract. With the development of deep learning, convolutional neural networks for single-image super-resolution have been proposed and achieved great success. However, most of these methods use L1 loss to guide network optimization, resulting in blurry restored images with sharp edges smoothed. This is because L1 loss limits the optimization goal of the network to the statistical average of all solutions within the solution space of that task. To solve this problem, this paper designs an image super-resolution algorithm based on second-order gradient loss. This algorithm imposes additional constraints on the optimization of the network from the high-order gradient level of the image so that the network can focus on the recovery of fine details such as texture during the learning process, and alleviate the problem of restored image texture over-smoothing to a certain extent. During network training, we extract the second-order gradient map of the generated image and the target image of the network and minimize the distance between them, this will guide the network to pay attention to the high-frequency detail information in the image to generate a high-resolution image with clearer edge and texture. It is worth mentioning that our proposed loss function has good embeddability, which can be easily integrated with existing image super-resolution networks. Experimental results show that the second-order gradient loss can significantly improve both Learned Perceptual Image Patch Similarity(LPIPS) and Frechet Inception Distance score(FID) performance of existing image super-resolution deep learning models.

Keywords: Single-image super-resolution · Gradient · Loss function

1 Introduction

Given a low-resolution (LR) image, the goal of Single-Image Super-Resolution (SISR) is to restore its corresponding high-resolution (HR) image, which has received great attention from researchers in recent years. As a fundamental visual task in computer vision, SISR has a wide range of practical applications, such

C. Cruz et al. (Eds.): IC 2023, CCIS 2036, pp. 17–28, 2024.
https://doi.org/10.1007/978-981-97-0065-3_2

as medical image enhancement [13,17,23], satellite imaging [20,21,29], etc. In addition, SISR can be combined with other high-level computer vision tasks to improve their performance, such as image classification [1,28], object detection [4,29], etc. However, SISR is inherently a highly challenging and ill-posed task, as LR images lose a significant amount of texture detail information compared to HR images, making it extremely difficult to generate HR images using only LR images. In addition, since a LR image may be obtained from multiple HR images through various degradations, the solution to the SR problem is not unique.

Recently, Convolutional Neural Networks (CNNs) have demonstrated outstanding performance in complex information recovery, achieving great success in many computer vision tasks, including SISR. However, existing CNN-based SISR methods often pursue high Peak Signal to Noise Ratio (PSNR) and Structural Similarity (SSIM), resulting in visually blurry restored images. This is because most of these methods ignore the structural prior knowledge within the image and only guide the optimization of the network by minimizing the mean absolute error between the recovered HR image and the ground truth image. As a result, the optimization objective of the network become the statistical mean of all possible solutions in this one-to-many problem, ultimately causing the reconstructed images to appear blurry.

Many previous works have demonstrated that introducing prior knowledge of images, such as total variation prior [26,27], sparse prior [3,33,34], gradient prior [22,35], can alleviate the ill-posedness of the super-resolution task to some extent. These prior knowledge can be regarded as additional constraints on the optimization objective of the network, which optimize the solution space of the task. Among all these prior knowledge, the gradient prior is one of the most effective ones, which can suppress noise and preserve edges in the process of image reconstruction. In fact, an image can be regarded as a two-dimensional discrete function, and the gradient of the image is actually the derivative of this two-dimensional discrete function, which measures the change rate of the pixel grayscale value of the image. Due to the fact that the pixel grayscale values of images often change significantly in some edge and texture regions, the gradient map of images can precisely represent the edge and texture details of images. In the field of mathematics, the derivative of a function contains information that can be used to depict the shape of the functional image, and the second-order derivative of the function contains more information than the first-order derivative, which has extremely important guiding significance for accurately modeling the functional image. Analogously, in the field of image processing, the second-order gradient map of images may also contain more informative prior knowledge than the first-order gradient map. To verify this idea, in this paper, we draw on the principles of function derivation to generate the second-order gradient map of images and visualize it for more intuitive comparison with the first-order gradient map. As shown in Fig. 1, the second-order gradient map shows more detailed information than the first-order gradient map. If it is fully utilized during network optimization, it can further compress the solution space of this task and reduce the difficulty of image restoration.

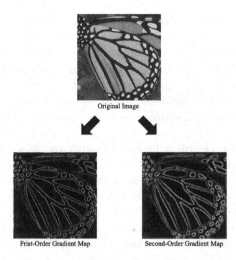

Original Image

Frist-Order Gradient Map   Second-Order Gradient Map

Fig. 1. Visualization of the first-order and second-order gradient maps of images

Based on the aforementioned discussion, this paper proposes an image super-resolution algorithm based on the second-order gradient (SG) loss. This algorithm replaces the loss function of the network with a combination of the SG loss and the L1 loss. The SG loss takes the second-order gradient map of the image as the starting point. To be specific, it first extracts the second-order gradient maps of the restored image and the HR image, and then minimizes the distance between them to fully exploit the high-frequency information contained in the second-order gradient map of the image. This forces the network to focus more on the restoration of high-frequency components such as textures and edges, improving the blurring of restored images caused by some existing methods that only use L1 loss as a constraint. Besides, it can be easily integrated into most existing SR methods without adding extra training parameters. This means that you can enhance the visual quality of the images restored by the network without increasing the computational cost. Experimental results on five widely used benchmark datasets demonstrate that, after integrating the proposed SG loss function into the network, it can achieve improved visual performance and recover fine details.

2   Related Works

2.1   Single Image Super-Resolution Methods

In the early days, researchers mostly adopted interpolation-based methods to tackle the SISR problem. The basic idea of these methods is to estimate the pixel value of the corresponding position in the HR image by weighted averaging the known pixel values around a certain position in the LR image. Considering that the common pixel variations in a local region of an image can be approximated by a continuous function, various weighting schemes have been designed for image interpolation. For example, bilinear interpolation proposed by using local linearity and bicubic [14] interpolation proposed by using high-order continuity. Although these methods are simple and fast to implement, the generated images always suffer from some unnatural artifacts and structural distortions. This is because many pixel variations inside an image are actually very complex and cannot be described by such simple predefined functions, especially for images with rich textures.

In recent years, deep learning-based methods have shown remarkable feature learning and extraction capabilities, enabling neural networks to simulate any function theoretically. By end-to-end model training, these networks can directly learn the mapping relationship between LR and HR images from data. Compared with early interpolation-based methods, these data-driven deep learning methods have brought significant improvements in performance. As a pioneer, Dong et al. [6] propose the first super-resolution convolutional neural network (SRCNN) consisting of three convolutional layers. However, this method lack the ability to upscale images, requiring preprocessing of the LR image by interpolation to the same size as the HR image before inputting it into the network. This lead to high computational complexity and information loss in the original LR image. To address this issue, Dong et al. [7] add a deconvolutional layer at the end of the network to implement image upscaling, achieving an end-to-end mapping from LR images to HR images. Shi et al. [30] design an efficient sub-pixel convolutional layer that can adjust the number of feature channels to enlarge image.

However, all the aforementioned methods only leverage shallow layers with limited receptive fields, which fail to fully exploit the contextual information from surrounding pixels. Therefore, many works employ deeper network for better performance. Kim et al. [15] attempt to increase the receptive field of the network by stacking more convolutional layers and introduce skip connections and gradient clipping to mitigate the problem of gradient vanishing that arises from deepening the network. Tai et al. [31] achieve deeper network without increasing the number of parameters by using recursive convolution and parameter sharing. Zhang et al. [39] design a residual dense connection mechanism to better exploit shallow features through feature reuse. However, all of these methods extract LR image features from a single scale, which limits their ability to adapt to noise and deformation. Li et al. [16] propose a multi-scale network that adaptively extracts image features of different scales to aid image reconstruction. Zhang et al. [38] find that these methods treat LR features equally, which hinders the performance of CNNs. Therefore, they incorporate attention mechanism into the network to focus on the more important parts of image reconstruction. Some of the latest works have also apply shift window-based self-attention mechanism to SISR field [18], achieving state-of-the-art performance. However, these methods overlook the effectiveness of the internal gradient prior of the image, producing blurry results, where edges and textures are smoothed.

2.2   Gradient Guided Super-Resolution Methods

Many traditional SISR methods use complex prior knowledge to limit the size of the solution space [8, 10, 26, 32, 40], which can help restore clearer high-frequency details. Inspired by the effectiveness of image priors in traditional methods, some recent deep learning-based methods have also attempted to combine image prior knowledge with neural networks [9, 22, 35]. Among them, the gradient prior knowledge is commonly used, which usually reflects the parts of the image where the pixel grayscale values change sharply, and this part is precisely the edge texture in the image. For example, Zhu et al. [40] propose a gradient-based SR method by collecting a dictionary of gradient patterns and modeling deformable gradient combinations. Fattal [10] designs a method based on image gradient edge statistics by learning the prior correlation of different resolutions. Yan et al. [32] propose a stochastic resonance method based on gradient contour sharpness. Yang et al. [35] use a pre-trained edge detector to extract image gradients and then use them to guide the deep network to reconstruct SR images with clearer edges. Ma et al. [22] design a dual-branch joint optimization network consisting of a main SR branch and a gradient-assisted branch, where the gradient-assisted branch uses the gradient map extracted from the LR image as input, and the optimization target becomes the gradient map of its corresponding HR image. Although these methods explore gradient prior knowledge to improve the visual quality of restored images, adding learnable parameters related to gradient information to the model greatly increases its complexity and reduces its computational efficiency. In contrast, the proposed method in this paper introduces a second-order gradient prior only during network optimization to provide additional supervision information, without adding any learnable parameters, and the computational cost can be neglected.

3   Our Method

3.1   Problem Definition

For the task of SISR, the goal is to predict a reasonable HR image I^{SR} from a LR input image I^{LR}, given its corresponding ground truth HR image I^{HR}, and ensure that the predicted HR image I^{SR} is as close as possible to the ground truth HR image I^{HR}. Therefore, in the actual training of the model, it is necessary to use already paired LR and HR image pairs (I^{LR}, I^{HR}). In fact, the LR image is obtained from the HR image through various types of degradation, but due to the complex and diverse forms of degradation and difficult modeling, for convenience of research, most works simply model the degradation process of the image as a bicubic interpolation downsampling operation. Therefore, the corresponding LR image can be generated from the HR image by the following formula:

$$I^{LR} = (I^{HR}) \downarrow_s \tag{1}$$

where $\downarrow_s$ represents a bicubic interpolation downsampling operation with a scaling factor of s. Usually, both LR and HR images are 3-channel RGB images, and the sizes of the two are $3 \times h \times w$ and $3 \times s \cdot h \times s \cdot w$, respectively, where h and w are the height and width of the LR image. If the SR network is represented as N with parameters θ, then the process of image SR can be represented as:

$$I^{SR} = N(I^{LR}; \theta) \tag{2}$$

If the network parameters can be optimized through the loss function f, then the optimization process of the SR network can be formulated as follows:

$$\widehat{\theta} = \underset{\theta}{argmin} f(N(I^{LR}; \theta), I^{HR}) \tag{3}$$

3.2 Second-Order Gradient Loss

Before introducing the SG loss, we need to discuss the L1 loss which used by most existing deep learning-based SR methods. This loss is obtained by calculating the mean absolute error between the image I^{SR} which predicted by the network and the ground truth HR image I^{HR} at each pixel, which can result in a high PSNR value for the recovered image. However, the visual results are often blurry and fail to preserve sharp edges in the original image. Nevertheless, due to its ability to accelerate convergence and improve SR performance, L1 loss remains the most widely used loss function in this field:

$$L_1 = \|I^{HR} - I^{SR}\|_1 \tag{4}$$

Considering that the L1 loss treats high-frequency and low-frequency information in the image equally, while ignoring the fact that high-frequency information is more difficult to recover, this paper proposes to use the second-order gradient map of the image as additional supervision information in the optimization process to encourage the network to pay more attention to high-frequency information during the recovery process and alleviate the problem of smoothing sharp edges in the image. The reason why this paper did not choose to use higher-order gradient maps of image is that studies have shown that as the order increases, the detail information in the gradient map becomes richer, which makes it more difficult for the network to learn and may even introduce additional errors. The method proposed in this paper requires extracting the second-order gradient map of the image. Specifically, we calculate the difference between each pixel and its diagonal neighbor in the image to obtain the first-order gradient map of the image. Then, the second-order gradient map of the image is obtained by calculating the difference between each diagonal neighbor pixels of the first-order gradient map. During actual training, an additional constraint is applied to the I^{SR} by minimizing the distance between the second-order gradient maps of I^{SR} and I^{HR}, in order to preserve high-frequency details in I^{SR}. The gradient map of the image I can be generated using the following formula:

$$\nabla x I(x,y) = I(x,y) - I(x-1, y-1) \tag{5}$$

$$\nabla y I(x,y) = I(x-1, y) - I(x, y-1) \tag{6}$$

$$\nabla I(x,y) = (\nabla x I(x,y), \nabla y I(x,y)) \tag{7}$$

$$G(I) = \|\nabla I\| \tag{8}$$

where (x, y) represents the coordinates of any point in the image, and $I(x, y)$ represents the pixel value of the image at (x, y). $G(\cdot)$ is the operation for extracting image gradients, which can be implemented by designing a convolution layer with fixed weight kernels. In this paper, we choose the Roberts filter weights as the convolution kernel weights. Compared with other edge detection filters, the Roberts filter is simple to implement and fast in computation. The second-order gradient map of the image can be obtained by applying $G(\cdot)$ twice. In summary, the proposed second-order gradient loss in this paper can be formulated as follows:

$$L_{SG} = \|G(G(I^{HR})) - G(G(I^{SR}))\|_1 \tag{9}$$

During the training of the network, since the second-order gradient map only contains high-frequency information of the image and lacks low-frequency information, it needs to be combined with L1 loss to jointly guide the optimization of the network. Therefore, the final loss function can be obtained by combining the SG loss and L1 loss as follows:

$$L_{total} = L_1 + L_{SG} \tag{10}$$

4   Experiments

4.1   DataSets and Metrics

The DIV2K [2] dataset is a high-quality visual dataset in the field of SISR, consisting of 800 training images, 100 validation images, and 100 testing images. For testing, this paper uses five standard public datasets: Set5 [5], Set14 [36], Urban100 [24], B100 [12], and Manga109 [25], which contain various scenes and can fully verify the effectiveness of the proposed method. Since paired HR and LR images are required for training, the corresponding LR images are obtained by downsampling the HR images with a scaling factor of 4 using bicubic interpolation before conducting experiments. Considering that evaluation metrics such as Peak Signal-to-Noise Ratio(PSNR) and Structural Similarity(SSIM) often contradict human perceptual quality, this paper uses perceptual metrics LPIPS [37] and FID [11], which are more consistent with human perception, as evaluation metrics to quantitatively compare the restoration results of the datasets. Lower LPIPS and FID values indicate better visual quality.

4.2   Implementation Details

For a fair comparison, we retrained several representative deep learning-based SR networks. Specifically, during the training process, data augmentation is performed on the training dataset by randomly cropping, rotating 90°, 180°, 270°, and flipping the original images to obtain approximately 32,000 HR images of size 480 × 480. In each training batch, 16 LR image patches with the size of 48 × 48 are used as the input. The proposed model is trained using the ADAM optimizer with default values of $\beta_1 = 0.9$, $\beta_2 = 0.999$ and $\epsilon = 1 \times 10^{-8}$. The initial learning rate is set to 1×10^{-4} and is subsequently halved every 2×10^{5} training iterations of back-propagation. The whole process is implemented in the PyTorch 2.0 platform with a Nvidia GeForce RTX 3090 24GB GPU.

Table 1. Quantitative comparisons of cnn-based SISR models with and without second-order gradient loss on five benchmark datasets for ×4 SR. best results are **highlighted**

DataSet	Metric	EDSR	EDSR+SG	RDN	RDN+SG	RCAN	RCAN+SG	SwinIR	SwinIR+SG
Set5	LPIPS ↓	0.1728	**0.1578**	0.1716	**0.1542**	0.1720	**0.1439**	0.1700	**0.1411**
	FID ↓	58.86	**51.96**	57.88	**47.74**	59.74	**54.14**	58.80	**46.92**
Set14	LPIPS ↓	0.2776	**0.2578**	0.2808	**0.2555**	0.2783	**0.2450**	0.2705	**0.2298**
	FID ↓	86.45	**82.60**	88.75	**82.08**	91.95	**83.24**	89.17	**83.47**
Urban100	LPIPS ↓	0.2037	**0.1943**	0.2107	**0.1964**	0.2047	**0.1888**	0.1923	**0.1719**
	FID ↓	25.56	**24.41**	26.12	**24.77**	25.71	**25.18**	24.54	**22.64**
B100	LPIPS ↓	0.3589	**0.3276**	0.3634	**0.3243**	0.3602	**0.3080**	0.3549	**0.2916**
	FID ↓	96.08	**86.54**	96.36	**81.68**	98.15	**82.41**	95.59	**82.68**
Manga109	LPIPS ↓	0.0997	**0.0935**	0.1018	**0.0932**	0.0991	**0.0881**	0.0938	**0.0829**
	FID ↓	12.58	**12.19**	13.25	**12.00**	12.48	**11.92**	11.82	**10.93**

4.3 Quantitative and Qualitative Comparisons

We choose several representative SR networks, including EDSR [19], RDN [39], RCAN [38], and SwinIR [18], to validate the effectiveness of our proposed SG loss function. These SR methods all employ the L1 loss function for optimization, and we made no modifications to their network architectures. Instead, we augmented the loss function with an SG loss term to provide additional supervision information. The additional computational burden brought by this operation can be ignored. The ×4 SR results on five benchmark datasets are shown in Table 1. The results marked with "+SG" indicate the outcomes obtained by adding the SG loss for auxiliary optimization to the original SR methods. From the data in the Table 1, it can be seen that for all models, the addition of the SG loss function as an auxiliary network optimization loss leads to lower LPIPS and FID values on all datasets compared to the original models. This conclusively illustrates that the second-order gradient map of the image, which contains high-frequency information, significantly aids the network in restoring images of superior perceptual quality. Especially on the more severely degraded B100 dataset, our method reduces the LPIPS values of the SwinIR and RCAN models by 0.0633 and 0.0522, respectively, which represents a significant improvement in terms of the LPIPS evaluation metric. Moreover, the FID values of these two models are also greatly improved compared to the original method.

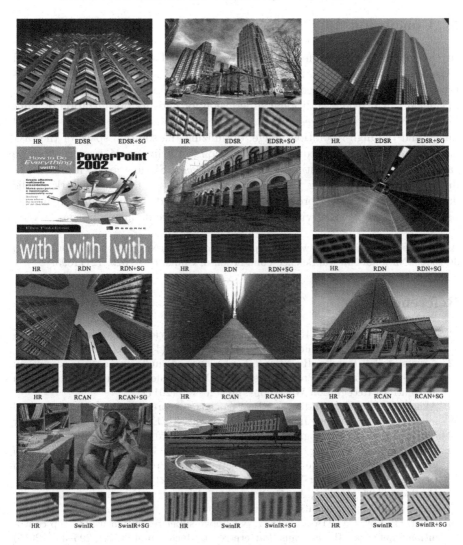

Fig. 2. Visual comparison of restoration results of different models on Set14, B100 and Urban100 datasets before and after integrating SG loss.

To further validate the effectiveness of our proposed SG loss, this section presents visual results of restored images on the Set14, B100 and Urban100 datasets. As shown in Fig. 2, before adding the proposed SG loss, although these methods which trained using only the L1 loss can restore the main contours of the objects, they can not accurately restore complex edges and textures, and even produce some distorted and deformed textures. In contrast, after adding our proposed SG loss as additional supervision, the network preserves the fine details in the image as much as possible, and the restored textures are more natural and realistic.

5 Conclusion

In this paper, a new high-frequency texture detail enhancement loss, called second-order gradient loss, is proposed to alleviate the problem of blurry high-resolution images generated by existing SISR methods trained with L1 loss. Specifically, it provides additional supervision for network optimization by minimizing the distance between the second-order gradient maps of the restored image and the high-resolution image, optimizing the solution space of the task and enhancing the high-frequency information in the image. Moreover, it can be easily integrated with existing deep-learning based SR methods without introducing additional training parameters. Evaluation on five benchmark public datasets shows that the proposed second-order gradient loss function in this paper is effective in preserving high-frequency texture details in images.

Acknowledgements. This work is supported by Beijing Natural Science Foundation: JQ20022; National Natural Science Foundation of China: 62072026, 72171093.

References

1. Agarwal, K., Macháň, R.: Multiple signal classification algorithm for super-resolution fluorescence microscopy. Nat. Commun. **7**(1), 13752 (2016)
2. Agustsson, E., Timofte, R.: Ntire 2017 challenge on single image super-resolution: Dataset and study. In: Proceedings of the IEEE Conference on Computer Vision and Pattern Recognition Workshops, pp. 126–135 (2017)
3. Akhtar, N., Shafait, F., Mian, A.: Bayesian sparse representation for hyperspectral image super resolution. In: Proceedings of the IEEE Conference on Computer Vision and Pattern Recognition, pp. 3631–3640 (2015)
4. Bai, Y., Zhang, Y., Ding, M., Ghanem, B.: SOD-MTGAN: small object detection via multi-task generative adversarial network. In: Proceedings of the European Conference on Computer Vision (ECCV), pp. 206–221 (2018)
5. Bevilacqua, M., Roumy, A., Guillemot, C., Alberi-Morel, M.L.: Low-complexity single-image super-resolution based on nonnegative neighbor embedding (2012)
6. Dong, C., Loy, C.C., He, K., Tang, X.: Image super-resolution using deep convolutional networks. IEEE Trans. Pattern Anal. Mach. Intell. **38**(2), 295–307 (2015)
7. Dong, C., Loy, C.C., Tang, X.: Accelerating the super-resolution convolutional neural network. In: Leibe, B., Matas, J., Sebe, N., Welling, M. (eds.) Accelerating the super-resolution convolutional neural network. LNCS, vol. 9906, pp. 391–407. Springer, Cham (2016). https://doi.org/10.1007/978-3-319-46475-6_25
8. Dong, W., Zhang, L., Shi, G., Wu, X.: Image deblurring and super-resolution by adaptive sparse domain selection and adaptive regularization. IEEE Trans. Image Process. **20**(7), 1838–1857 (2011)
9. Fang, F., Li, J., Zeng, T.: Soft-edge assisted network for single image super-resolution. IEEE Trans. Image Process. **29**, 4656–4668 (2020)
10. Fattal, R.: Image upsampling via imposed edge statistics. In: ACM SIGGRAPH 2007 papers, p. 95-es (2007)
11. Heusel, M., Ramsauer, H., Unterthiner, T., Nessler, B., Hochreiter, S.: GANs trained by a two time-scale update rule converge to a local nash equilibrium. In: Advances in Neural Information Processing Systems 30 (2017)

12. Huang, J.B., Singh, A., Ahuja, N.: Single image super-resolution from transformed self-exemplars. In: Proceedings of the IEEE Conference on Computer Vision and Pattern Recognition, pp. 5197–5206 (2015)
13. Huang, Y., Shao, L., Frangi, A.F.: Simultaneous super-resolution and cross-modality synthesis of 3D medical images using weakly-supervised joint convolutional sparse coding. In: Proceedings of the IEEE Conference on Computer Vision and Pattern Recognition, pp. 6070–6079 (2017)
14. Keys, R.: Cubic convolution interpolation for digital image processing. IEEE Trans. Acoust. Speech Signal Process. **29**(6), 1153–1160 (1981)
15. Kim, J., Lee, J.K., Lee, K.M.: Accurate image super-resolution using very deep convolutional networks. In: Proceedings of the IEEE Conference on Computer Vision and Pattern Recognition, pp. 1646–1654 (2016)
16. Li, J., Fang, F., Mei, K., Zhang, G.: Multi-scale residual network for image super-resolution. In: Proceedings of the European Conference on Computer Vision (ECCV), pp. 517–532 (2018)
17. Li, Y., Sixou, B., Peyrin, F.: A review of the deep learning methods for medical images super resolution problems. IRBM **42**(2), 120–133 (2021)
18. Liang, J., Cao, J., Sun, G., Zhang, K., Van Gool, L., Timofte, R.: SwinIR: image restoration using swin transformer. In: Proceedings of the IEEE/CVF International Conference on Computer Vision, pp. 1833–1844 (2021)
19. Lim, B., Son, S., Kim, H., Nah, S., Mu Lee, K.: Enhanced deep residual networks for single image super-resolution. In: Proceedings of the IEEE Conference on Computer Vision and Pattern Recognition Workshops, pp. 136–144 (2017)
20. Lu, T., Wang, J., Zhang, Y., Wang, Z., Jiang, J.: Satellite image super-resolution via multi-scale residual deep neural network. Remote Sens. **11**(13), 1588 (2019)
21. Luo, Y., Zhou, L., Wang, S., Wang, Z.: Video satellite imagery super resolution via convolutional neural networks. IEEE Geosci. Remote Sens. Lett. **14**(12), 2398–2402 (2017)
22. Ma, C., Rao, Y., Cheng, Y., Chen, C., Lu, J., Zhou, J.: Structure-preserving super resolution with gradient guidance. In: Proceedings of the IEEE/CVF Conference on Computer Vision and Pattern Recognition, pp. 7769–7778 (2020)
23. Mahapatra, D., Bozorgtabar, B., Garnavi, R.: Image super-resolution using progressive generative adversarial networks for medical image analysis. Comput. Med. Imaging Graph. **71**, 30–39 (2019)
24. Martin, D.R., Fowlkes, C.C., Tal, D., Malik, J.: A database of human segmented natural images and its application to evaluating segmentation algorithms and measuring ecological statistics. In: Proceedings of the Eighth International Conference On Computer Vision (ICCV-01), Vancouver, British Columbia, Canada, July 7–14, 2001 -Volume 2, pp. 416–425. IEEE Computer Society (2001). https://doi.org/10.1109/ICCV.2001.937655
25. Matsui, Y., et al.: Sketch-based manga retrieval using manga109 dataset. Multimedia Tools Appl. **76**, 21811–21838 (2017)
26. Ng, M.K., Shen, H., Chaudhuri, S., Yau, A.C.: Zoom-based super-resolution reconstruction approach using prior total variation. Opt. Eng. **46**(12), 127003–127003 (2007)
27. Ng, M.K., Shen, H., Lam, E.Y., Zhang, L.: A total variation regularization based super-resolution reconstruction algorithm for digital video. EURASIP J. Adv. Signal Process. **2007**, 1–16 (2007)
28. Qin, B., Li, D.: Identifying facemask-wearing condition using image super-resolution with classification network to prevent COVID-19. Sensors **20**(18), 5236 (2020)

29. Shermeyer, J., Van Etten, A.: The effects of super-resolution on object detection performance in satellite imagery. In: Proceedings of the IEEE/CVF Conference on Computer Vision and Pattern Recognition Workshops (2019)
30. Shi, W., et al.: Real-time single image and video super-resolution using an efficient sub-pixel convolutional neural network. In: Proceedings of the IEEE Conference on Computer Vision and Pattern Recognition, pp. 1874–1883 (2016)
31. Tai, Y., Yang, J., Liu, X.: Image super-resolution via deep recursive residual network. In: Proceedings of the IEEE Conference on Computer Vision and Pattern Recognition, pp. 3147–3155 (2017)
32. Yan, Q., Xu, Y., Yang, X., Nguyen, T.Q.: Single image superresolution based on gradient profile sharpness. IEEE Trans. Image Process. **24**(10), 3187–3202 (2015)
33. Yang, J., Wright, J., Huang, T., Ma, Y.: Image super-resolution as sparse representation of raw image patches. In: 2008 IEEE Conference on Computer Vision and Pattern Recognition, pp. 1–8. IEEE (2008)
34. Yang, J., Wright, J., Huang, T.S., Ma, Y.: Image super-resolution via sparse representation. IEEE Trans. Image Process. **19**(11), 2861–2873 (2010)
35. Yang, W., et al.: Deep edge guided recurrent residual learning for image super-resolution. IEEE Trans. Image Process. **26**(12), 5895–5907 (2017)
36. Zeyde, R., Elad, M., Protter, M.: On single image scale-up using sparse-representations. In: Boissonnat, J.-D., et al. (eds.) Curves and Surfaces 2010. LNCS, vol. 6920, pp. 711–730. Springer, Heidelberg (2012). https://doi.org/10.1007/978-3-642-27413-8_47
37. Zhang, R., Isola, P., Efros, A.A., Shechtman, E., Wang, O.: The unreasonable effectiveness of deep features as a perceptual metric. In: Proceedings of the IEEE Conference on Computer Vision and Pattern Recognition, pp. 586–595 (2018)
38. Zhang, Y., Li, K., Li, K., Wang, L., Zhong, B., Fu, Y.: Image super-resolution using very deep residual channel attention networks. In: Proceedings of the European conference on computer Vision (ECCV), pp. 286–301 (2018)
39. Zhang, Y., Tian, Y., Kong, Y., Zhong, B., Fu, Y.: Residual dense network for image super-resolution. In: Proceedings of the IEEE Conference on Computer Vision and Pattern Recognition, pp. 2472–2481 (2018)
40. Zhu, Y., Zhang, Y., Bonev, B., Yuille, A.L.: Modeling deformable gradient compositions for single-image super-resolution. In: Proceedings of the IEEE Conference on Computer Vision and Pattern Recognition, pp. 5417–5425 (2015)

The Implementation and Optimization of FFT Calculation Based on the MT-3000 Chip

Dong Cheng[1], Guilan Li[1], Aochuang Song[2], and Bangjian Xu[2(✉)]

[1] College of Electrical and Information Engineering, Hunan University,
Changsha, China
[2] College of Computer Science and Electronic Engineering, Hunan University,
Changsha, China
bjxu@hnu.edu.cn

Abstract. Based on the accelerator chip MT-3000, the FFT algorithm in SAR imaging has been implemented and optimized. The optimization includes vectorization and MPI, DMA transmission and dual buffer transmission, and linear assembly. The experimental results show that on the platform, the performance increased by more than 99.2% after the FFT function was optimized.

Keywords: MT-3000 · FFT · Vector · MPI · DMA transmission · Dual buffer transmission · Linear assembly

1 Introduction

With the rapid development of integrated circuits and communication technology, digital signal processing is widely used in communication and other areas [1]. FFT is a basic algorithm commonly used in signal processing. In SAR radar imaging and other algorithms, FFT has a vital role [2]. Base 2FFT can significantly reduce the amount of computing, but it requires a large number of iterative operations. The requirements for the number of input sequences must be a power of 2 [3]. With the deepening of the FFT theoretical algorithm, the split radix FFT [4] [5], base 4FFT [6], base 8FFT [7], mixed-radix FFT [8] and other high-efficiency algorithms were proposed one after another. High performance computing (HPC) refers to a technology that uses powerful processor clusters working in parallel to process massive multidimensional datasets and solve complex problems at extremely high speeds [9]. In recent years, due to high performance computing shows many advantages, more and more high-performance computing platforms have been used for radar signal processing [10] [11]. Internationally, commercial high-performance calculation accelerators are mainly represented by NVIDIA, AMD and Intel [12]. In 2021, AMD launched the Instant Mi250X processor dual-precision floating-point computing capacity of up to 95.7TFLOPS (Tera Floating Point Operations Per Second) [13].

C. Cruz et al. (Eds.): IC 2023, CCIS 2036, pp. 29–40, 2024.
https://doi.org/10.1007/978-981-97-0065-3_3

NVIDIA released the "H100" with a new Hopper architecture in March 2022. H100 increases the floating-point operations per second (FLOPS) of the Tensor Core by three times, and provides HPC with a FP64 floating-point operation of 60 teraFLOPS [14]. In the field of parallel computing research, using parallel programming models to develop parallel programs is an effective method. There are three common parallel programming models nowadays: shared variable programming model, data parallel programming model, and message passing programming model [15]. The message passing programming model can be used in distributed memory parallel architectures, and is also very applicable in shared memory architectures. The representative implementation of the current message passing programming model is MPI [16]. This article is based on the accelerator MT-3000 and optimizes the FFT algorithm for rows and columns of a matrix with a size of 32768 * 8192, and compares its performance. This chip supports vector computing and MPI programming, providing a good research platform for multi-core programming. The research content of this article mainly includes the following aspects: (1) Analyzing the characteristics of the chip architecture; (2) Implement FFT algorithm for the rows and columns corresponding to a matrix of size 32768 * 8192; (3) Optimize the FFT algorithm for rows and columns in (2) on this platform, using optimization methods such as DMA dual channel transfer, vectorization and MPI multi-core parallelization, double buffering transfer, and linear assembly; (4) Performance analysis was conducted on the experimental results before and after the optimization of the FFT algorithm.

2   Chip Overall Structure

The chip uses a heterogeneous fusion architecture composed of multi-core CPU and 4 GPDSP_Cluster. Among them, the multi-core CPU contains 32 FT-C662 CPU cores, and each GPDSP_Cluster contains 24 FT-M64DSP cores. The DSP kernel architecture of the chip is shown in Fig. 1, using the scalar/vector coordinated architecture based on the Very Long Instruction Word (VLIW). The LIP/Fetch/Dispatch unit extracts the instructions that need to be executed from the instruction package and sends them to the scalar unit and vector unit for execution. The scalar memory is a local memory corresponding to the scalar component. The vector parts consist of 16 homogeneous Vector Process Element (VPE). Array Memory (AM) is a vector data access to the 16-way SIMD width, and provides a higher access bandwidth for vector parts. The DMA (Direct Memory Access) component initiates access to specific storage resources by configuring transmission parameters, supporting flexible and efficient data transmission methods (such as point-to-point, broadcast, segmentation, SuperGather, etc.) to fully adapt to different application requirements [17].

3   FFT Algorithm Implementation

3.1   Base 2FFT

The implementation, optimization and debugging of all FFT functions in this article are based on MT-3000IDE. Depending on the sequence decomposition or

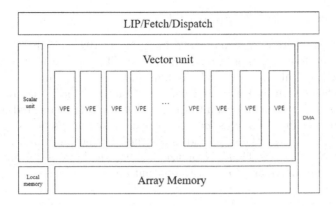

Fig. 1. DSP kernel architecture.

selection method, the base 2FFT is divided into: Decimation-in-time(DIT) and
Decimation-in-frequency(DIF). This article uses the DIT-FFT. The pseudocode
of the program is shown in Algorithm 1. The number of input data is n $= 2^k$,
and the element type is Double type. When you enter L=0, it means FFT, and
L $= 1$ is used for IFFT.

Algorithm 1. FFT

Input: pr, pi, n, k, fr, fi, l
Output: fr, fi
1: $fr = BitReverse(pr); fi = BitReverse(pi);$
2: $pr = cos(2 * PI/(1.0 * n)); pi = -sin(2 * PI/(1.0 * n));$
3: **if** $l! = 0$ **then**
4:    $pi = -pi;$
5: **end if**
6: **for each** $l0 \in [k - 2, 0]$ **do**
7:    $m = m/2; nv = 2 * nv;$
8:    **for each** $it \in [0, (m - 1) * nv]$ **do**
9:       **for each** $j \in [0, (nv/2) - 1]$ **do**
10:          $bxcr = pr[m * j] * fr[it + j + nv/2] - pi[m * j] * fi[it + j + nv/2];$
11:          $bxci = pr[m * j] * fi[it + j + nv/2] + pi[m * j] * fr[it + j + nv/2];$
12:          $fr[it + j + nv/2] = fr[it + j] - bxcr; fi[it + j + nv/2] = fi[it + j] - bxci;$
13:          $fr[it + j] = fr[it + j] + bxcr; fi[it + j] = fi[it + j] + bxci;$
14:       **end for**
15:    **end for**
16: **end for**
17: **if** $l! = 0$ **then**
18:    **for each** $i \in [0, n - 1]$ **do**
19:       $fr[i] = fr[i]/(1.0 * n); fi[i] = fi[i]/(1.0 * n);$
20:    **end for**
21: **end if**

3.2　Matrix Line and Column FFT

The matrix size entered in this article is 32768 * 8192, and the element type is double. The four functions FFTY, FFTX, IFFTX, and IFFTY sequentially perform FFT calculations on each column and row of the matrix, as well as IFFT calculations on each row and column. Complete the FFT calculation of matrix rows and columns by calling the FFT function implemented in the previous section. Taking FFTX as an example, the program pseudocode is shown in Algorithm 2. Among them, pr and pi are the real and imaginary parts of the input data, Nrow size is 32768, Ncol size is 8192, Mcol size is 13, and fr and fi are the real and imaginary parts of the output data.

Algorithm 2. FFTX

Input: pr, pi, fr, fi
Output: fr, fi
1: **for** each $i \in [0, Nrow]$ **do**
2:　$FFT(\&pr[i * Ncol], \&pi[i * Ncol], Ncol, Mcol, \&fr[i * Ncol], \&fi[i * Ncol], 0);$
3: **end for**

4　Optimization of FFT Algorithm

4.1　Data Processing

Generate the Rotation Factor for Each Level of FFT Butterfly. The second, third, and fourth levels of butterfly calculation have 2, 4, and 8 rotation factors, respectively. In order to facilitate vector calculation, it is necessary to expand these three rotation factors to 16. The rotation factor of the fifth level and above butterfly is a multiple of 16, so there is no need to expand it further.

Generate the Index of the Bit Reversal. Before FFT operation, the input data needs to be bitwise reversed first. The index transmission of this platform supports discontinuous retrieval, therefore, corresponding index values are generated based on the law of code bit reversal, and DMA index transmission is used to retrieve data and achieve code bit reversal.

Generate the Index of the Matrix Transposition. After FFTY is calculated, the matrix has changed from 32768*8192 to 8192*32768, and the matrix needs to be restored to the original scale. The Matrix transposition also uses index transmission to complete.

4.2   DMA Dual Channel Transmission

In this chip, each DMA is set with 20 independent logical channels, which are numbered 0–19. The DMA logic channel is an interface between the DSP kernel and DMA. It completes the configuration and maintenance of the DMA parameter and the startup of DMA transactions. DMA data transmission mode includes point-to-point transmission, segmented data transmission, broadcast data transmission, Super Gather (SG) data transmission, etc. The platform provides the Super Gather (SG) data transmission that is an index transmission. It is characterized by "two reading and one writing". The specific description is as follows: the source address (Src_addr) of the data is not regular. The DMA cannot be produced. It must be generated through the source base address and source index (addr). Each Src_addr corresponds to data of one word width (64 bits). The direction of data transmission, where the space for storing the source address index is in DDR or GSM, and the source data is moved to AM or SM in DDR or GSM. The destination address is regular, and the user can generate the start address and frame index by configuring the DMA destination. This article uses index transmission to implement the bit reversal of FFT functions, that is, taking a column or row of data based on the index value and performing bit reversal operation to transfer it to vector space, in order to save time.

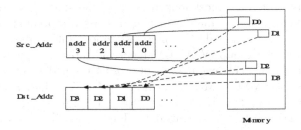

Fig. 2. Super Gather data transmission.

The DMA parameter consists of 8 words, each of which is 64 bit. Transmission control word 1 is mainly used to control the transmission type and the transmission end of the interrupt code [18]. To achieve DMA dual channel transmission, it is necessary to configure the transmission end interrupt parameter for the corresponding bit in transmission control word 1, and the event enable register (EER) is set. If it is SG transmission, you need to configure the SG data transmission control register (SGTC). The SG data transmission control register (SGTC) has a total of 64 bits. If a bit from 0 to 19 is written as 1, it indicates that the corresponding logical channel parameter is for SG data transmission. According to the DMA transmission process, combined with the multi-core parallel optimization in the following text, the multi-core SG dual-channel transmission function and the multi-core point-to-point dual channel transmission function were finally realized. After calling the multi-core SG dual-channel transmission function, the corresponding bit of Event Set Register (ESR) is set to 1, starting the first DMA logical channel. After the DMA channel starts, read the transmission parameters from the parameter RAM and submit them to the DMA host channel for processing. After the current DMA transaction transfer ends, an interrupt event corresponding to the TCC value will be generated to start the next logical channel and complete the transfer of the second array. When the transfer of two arrays ends, detect the corresponding bit fields of the register CIPR, where 1 indicates the end of the transfer [19].

4.3   Vectorization

Using the vector C interface provided by the platform, the FFT correlation function was vectorized. The FFT of 8192 points in this article uses a function file "ACSP_fft_rsp_2_8192" to achieve 13 level butterfly calculations at once. The butterfly operation process at each level is: first take out the input data, then call the vector "vec_muli" interface to multiply the input data and the rotation factor, and finally call the vector "vec_mula" and vector "vec_mulb" interfaces to output data. The first to fifth level butterfly calculations are implemented separately. The sixth to thirteenth level butterfly calculations use the same set of core calculations. The core pseudocode after vector optimization is shown in Algorithm 3.

Among them, the outermost loop represents the sixth to thirteenth levels of butterfly operations, the middle loop represents the number of all butterfly groups in the L-level, and the innermost loop implements the operations within each butterfly group. VNUM_Member indicates the number of vectors of the butterfly group. For example, a total of 64 input data of the sixth-level butterfly group, the VNUM_Member value is 4. NUM_TwiddleFactor indicates the number of rotation factor, and the vector space InputR and InputI are the real and imaginary parts of the input data. The vector spaces CoefficientR and CoefficientI are storage spaces for the real and imaginary parts of rotation factors. Vector Processing Element Src_Upper_Real, Src_Upper_Image is the real and imaginary parts of the input data in the upper half of the butterfly shape. Src_Lower_Real, Src_Lower_Imag is the real and imaginary parts of the input data in the lower half of the butterfly shape. Num1 data is all 1. Num2 data is all -1. Vector Processing Element Rotation_Real and Rotation_Imag reads data from vector spaces CoefficientR and CoefficientI. First take out the butterfly input data, calculate the real and imaginary parts of the product of the lower part of the butterfly input data and the rotation factor, and temporarily store them in the Vector Processing Element Dst_Lower_Real and Dst_Lower_Imag; then call the "vec_mula" interface to calculate the output results of the real and imaginary parts of the upper half of the butterfly shape, using Dst_Upper_Real, Dst_Upper_Imag to temporarily storage; Calling the "vec_mulb" interface to calculate the output results of the real and imaginary parts of the lower half of the butterfly shape, and use Dst_Upper_Real_1, Dst_Upper_Imag_1 temporary storage. After the operation is over, the result of the vector register is stored to the specified AM space until the cycle ends.

Algorithm 3. ACSP_FFT_RSP_2_8192

Input: *InputR, InputI, CoefficientR, CoefficientI, L0, N*
Output: *InputR, InputI*
 1: *Corecalculation*;
 2: **for** each $L \in [5, 13]$ **do**
 3:     $NUM\_Group = NUM\_Group/2$;
 4:     $NUM\_TwiddleFactor = NUM\_TwiddleFactor * 2$;
 5:     $VNUM\_Member = VNUM\_Member * 2$;
 6:     $VNUM\_Member\_Half = VNUM\_Member\_Half * 2$;
 7:     **for** each $j \in [0, NUM\_Group]$ **do**
 8:       **for** each $k \in [0, VNUM\_Member\_Half]$ **do**
 9:         **if** $L0! = 0$ **then**
10:           $Rotation\_Imag = vec\_muli(Rotation\_Imag, Num2)$;
11:         **end if**
12:         $temp1 = vec\_muli(Src\_Lower\_Imag, Rotation\_Imag)$;
13:         $temp2 = vec\_muli(Src\_Lower\_Imag, Rotation\_Real)$;
14:         $Dst\_Lower\_Real = vec\_mulb(Src\_Lower\_Real, Rotation\_Real, temp1)$;
15:         $Dst\_Lower\_Imag = vec\_mula(Src\_Lower\_Real, Rotation\_Imag, temp2)$;
16:         $Dst\_Upper\_Real = vec\_mula(Src\_Upper\_Real, Num1, Dst\_Lower\_Real)$;
17:         $Dst\_Upper\_Imag = vec\_mula(Src\_Upper\_Imag, Num1, Dst\_Lower\_Imag)$;
18:         $Dst\_Upper\_Real\_1 = vec\_mulb(Src\_Upper\_Real, Num1, Dst\_Lower\_Real)$;
19:         $Dst\_Upper\_Imag\_1 = vec\_mulb(Src\_Upper\_Imag, Num1, Dst\_Lower\_Imag)$;
20:       **end for**
21:     **end for**
22: **end for**
23: **if** $L0! = 0$ **then**
24:     **for** each $j \in [0, N]$ **do**
25:       $temp1 = vec\_ld(j, \&InputR[0])$; $temp2 = vec\_ld(j, \&InputI[0])$;
26:       $temp1 = vec\_muli(temp1, tempN)$; $temp2 = vec\_muli(temp2, tempN)$;
27:       $vec\_st(temp1, j, InputR)$; $vec\_st(temp2, j, InputI)$;
28:     **end for**
29: **end if**

4.4 Double Buffer Transmission

Figure 3 is a schematic diagram of the dual buffer mechanism. During operation, the input data is first initialized and configured, and data is sequentially transmitted to Buffer0 and Buffer1. After entering the loop, the first step is to check whether the data input of Buffer0 is complete. If the transmission has already been completed, a vector calculation operation is performed on it. At the same time, DMA starts transmitting data to Buffer1. If the data transmission in Buffer1 is completed and written back to DDR, the newly read data is calculated and the data in Buffer0 is updated. This article mainly utilizes dual channel DMA transmission to achieve dual buffering transmission. Before iteratively calling the FFT function to process matrix rows or columns, the data from the first row or column is transferred to the first buffer space of AM. After entering the loop, the data from the second row or column is transferred to the second buffer space, and the FFT function is called to process the data from the first row or column. The calculated data is returned to the DDR and the next loop begins. The third row or column of data is fed into the first buffer space, and the FFT function is called to process the data in the second buffer. And so on. See the next section for specific implementation.

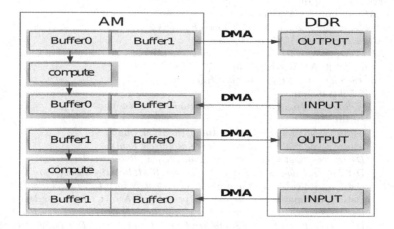

Fig. 3. Double buffer mechanism schematic diagram.

4.5 MPI

The MPI (Message Passing Interface) parallel programming environment of the platform transplanted is a parallel library MPI.LIB, which can be called directly through C/C++. The FFT calculation for each column and row of the matrix can be evenly distributed to each core of the platform. For example, for the FFTX function, multiple DSP cores can be used to perform FFT operations on a certain row of the matrix simultaneously, saving computational time and improving program performance. The core pseudocode after FFTX parallelization optimization is shown in Algorithm 4.

Among them, AM_R, AM_I, and AM_1, AM_2 alternately point to the two buffer spaces of the real and imaginary parts, respectively. First, transfer the real and imaginary data of the first row to the first buffer space of AM through SG dual channels. After entering the loop, start the SG dual channel transmission to transfer the real and imaginary data of the second row to the second buffer space. At the same time, call the FFT function to process the data of the first row, and the calculated data is sent back to the DDR through point-to-point dual channel transmission.

4.6 Linear Assembly

The platform provides model 2 and model 4 vector Load instructions, and model 16 vector Store instruction. Vector Load/Store instruction can be called by writing linearly assembly sentences. According to the previous "ACSP_FFT_RSP_

Algorithm 4. FFTX

Input: Din_R, Din_I, $Dout\_R$, $Dout\_I$, $myid$, $numprocs$
Output: $Dout\_R$, $Dout\_I$
1: $DMA\_SG\_Link(\&Din\_R[myid * Ncol], \&bitrev[0], \&AM[0], Ncol, \&Din\_I[myid * Ncol], \&bitrev[0], \&AMP[0], Ncol)$;
2: $n = 0$;
3: **for** each $i \in [myid, Nrow]$ **do**
4:    $m = n + 1$;
5:    $AM\_R = AM + (m\%2) * 1040$;
6:    $AM\_I = AMP + (m\%2) * 1040$;
7:    $AM\_1 = AM + (n\%2) * 1040$;
8:    $AM\_2 = AMP + (n\%2) * 1040$;
9:    $DMA\_SG\_Link(\&Din\_R[(i + numprocs) * Ncol], \&bitrev[0], \&AM\_R[0], Ncol, \&Din\_I[(i + numprocs) * Ncol], \&bitrev[0], \&AM\_I[0], Ncol)$;
10:    $ACSP\_fft\_rsp\_2\_8192(\&AM\_1[0], \&AM\_2[0], \&xwr[0], \&xwi[0], 0, Ncol)$;
11:    $DMA\_trans\_link(\&AM\_1[0], \&Dout\_R[i*Ncol], Ncol, \&AM\_2[0], \&Dout\_I[i*Ncol], Ncol)$;

12:    $n = n + 1$;
13:    $i = i + numprocs$;
14: **end for**

2_8192" function code, the corresponding linear assembly statement was written. Based on the results of data shuffling operations in model 2 and 4, the corresponding Vector Load/Store instructions have been added to optimize the first and second level butterfly operations of the function. Finally, perform soft pipeline optimization on some loops of the function. Implemented assembly code for 8192 point FFT.

5 Experiment and Analysis

The experimental data in this article is a 32768 * 8192 point complex number randomly generated on the platform. The real and imaginary parts of the complex number are stored separately, and the data type is double. The experiment was conducted on this platform, and the number of DSP running cycles was calculated based on the development program. This section mainly analyzes the experiment from three aspects: experimental environment, correctness analysis, and performance analysis.

5.1 Lab Environment

The experimental environment in this article is based on the MT-3000 accelerator chip and the corresponding CUDA development and commissioning operation environment that supports the chip heterogeneous multi-core. The chip software development environment consists of a series of software tools, which mainly include compilers, compilations, linkors, and other binary tools. The compiler process of the compiler can be controlled by configure various options of the compiler. The specific experimental environmental parameters are shown in Table 1.

Table 1. Experimental environmental parameters.

Parameter	MT-3000
Compiler (centered)	**MT-3000**
Compiler optimization level	**O2**
Register length/b	**64**
Frequency	**1.8 GHz**
TFLOPS	**10**
AM	**768 KB**

5.2   Correctness Analysis

On this platform, the data results before and after the optimization of FFTY, FFTX, IFFTX, and IFFTY are compared respectively. Take FFTX as an example, Table 2 is part of the experimental results before and after fftx optimization. For reference, the optimized FFT results can maintain accuracy of more than ten digits with the results before the optimization. The experimental results show that within a certain accuracy range, the optimized FFT can output the correct results.

Table 2. Results before and after FFTX optimization.

Before optimization		After optimization	
335.50336000000004	−213.65509814565877	335.50336000000004	−213.65509814565928
−106.84801336486535	−71.245644789948060	−106.84801336486561	−71.245644789949050
−53.444455266499986	−42.763737363637105	−53.444455266500100	−42.763737363636540
−35.643255271063440	−30.557193641510835	−35.643255271063936	−30.557193641512598
−26.742644801347648	−23.775771154105502	−26.742644801347700	−23.775771154105016
−21.402270141905962	−19.460312864281114	−21.402270141905674	−19.460312864281370
−17.842013387585787	−16.472681450064478	−17.842013387586032	−16.472681450064677
−15.298966864748415	−14.281746161194397	−15.298966864749300	−14.281746161196530
−13.391676736573420	−12.606320129890669	−13.391676736573440	−12.606320129890353
−11.908224204821977	−11.283610959005038	−11.908224204821725	−11.283610959004978

5.3   Performance Analysis

This article optimizes the four functions of FFTY, FFTX, IFFTX, and IFFTY on this platform. The optimized function can be divided into two versions: one is the vectorized and parallelized version, and the other is the version that adds linear assembly for optimization on top of the former. Call the timer's timing function inside the platform to record the algorithm execution time before and after function optimization, and analyze the performance of the experimental results. The optimized program was run using 22 cores, and the test time results are shown in Table 3, which lists the running time of the relevant FFT functions before and after optimization. Because IFFT operations require the final result to be multiplied by a constant of 1/N, the operation time is slower than FFT. Model 2 and Model 4 vector load instructions, and Model 16 vector store instructions are a type of doubleword vector load/store instruction. Compared to regular vector load/store instructions, the same execution cycle can read and write more data at once. Therefore, the linear assembly version will have a shorter runtime than the first vectorized version. It can be seen that after FFTY optimization, the average performance has improved by 99.284% 99.290%; The average performance improvement of FFTX after optimization was 99.926% 99.933%; The average performance improvement after IFFTX optimization is 99.927% 99.935%; After IFFTY optimization, the average performance improved by 99.213% 99.261%; It can be seen that all related FFT functions have achieved a performance improvement of over 99%.

Table 3. Comparison of running time before and after optimization (unit: s).

Function	Before optimization	Optimized version 1	Optimized version 2
FFTY	2008.903	14.375	14.260
FFTX	1729.951	1.288	1.158
IFFTX	1790.572	1.312	1.160
IFFTY	2089.295	16.438	15.449

6 Conclusion

Based on the accelerator chip, this article has completed the optimization of the FFT algorithm for a matrix size of 32768 * 8192. The experimental results show that the performance of the FFT algorithm optimized on this platform has been greatly improved, verifying the effectiveness of the optimization method proposed in this paper and the high-performance computing advantages of the chip.

Acknowledgements. Thanks a lot for the research help from National University of Defense Technology.

References

1. Huang, J.: FFT based on YHFT-Matrix design and realize. National University of Defense Technology (2012)
2. Xiang, H.: FFT algorithm design and implementation. National University of Defense Technology (2014)

3. Cooley, J.W., Tukey, J.W.: An algorithm for the machine calculation of complex Fourier series. Mathem. Comput. **19**, 297–301 (1965)
4. Duhamel, P., Hollmann, H.: Split radix FFT algorithm. Electron. Lett. **20**(1), 14–16 (1984)
5. Liu, H., Xie, Z.: Discussion and improvement of the FFT algorithm of the split foundation. Commun. Technol. **03**, 124–125 (2008)
6. Swartzlander, E.E., Young, W.K.W., Joseph, S.J.: A radix 4 delay commutator for fast Fourier transform processor implementation. IEEE J. Solid-State Circ. **19**(5), 702–709 (1984)
7. Bouguezel, S., Ahmad, M.O., Swamy, M.N.S.: Improved radix-4 and radix-8 FFT algorithms. In: 2004 IEEE International Symposium on Circuits and Systems (IEEE Cat. No. 04CH37512), vol. 3, p. III-561. IEEE (2004)
8. Singleton, R.C.: An algorithm for computing the mixed radix fast Fourier transform. IEEE Trans. Audio Electroac. AU **17**(2), 93–103 (1969)
9. Bai, X.: Research on the development of high performance computing. Military Civilian Dual-use Technol. Products (02), 26–29 (2023). https://doi.org/10.19385/J.CNKI.1009-8119.2023.02.001
10. Denham, M., Lamperti, E., Areta, J.: Weather radar data processing on graphic cards. J. Supercomput. **74**(2), 868–885 (2018)
11. Cui, Z., Quan, H., Cao, Z., Xu, S., Ding, C., Wu, J.: SAR Target CFAR detection via GPU parallel operation. IEEE J. Selected Topics Appli. Earth Observ. Remote Sensing **11**(12), 4884–4894 (2018)
12. Tiebin, W., Feng, G., Di, W.: Research on the core operation architecture of high-performance processor computing for E-Class. Comput. Eng. Sci. **45**(05), 761–771 (2023)
13. AMD Instinct MI200 datasheet [EB/OL]. https://www.amd.com/content/dam/amd/en/documents/instinct-tech-docs/instinct-mi200-datasheet.pdf
14. NVIDIA Hopper architecture whitepaper (H100 tensor coreGPU architecture)[EB/OL]. https://resources.nvidia.com/en-us-tensor-core
15. Liao, K.: MPI transplantation based on FT-C6XX multi-core DSP realizes and optimization. National University of Defense Technology (2015)
16. Petersen, W., Arbenz, P.: Introduction to Parallel Computing: A practical guide with examples in C. OUP Oxford (2004)
17. Liu, S., et al.: A self-designed heterogeneous integration accelerator for E-class high performance computing. Computer Res. Develop. **58**(06), 1234–1237 (2021)
18. Hu, Y.:32-bit high-performance M-DSP DMA design and verification of high-efficiency data transmission. National University of Defense Technology (2015)
19. Guo, P., Chen, M., Liang, Z., Ma, X., Xu, B.: Optimization and implementation of the accumulation algorithm for the FT-M7002 platform. Computer Eng. Sci. **44**(11), 1909–1917 (2022)

EDFI: Endogenous Database Fault Injection with a Fine-Grained and Controllable Method

Haojia Huang, Pengfei Chen$^{(\boxtimes)}$, and GuangBa Yu

Sun Yat-sen University, Guangzhou, China
{huanghj78,yugb5}@mail2.sysu.edu.cn, chenpf7@mail.sysu.edu.cn

Abstract. Database reliability is an essential issue in many applications that rely on databases, especially in modern artificial intelligence (AI) applications. One common method to uncover the reliability weakness of databases is fault injection, which can introduce faults into the running database to observe and evaluate its reaction on reliability and performance after the occurrence of faults. Moreover, the fault injection can generate a large and diverse amount of realistic data for training and evaluating anomaly detection algorithms, which can also enhance database reliability. However, existing fault injection tools for testing database reliability are either coarse-grained or imprecise to mimic real-world faults, which limits their applicability and effectiveness. In this paper, we present EDFI, a fine-grained and controllable fault injection framework for endogenous database fault. EDFI can inject endogenous database faults for specific SQL statements from specific user connections, and support extensible fault types and scenarios. We demonstrate the effectiveness of EDFI by generating large and diverse training data to validate commonly used anomaly detection algorithms. The results show that EDFI can effectively simulate realistic endogenous database faults and provide valuable insights to improve anomaly detection algorithms.

Keywords: Database Reliability · Fault Injection · Endogenous Fault · Anomaly Detection

1 Introduction

As a core component of most systems especially AI systems, databases carry a lot of business data and sensitive data. As large language models (LLM) emerge, the storage of massive training and inference data poses a new challenge for databases. If the database behaves anomalously, it will have a serious impact on the functionality and performance of the database, and even lead to data loss, leakage, or tampering [2,6,8,15]. At present, the design and implementation of databases are becoming more and more complex. As a result, databases are facing a variety of internal and external faults, such as hardware failures, software defects, network failures and so on. Therefore, ensuring the reliability of databases is an important and challenging task.

C. Cruz et al. (Eds.): IC 2023, CCIS 2036, pp. 41–60, 2024.
https://doi.org/10.1007/978-981-97-0065-3_4

One common method to improve the reliability of databases is fault injection, which introduces faults to observe and evaluate the behavior and performance of the system after the occurrence of faults, thereby testing the reliability and robustness of the system. Moreover, fault injection can simulate a large and diverse amount of realistic fault scenarios to provide sufficient training data for anomaly detection algorithms and evaluate the effectiveness of these algorithms, which are another important method for improving the reliability of databases.

Currently, there are lots of fault injection tools, such as ChaosBlade [9], an open source experimental injection tool, which can inject faults for many application areas, such as Java applications, C++ applications, Docker containers, cloud native platforms, and etc. Besides, Amazon Aurora database has its own fault injection function specifically for databases [28].

However, after conducting a thorough analysis of most current fault injection tools, we identify three primary limitations of existing fault injection tools when applying them into database domain.

- **Uncontrollable blast radius.** Most of the fault injection tools cannot control the blast radius [49] of faults well. The blast radius indicates how widely fault injection may affect the system or the application. For example, Amazon Aurora database supports faults simulation using fault injection queries. A fault injection query will force a faulty occurrence of the Aurora MySQL DB instance, which will inevitably affect other normal queries.
- **Interference with programs from outside.** Most tools affect the program externally when injecting faults. For example, Chaosblade-exec-os [10] simulates a disk burn fault by creating a new process to execute a large number of read and write operations. However, breaking the program from the outside is to test the performance of the program when it is subjected to the external environment, and it cannot well simulate the real scenarios when the program fails internally.
- **Excessive source code intrusion.** Most tools change the behavior of program through code instrumentation. For example, WAFFLE [46] injects delay into threads to detect MemOrder bugs by instrumenting sleep operation into the program. On one hand, this method cannot achieve runtime fault injection, on the other hand, it requires recompiling and redeploying the target system, which increases testing time and cost.

In this paper, we present **EDFI**, a fine-grained and controllable fault injection framework for endogenous database fault. EDFI attaches to a database process with symbol tables by using GDB [14], and filters the execution flow of the target SQL statement by setting breakpoints on the SQL parser function. After that, EDFI can achieve fault injection by injecting specified logic through dynamic link libraries or by using GDB instructions to modify variables or to change the execution flow.

EDFI has the following advantages.

- EDFI can accurately control the blast radius of faults because it is capable of injecting faults into a specific SQL statement from specific a user connection without affecting other users or even other SQL statements from the same user.

- EDFI can more realistically simulate internal faults of the system because it can cause a specific process faulty.
- EDFI is extensible because it injects custom logic into a process by opening a dynamic link library in the process, which makes it easy to extend to support new faults.
- EDFI's fault injection methodology is white-box because it allows runtime fault injection at the source code level.

Contributions. To sum up, the paper makes two main contributions.

- We present EDFI, a fine-grained and controllable fault injection framework for endogenous database faults. EDFI is suitable for most databases implemented in C or C++ such as PostgreSQL [24], MySQL [20], openGauss [23], etc.
- We evaluate the effectiveness of EDFI by applying some common anomaly detection algorithms and comparing their effectiveness on the fault scenarios generated by EDFI. The results also provide some insights to improve anomaly detection algorithms.

2   Background and Related Work

Database Reliability. With the development and application of information technology, data has become an important resource and driving force for social and economic activities. The collection, storage, processing and analysis of data are of great significance to improve production efficiency, optimizing management decisions, and innovating business models. As such, the reliability of database is essential in ensuring the integrity, security and availability of data. There are numerous methods for improving database reliability, among which performance monitoring and fault injection are two significant approaches.

Database performance monitoring includes observability and anomaly detection, which tracks the database KPI(Key Performance Indicator) data and uses anomaly detection algorithms such as kNN [31,42], IForest [40], LOF [33] and so on, to help database administrators get deep insights into database health. Tools like SolarWinds ® Database Performance Analyzer [11] enables deep visibility into database performance and expert advice for performance optimization. However, even with a monitoring system in place, there remain two issues that must be addressed. The first is how to verify the effectiveness of the monitoring system, and the second is how to respond when an anomaly is detected. This is where fault injection comes into play. With fault injection, we can test the effectiveness of anomaly detection algorithms and the reliability of databases.

Fault Injection. Fault injection is a testing technique that actively introduces faults into the system to observe its operation under abnormal conditions. Recently, numerous excellent tools have been proposed to inject faults to achieve automatic testing for application, carry out chaos engineering experiments and generate anomaly data for training anomaly detection algorithms. Fault injection tools possess various capabilities in many aspects, such as emulating hardware faults like CPU, memory, and bus faults [37,45], injecting faults of system resource exhaustion [9,25], causing instances in distributed systems to crash [18,21], injecting errors or delays into service requests [30,34,47], and so on.

Gradually, with the introduction of Chaos Engineering [32], fault injection has become a component of chaos engineering experimental tools. Chaos Engineering is the discipline of experimenting on a system in order to build confidence in the system's capability to withstand turbulent conditions in production [4]. For example, ChaosMonkey [21] randomly terminates virtual machine instances and containers that run inside of the production environment. There are also studies on chaos engineering for databases. For instance, databases rely on the main memory to perform calculations and present accurate results while it is not well understood how a database system can handle bit-flips in memory. Hence, it is valuable to apply chaos engineering to database. One such study [48] utilized ptrace(2) [26] which can access and modify the internal state of the database process [50] to achieve fault injection.

ptrace and GDB. ptrace is the Linux kernel's interface which gives users access to read and write another process states, such as memory and registers. Several well-known tools such as GDB [14] and strace [17] are implemented based on ptrace.

ptrace is a critical technology for implementing runtime tracking and control of processes. At present, there are some works based on ptrace to inject logic into processes [1]. However, it remains challenging for developers to achieve runtime logic injection into a process and accurately select injection points, as ptrace only provides a set of read and write interfaces for the process.

GDB is a program debugger based on ptrace. It is encapsulated on the basis of ptrace interface and provides some more user-friendly APIs. It allows users to easily set breakpoints, modify variables and perform other operations with process symbol table. As such, GDB reduces the difficulty of injecting logic into the process at run time and enables accurate selection of fault injection points. Therefore, the method proposed in this paper is an accurate fault injection method for processes during runtime based on GDB, which achieves automatic interaction with GDB by means of GDB Python API [27] and Tcl scripts [16].

3   Overview of EDFI

In this section, we provide an overview of EDFI based on its two characteristics: **endogenous fault** and **fine-grained and controllable method**.

Endogenous Fault. An endogenous fault refers to a fault caused by internal factors in the system, as opposed to an exogenous fault, which refers to a fault caused by external factors. Taking database as an example, endogenous faults include decreased database process performance and erroneous queries, while exogenous faults encompass hardware failures and network outages.

We implement EDFI by attaching to the database process via GDB to change its execution logic. In other words, we can cause the database processes to encounter issues such as resource hog, slow SQL and deadlocks, which are all endogenous faults. The implementation of faults can be divided into two strategies: one is to affect the original execution flow using GDB, and the other is to insert a new execution flow through dynamic link libraries.

For the first strategy, we use GDB to block the process or modify related variables to cause slow SQL or deadlocks. For the second strategy, we use the `dlopen` [12] or `__libc_dlopen_mode` [29] functions to open dynamic link libraries to simulate resource hog. The former is a standard library function exported by libdl.so, while the latter is an underlying libc function exported by libc.so. We use both to ensure a higher success rate, as some platforms may not load libdl.so.

Fine-Grained and Controllable Method. We achieve EDFI with a fine-grained and controllable method since we use GDB to filter the specific SQL statement from the specific user connection. As shown in Fig. 1, we set a breakpoint on the function related to SQL statement processing within the process to capture SQL statements. This enables us to determine whether to inject faults based on the current SQL statement, providing us with a high degree of control over the fault injection process. The precise setting of breakpoints depends on the process symbol table, which allows us to identify the exact functions and locations in the code where SQL statements are processed. By leveraging this information, we can place breakpoints strategically at the key points within the code where SQL statements are parsed and executed. These breakpoints act as trigger points for our fault injection mechanism.

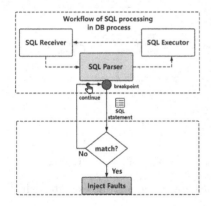

Fig. 1. Workflow of filtering SQL.

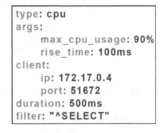

Fig. 2. An example configuration file for injecting CPU utilization anomaly.

4   EDFI's Design and Implementation

EDFI is a fine-grained and controllable fault injection framework for endogenous database fault. Its design and implementation bring many advantages, as we have mentioned in §1. The most prominent advantage is its ability to simulate the internal faults of the system more realistically, that is, it injects endogenous faults into the database process.

Most fault injection tools simulate external faults which can simulate faults caused by the system environments. When using a common fault injection tool to inject CPU utilization anomaly into a multi-process database like PostgreSQL [24], so as to test the anomaly detection algorithms working for it, the period of abnormal CPU utilization is likely to be detected. However, if the algorithms only collect overall CPU utilization, they cannot locate the faulty process. Even if the algorithms collect the CPU utilization of each process, the located process may have no practical significance. In contrast, EDFI can make the system itself faulty, allowing for more realistic simulation of the system's status to test fault tolerance and anomaly detection mechanisms.

To use EDFI, it is necessary to provide a configuration file in yaml [22] format, which is a fault injection plan with necessary information. Figure 2 shows an example configuration file for injecting CPU utilization anomaly.

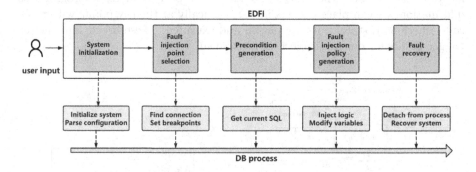

Fig. 3. Workflow of EDFI.

As shown in Fig. 3, EDFI's workflow consists of the following five parts:

- System initialization (Sect. 4.1). EDFI starts by launching the executor which undertakes most of the work of EDFI.
- Fault injection point selection (Sect. 4.2). EDFI then finds the specific worker process or thread and sets a breakpoint on its critical path.
- Fault injection precondition generation(Sect. 4.3). After obtaining the SQL statement matching policy from the configuration file, EDFI generates preconditions for fault injection based on it, and performs fault injection when the preconditions are met.

– Fault injection policy generation (Sect. 4.4). In this part, EDFI will convert the information such as the fault type and related parameters from the configuration file into the corresponding fault policy and inject it into the database process.
– Fault recovery (Sect. 4.5). After completing the fault injection, EDFI will restore the normal operation of the system.

4.1 System Initialization

The architecture of EDFI is shown in Fig. 4 and consists of an executor implemented using the GDB Python API, an entry script, and GDB. The executor is responsible for executing main logic of EDFI. The GDB Python API provides a simple way to extend the functionality of GDB, making it programmable and meeting more of our needs. For example, by inheriting the basic breakpoint class, we can inject custom logic when the program reaches the breakpoint.

As we all know, GDB is an interactive command-line tool. Therefore, a crucial aspect of implementing EDFI is to enable automated interaction with GDB and fully utilize GDB's capabilities. To achieve this, the first step during the initialization phase is to launch the executor through the entry script and establish its connection with GDB. As shown in Fig. 4, ① EDFI launches GDB via an entry script to prepare for controlling the database process. The entry script interacts with GDB by means of Expect tool [3] which is implemented based on Tcl. ② After launching GDB, the entry script sends a request to GDB to execute command **source file_name**. Then, GDB loads the executor script and executes it. ③ The executor calls GDB through the GDB Python API. For example, it uses the **parse_and_eval** function to evaluate the values of arguments. ④ Ultimately, GDB, under the control of the executor, is able to control the database process.

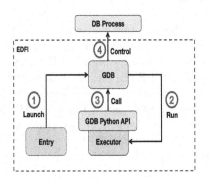

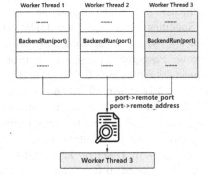

Fig. 4. The process of initializing the system.

Fig. 5. Workflow of searching target worker thread.

After the executor is launched, it will parse a configuration file in yaml format. The configuration file contains the necessary information for the fault injection such as the target connection information, fault type and SQL statement

matching policy. The connection information includes the host address and port
number of the target client connection. The fault type indicates the fault to be
injected such as resource hog(CPU, disk, memory), slow SQL and deadlocks. The
SQL statement matching policy specifies a series of SQL statements with regular
matching form that will trigger faults. For example, `^SELECT` specifies a series of
SELECT statements for subsequent generation of fault injection preconditions.

4.2 Fault Injection Point Selection

Database systems typically handle a large number of SQL requests from different
users, resulting in a vast search space for fault injection points. Fortunately, most
databases allocate a worker process or thread, depending on the connection
model of the database, to each client to handle requests, and the processing flow
of SQL requests is fixed. Therefore, we select fault injection points in two steps.

Select Worker Process or Thread. Since the database server allocates a
worker process or thread to handle requests on that connection, EDFI can identify the corresponding process or thread identifier according to the host address
and port number. For databases with a multi-process connection model like
PostgreSQL, it is easy to find the target process by using the `netstat` command
[5]. For databases with a multi-thread connection model like openGauss, one
possible approach is to attach to the database process with GDB firstly. Then,
traverse each connection thread. As shown in Fig. 5, for each of them, switch to
the frame of the function with connection parameters, print the function parameters to obtain the host address and port number of the connection, and compare
with the configuration to find the target thread.

Select SQL Statement. After the target process or thread is found, the next
step is to set breakpoints on functions related to SQL statement processing
within the process to capture SQL statements.

Specifically, we first use GDB to analyze the function call chain of SQL
requests, so as to analyze the parsing functions that most normal sql call.
Taking PostgreSQL as an example, we find that all simple SQL query
requests, such as SELECT, INSERT, UPDATE, DELETE, etc. are handled by
`exec_simple_query` function, so this is a breakpoint that can be selected.

Then, EDFI will set a custom breakpoint on the SQL parser function. After
that, every time the target process parses the SQL statement, it will reach the
breakpoint and then deliver the current SQL statements to EDFI.

As show in **Listing 1.1**, EDFI captures and matches SQL statements by
overriding the `stop` function. If the `stop` function returns True, the process will
be stopped at the location of the breakpoint, otherwise the process will continue.

Listing 1.1. Critical code for implementing breakpoints.

```
import gdb
class SQLFilterBreakpoint(gdb.Breakpoint):
    def __init__(self):
```

```
super(SQLFilterBreakpoint, self).__init__(SQL_PARSER_NAME,
    gdb.BP_BREAKPOINT)

def stop(self):
    # Capture the current SQL statement
    sql = gdb.parse_and_eval(SQL_PARSER_PARAM).string()
    if match(sql):
        return inject_fault()
    return False
```

4.3   Fault Injection Precondition Generation

After obtaining the SQL statement matching policy from the configuration file, EDFI generates preconditions for fault injection based on it. The matching policy specifies how to select the SQL statements that should trigger the fault injection by regular expression match.

As mentioned above, EDFI captures the current executed SQL statement when the process reaches the breakpoint. Therefore, it will continuously determine whether the currently executed SQL statement meets the precondition and fault injection will occur when the precondition is met. For example, when a user executes a SELECT statement, the SQL parser reaches the breakpoint when parsing this statement. At this point, EDFI will obtain the SELECT statement with the help of GDB and inject fault if the matching policy is a regular expression such as "^SELECT".

EDFI also supports injecting faults without precondition if the **filter** field in the configuration file is empty. In this case, EDFI will immediately inject faults upon being launched.

Table 1. List of fault types supported by EDFI.

Fault Type	Description
Resource hog	Cause abnormal utilization of CPU, disk or virtual memory
Slow SQL	Inject delay into a specific SQL
Deadlock	Cause a specified database object to deadlock

4.4   Fault Injection Policy Generation

EDFI will convert the fault type and related parameters from the configuration file into the corresponding fault policy and inject it into the database. As shown in Table 1, it currently supports injection of the following types of faults.

Resource Hog (CPU, Disk, Memory). This type of fault is achieved by injecting a custom logic that can change the utilization of the resource. Figure 6 shows the process of injecting a resource hog fault. When the target process reaches the breakpoint, EDFI causes the process to execute `dlopen` function to load a dynamic link library. The logic in the library is customized, so by defining some computationally or IO intensive logic, the target process can experience resource exhaustion faults. This also indicates the extensibility of EDFI.

Taking CPU utilization as an example, users can set the expected abnormal CPU utilization, the rise time of CPU utilization, and the duration of the entire CPU utilization anomaly in the configuration file like Fig. 2. EDFI will fill in these parameters to the template file prepared in advance. Then EDFI compiles the file into a dynamic link library and invokes it when triggered by the specific SQL statement.

In the template file, we use `__attribute__((constructor))` to declare the function that contains the logic to be executed, so that it will be executed when the dynamic link library is opened.

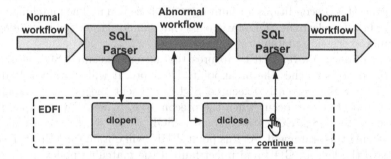

Fig. 6. Workflow of injecting resource hog fault.

Slow SQL. Due to the fact that a worker process or thread will be blocked and be controlled by EDFI when a breakpoint is reached, EDFI can resume the execution of the specific SQL statement after a certain period of blocking so as to simulate slow SQL as shown in Fig. 7. This means that EDFI can delay the response time of the target database process by pausing the execution of the SQL statement at the breakpoint and waiting for a predefined duration before resuming it.

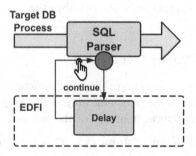

Fig. 7. Workflow of injecting delay fault.

As shown in **Listing 1.1**, when the process reaches the breakpoint, it will be blocked. At the same time, the `stop` function of the breakpoint object will be called. Since the `stop` function will invoke `inject` function if the SQL statement meets the precondition, we can call a `sleep` function in the `inject` function to inject delay fault.

Deadlock. With GDB, EDFI can modify the runtime logic anywhere in the source code, making it easy to dynamically modify the process's runtime logic. For example, in a PostgreSQL database, the worker process will call a specific function to acquire a corresponding lock before accessing a database object. Figure 8 shows the process of injecting a deadlock fault. EDFI will set a breakpoint on that function to change the lock type to **exclusive lock** before the target process access the specific object. As a result, the object is locked with an **exclusive lock**. Then EDFI delays the execution of releasing the lock for a while. In this way, other process accessing the specific object will become waiting status until the lock is released.

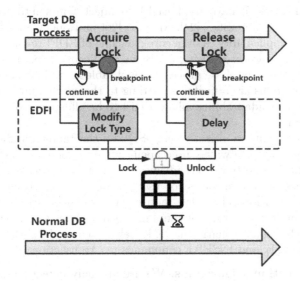

Fig. 8. Workflow of injecting deadlock fault.

4.5   Fault Recovery

After fault injection, EDFI will ensure the fault is completed and removed within the specified time. For resource hog, EDFI will call corresponding functions to close the dynamic link library. The function that closes a dynamic link library opened by `dlopen` is `dlclose`, while the function that closes a dynamic link library opened by `__libc_dlopen_mode` is `__libc_dlclose`. After that, EDFI uses the GDB command `detach` to relinquish the control over the database process and detach itself from the process so that the process can run normally.

5   Evaluation

In this section, we first present the experimental setup for EDFI. Then we conduct experiments to answer

- **Q1** Can EDFI indeed inject the expected faults?
- **Q2** Can the injected fault be detected by the anomaly detection algorithms?
- **Q3** Does different degrees of faults have an impact on anomaly detection algorithms?

5.1   Experimental Setup

Target Database. We have used EDFI to inject faults into some popular databases such as PostgreSQL [24], MySQL [20] and openGauss [23], demonstrating its high applicability. As the core principle of EDFI revolves around the control of database processes using GDB, its applicability extends to different databases as long as a database process symbol table is available, and the breakpoint setting positions are adjusted according to the specific implementation of each database. This adaptability allows EDFI to be effective across a variety of database systems.

In this experiment, we select PostgreSQL as the target database for fault injection. PostgreSQL is a widely used open-source relational database management system that supports various advanced features. For example, PostgreSQL provides a series of system views for users to observe performance statistics, which makes it easier for us to collect metrics. As for benchmark, we use pgbench [13] which is a built-in program for running benchmark tests on PostgreSQL. By default, pgbench tests a scenario that is loosely based on TPC-B, involving five SELECT, UPDATE, and INSERT commands per transaction.

Anomaly Detection Algorithms. We use anomaly detection algorithms to verify the effectiveness of fault data generated by EDFI. Specifically, we use the unified interface provided by pyod [51]. We pick out several representative detection algorithms from pyod toolkits and use them with default parameters. The algorithms we use are shown in Table 2. Most of these algorithms have been proven to perform well on many benchmarks [35].

Table 2. The algorithms used in our experimentation.

Ref	Abbr	Full name	Type	Year
[39]	ECOD	Empirical Cumulative Distribution	Probabilistic	2022
[36,43]	MCD	Minimum Covariance Determinant	Linear Model	1999
[33]	LOF	Local Outlier Factor	Proximity-Based	2000
[31,42]	kNN	k Nearest Neighbors	Proximity-Based	2000
[40]	IForest	Isolation Forest	Outlier Ensembles	2008
[41]	LODA	Lightweight On-line Detector	Outlier Ensembles	2016
[38]	VAE	Variational AutoEncoder	Neural Networks	2013
[44]	DeepSVDD	Deep One-Class Classification	Neural Networks	2018

Evaluation Metrics. We use Precision, Recall, F1 Score, and Area Under the Curve(AUC) to compare the performance of each algorithm. AUC provides a summary of the overall performance of the classifiers. The closer the AUC value is to 1, the better the algorithm's performance is. AUC values above 0.5 are preferable since a value equal to 0.5 indicates that the classifier's performance is random guessing and lower than 0.5 means that the classifier's classification scheme should be inverted.

5.2 Fault Injection Effectiveness of EDFI

In this part, we use EDFI to inject faults into PostgreSQL and collect relevant metrics to check its effectiveness.

Resource Hog. For our experiments, we inject faults related to abnormal CPU, disk, and virtual memory utilization to assess the effects of resource hog faults. We collect corresponding metrics, including the CPU utilization, disk utilization and virtual memory utilization of the target process.

Taking CPU utilization as an example, we can easily observe the effect of the fault by using the shell command `top`. With `top` command, we can clearly see that the CPU utilization of the database process is gradually increasing after injecting the fault.

Further, we can collect the corresponding utilization of the target process throughout the entire process of running pgbench to observe the trend before and after fault injection. As shown in Fig. 9, the x-axis represents the timestamps and the y-axis represents the metrics normalized to the range 0–1. The grey background indicates the execution period of the fault.

– Fig. 9(a) shows the trend of CPU utilization before and after the injection of CPU utilization anomaly. We inject intensive computations to increase the CPU utilization of the process. We can see that the utilization will drop to close to 0 for a short period of time before rising, which is about 2 s (as can also be seen in Fig. 9(b)). This is due to EDFI loading symbols after attaching to the target process using GDB. This phenomenon can be ignored in the case of coarse-grained or long-duration metric collection. Alternatively, we can eliminate this time period by loading the necessary symbol table before attaching the process.

- Fig. 9(b) shows the trend of disk utilization before and after the injection of disk utilization anomaly. We inject logic that performs frequent file read and write operations to increase the disk utilization of the process. We can see that the disk utilization remains high for about 20 s after the execution period of the fault, which is caused by the operating system's inability to finish the read and write operations that we injected in a timely manner. This phenomenon is related to the size of the content that is read from or written to the file in the injected logic and the specifications of the machine.
- Fig. 9(c) shows the trend of virtual memory utilization before and after the injection of memory utilization anomaly. We inject logic that calls `malloc` [7] function so as to increase the virtual memory utilization of the process.

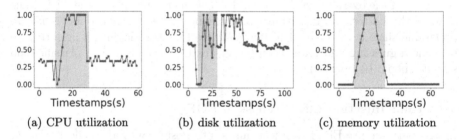

(a) CPU utilization   (b) disk utilization   (c) memory utilization

Fig. 9. The trend of change for each metric

Slow SQL. We use pgbench to generate transactions including SELECT, UPDATE, and INSERT commands. After running, pgbench will report related statistics such as the number of transactions actually processed, tps and so on. We can check the effectiveness by comparing these statistics before and after fault injection.

Specifically, we use pgbench to run transactions twice, setting the duration of each run to five minutes with the option -T. In the first run, we let it run normally, while in the second run, we inject an 100-ms delay fault into the INSERT statements with EDFI. Table 3 shows the statistical results of two runs. By comparison, we can see that the tps of the second run is significantly lower than that of the first run, and the latency of INSERT statements is around 100 ms. Besides, in the abnormal scenario, both SELECT and UPDATE statements have about 0.3 ms more execution delay than in the normal scenario, even without injected delay. This extra time is the overhead of capturing and matching SQL statements. However, this extra time only occurs on the worker process where the fault is injected, which reflects the ability of EDFI to accurately control the range of fault impact.

Table 3. The average latency and tps of two scenarios: normal execution and execution with injected INSERT statements delay.

	SELECT(ms)	UPDATE(ms)	INSERT(ms)	TPS
Normal scenario	0.187	0.223	0.158	334.367
Abnormal scenario	0.553	0.563	101.066	10.177

Deadlock. We run pgbench to perform UPDATE operations on table pgbench_accounts. Then we inject a deadlock fault into another worker process which is also operating on the table pgbench_accounts. As a result, the process running pgbench is blocked due to the deadlock. At this time, if we query the table pg_locks, we will find that table pgbench_accounts has been acquired with an exclusive lock.

Figure 10 shows the CPU utilization of the pgbench process when injecting a deadlock fault into another worker process to block the pgbench process for a while. The grey background indicates the execution period of the fault. Four different lines represent the four scenarios where deadlocks last for four different times. Figure 10 indicates that the CPU utilization of the process significantly decreases when a deadlock occurs, and the longer the duration of the deadlock, the lower the CPU utilization.

5.3 Further Validate the Effectiveness of EDFI Through Anomaly Detection Algorithms

In this section, we supply the fault data generated by EDFI to the anomaly detection algorithm and subsequently assess the efficacy of the faults injected by EDFI through the analysis of the algorithm's outcomes.

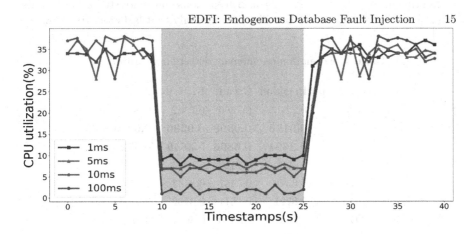

Fig. 10. The CPU utilization of the pgbench process when injecting a deadlock fault.

Firstly, we use pgbench to run 18,000 transactions and inject disk utilization anomaly lasting for 20 s every minute during the run. Then, we collect the CPU and disk utilization of the pgbench worker process throughout the run and use these metrics as training datasets for various anomaly detection algorithms. Finally, we calculate the metrics presented in Table 4 to evaluate their performance.

From Table 4, we can see that MCD algorithm performs the best, while LOF algorithm performs the worst. MCD algorithm estimates data mean and covariance using a subset of normal data, then identifies outliers using Mahalanobis distance [19], which reflects the relationship between data points and the overall dataset. In contrast, LOF algorithm identifies outliers by calculating local density deviation, reflecting the relationship between data points and their surroundings. Hence, the structure of experimental data, consisting of alternating segments of normal and abnormal data, determines MCD algorithm's superior performance.

The majority of algorithms achieve F1-scores and AUC greater than 0.7, indicating that the faults we inject can be detected well by anomaly detection algorithms. In addition, different anomaly detection algorithms may perform significantly differently on the same set of data. This indicates that different anomaly detection algorithms may have different sensitivities to the same fault. Hence, when choosing an anomaly detection algorithm, it is necessary to consider the type of fault that the algorithm is suitable for.

5.4   Sensitivity Analysis of Anomaly Detection Algorithms

In this part, we attempt to use EDFI to analyze the sensitivity of anomaly detection algorithms and provide some insights. Since EDFI can inject any specified values of resource hog into processes, we can use it to compare the performance of anomaly detection algorithms under different degrees of faults. We conduct our experiment as follows. First, we run transactions continuously for one minute using pgbench, and inject a CPU utilization anomaly that lasts for 20 s. Then we

Table 4. Evaluation on anomaly detection algorithms

Algorithm	Precision	Recall	F1 Score	AUC
ECOD	0.7216	0.7390	0.7302	0.8645
MCD	**0.9176**	**0.9398**	**0.9286**	**0.9643**
LOF	**0.3804**	**0.3896**	**0.3849**	**0.5720**
kNN	0.6980	0.7149	0.7063	0.6997
IForest	0.8578	0.7751	0.8143	0.9462
LODA	0.8721	0.7671	0.8162	0.9496
VAE	0.8651	0.7470	0.8017	0.9266
DeepSVDD	0.8356	0.7550	0.7932	0.9223

capture a series of CPU utilization of the process by changing two variables: the maximum abnormal CPU utilization and the duration of CPU utilization reaches the maximum value. Finally, we choose the MCD algorithm which performs best in the previous experiments to conduct this experiment.

We calculate the AUC values of the MCD algorithm's results across different degrees of faults. As shown in Fig. 11, the x-axis represents the maximum CPU utilization and the y-axis represents the AUC value of the MCD algorithm. The four lines represent scenarios with different times required to reach maximum CPU utilization.

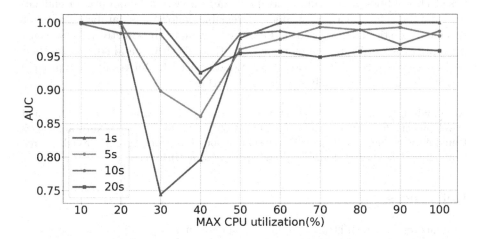

Fig. 11. The performance of MCD algorithm under different degrees of faults.

Figure 11 shows that anomaly detection performance varies with different fault degrees. AUC drops significantly when maximum CPU utilization is between 30–40%, as normal CPU utilization is around 30%, making anomalous data less distinguishable. This effect decreases as the time to reach maximum CPU utilization increases, as data is less concentrated at maximum CPU utilization, resulting in more detectable anomalous data. Therefore, in normal scenarios where there is high CPU utilization, we can inject low CPU utilization fault with the help of EDFI's ability to injectt arbitrary CPU utilization values, so as to simulate some application logic errors or disk IO timeout faults. If high CPU utilization faults are inevitably injected, in order to improve the performance of anomaly detection algorithms, it is necessary to add more features, such as memory usage, IO utilization, etc.

6   Limitations

Despite its many advantages, EDFI has some limitations. For one thing, EDFI relies on the symbol table of the target process to set breakpoints for filtering

SQL statements and selecting fault injection points. For another, EDFI introduces additional overhead to the target process when filtering SQL statements, but this overhead only affects the target client connection being injected.

7 Conclusion

Reliability research in databases necessitates the use of controllable fault injection tools for endogenous faults and flexible fault injection tools to generate rich training data for anomaly detection algorithms. In response to this need, we propose EDFI, a fine-grained and controllable fault injection framework for endogenous database fault. EDFI can inject extensible endogenous database faults for specific SQL statements based on GDB. We evaluate the effectiveness of EDFI by applying several common anomaly detection algorithms and comparing their performance on the fault scenarios generated by EDFI. Our results also provide insights to improve anomaly detection algorithms.

Acknowledgements. The research is supported by the National Key Research and Development Program of China (2019YFB1804002), the National Natural Science Foundation of China (No.62272495), the Guangdong Basic and Applied Basic Research Foundation (No.2023B1515020054), and sponsored by Huawei.

References

1. linux-inject (2016). https://github.com/gaffe23/linux-inject
2. Postmortem of database outage of january 31 (2017). https://about.gitlab.com/blog/2017/02/10/postmortem-of-database-outage-of-january-31/
3. Expect - tcl/tk (2018). https://core.tcl-lang.org/expect/index
4. Principles of chaos engineering (2019). https://principlesofchaos.org/
5. View your linux server's network connections with netstat (2020). https://opensource.com/article/22/2/linux-network-security-netstat
6. Github availability report: November 2021 (2021). https://github.blog/2021-12-01-github-availability-report-november-2021/
7. malloc - cppreference.com - c++ reference (2022). https://en.cppreference.com/w/c/memory/malloc
8. 2023 faa system outage (2023). https://en.wikipedia.org/wiki/2023_FAA_system_outage
9. Chaosblade: An easy to use and powerful chaos engineering toolkit (2023). https://github.com/chaosblade-io/chaosblade
10. Chaosblade-exec-os: Basic resources chaos experiment executor (2023). https://github.com/chaosblade-io/chaosblade-exec-os
11. Database performance tuning to improve dbms performance (2023). https://www.solarwinds.com/database-performance-analyzer/use-cases/database-performance-tuning
12. dlopen(3) - linux manual page (2023). https://man7.org/linux/man-pages/man3/dlopen.3.html
13. Documentation: 15: pgbench - postgresql (2023). https://www.postgresql.org/docs/current/pgbench.html

14. Gdb: The gnu project debugger (2023). https://www.sourceware.org/gdb/
15. Issues with heroku api, dashboard, and cli (2023). https://status.heroku.com/incidents/2558
16. Language - tcl/tk (2023). https://www.tcl.tk/about/language.html
17. linux syscall tracer (2023). https://strace.io/
18. litmuschaos/litmus (2023). https://github.com/litmuschaos/litmus
19. Mahalanobis distance (2023). https://en.wikipedia.org/wiki/Mahalanobis_distance
20. Mysql (2023). https://www.mysql.com/
21. Netflix/chaosmonkey (2023). https://github.com/Netflix/chaosmonkey
22. The official yaml web site (2023). https://yaml.org/
23. opengauss (2023). https://opengauss.org/en/
24. Postgresql: The world's most advanced open source relational database (2023). https://www.postgresql.org/
25. A powerful chaos engineering platform for kubernetes (2023). https://chaos-mesh.org/
26. ptrace(2) - linux manual page (2023). https://man7.org/linux/man-pages/man2/ptrace.2.html
27. Python api (debugging with gdb) - sourceware.org (2023). https://sourceware.org/gdb/onlinedocs/gdb/Python-API.html
28. Relational database - amazon aurora mysql postgresql (2023). https://aws.amazon.com/rds/aurora/
29. Source code of dlfcn (2023). https://codebrowser.dev/glibc/glibc/include/dlfcn.h.html
30. Alvaro, P., Andrus, K., Sanden, C., Rosenthal, C., Basiri, A., Hochstein, L.: Automating failure testing research at internet scale. In: Proceedings of the Seventh ACM Symposium on Cloud Computing, pp. 17–28 (2016)
31. Angiulli, F., Pizzuti, C.: Fast outlier detection in high dimensional spaces. In: Elomaa, T., Mannila, H., Toivonen, H. (eds.) PKDD 2002. LNCS, vol. 2431, pp. 15–27. Springer, Heidelberg (2002). https://doi.org/10.1007/3-540-45681-3_2
32. Basiri, A., et al.: Chaos engineering. IEEE Softw. **33**(3), 35–41 (2016). https://doi.org/10.1109/MS.2016.60
33. Breunig, M.M., Kriegel, H.P., Ng, R.T., Sander, J.: Lof: identifying density-based local outliers. In: Proceedings of the 2000 ACM SIGMOD International Conference on Management of Data, SIGMOD 2000, pp. 93–104. Association for Computing Machinery, New York (2000). https://doi.org/10.1145/342009.335388
34. Chen, Y., Sun, X., Nath, S., Yang, Z., Xu, T.: Push-Button reliability testing for Cloud-Backed applications with rainmaker. In: 20th USENIX Symposium on Networked Systems Design and Implementation (NSDI 2023), pp. 1701–1716. USENIX Association, Boston, MA (Apr 2023). https://www.usenix.org/conference/nsdi23/presentation/chen-yinfang
35. Han, S., Hu, X., Huang, H., Jiang, M., Zhao, Y.: Adbench: anomaly detection benchmark. In: Neural Information Processing Systems (NeurIPS)
36. Hardin, J., Rocke, D.M.: Outlier detection in the multiple cluster setting using the minimum covariance determinant estimator. Comput. Stat. & Data Anal. **44**(4), 625–638 (2004). https://doi.org/10.1016/S0167-9473(02)00280-3, https://www.sciencedirect.com/science/article/pii/S0167947302002803
37. Kanawati, G.A., Kanawati, N.A., Abraham, J.A.: Ferrari: a tool for the validation of system dependability properties. In: FTCS, pp. 336–344 (1992)
38. Kingma, D.P., Welling, M.: Auto-encoding variational bayes. arXiv preprint arXiv:1312.6114 (2013)

39. Li, Z., Zhao, Y., Hu, X., Botta, N., Ionescu, C., Chen, G.: Ecod: unsupervised outlier detection using empirical cumulative distribution functions. IEEE Trans. Knowl. Data Eng. (2022). https://doi.org/10.1109/TKDE.2022.3159580

40. Liu, F.T., Ting, K.M., Zhou, Z.H.: Isolation-based anomaly detection. ACM Trans. Knowl. Discov. Data **6**(1) (2012). https://doi.org/10.1145/2133360.2133363

41. Pevný, T.: Loda: Lightweight on-line detector of anomalies. Mach. Learn. **102**(2), 275–304 (2016). https://doi.org/10.1007/s10994-015-5521-0

42. Ramaswamy, S., Rastogi, R., Shim, K.: Efficient algorithms for mining outliers from large data sets. In: Proceedings of the 2000 ACM SIGMOD International Conference on Management of Data, SIGMOD 2000, pp. 427–438. Association for Computing Machinery, New York (2000). https://doi.org/10.1145/342009.335437

43. Rousseeuw, P.J., Driessen, K.V.: A fast algorithm for the minimum covariance determinant estimator. Technometrics **41**(3), 212–223 (1999). https://doi.org/10.1080/00401706.1999.10485670, https://www.tandfonline.com/doi/abs/10.1080/00401706.1999.10485670

44. Ruff, L., et al.: Deep one-class classification. In: International Conference on Machine Learning, pp. 4393–4402. PMLR (2018)

45. Schlichting, R.D., Schneider, F.B.: Fail-stop processors: an approach to designing fault-tolerant computing systems. ACM Trans. Comput. Syst. (TOCS) **1**(3), 222–238 (1983)

46. Stoica, B.A., Lu, S., Musuvathi, M., Nath, S.: Waffle: exposing memory ordering bugs efficiently with active delay injection. In: Proceedings of the Eighteenth European Conference on Computer Systems, pp. 111–126 (2023)

47. Sun, X., et al.: Automatic reliability testing for cluster management controllers. In: 16th USENIX Symposium on Operating Systems Design and Implementation (OSDI 2022), pp. 143–159. USENIX Association, Carlsbad, CA (Jul 2022). https://www.usenix.org/conference/osdi22/presentation/sun

48. Tran, T.L.: Chaos Engineering for Databases. Ph.D. thesis, Universiteit van Amsterdam (2020)

49. Tucker, H., Hochstein, L., Jones, N., Basiri, A., Rosenthal, C.: The business case for chaos engineering. IEEE Cloud Comput. **5**(3), 45–54 (2018). https://doi.org/10.1109/MCC.2018.032591616

50. Volkmar, S.: Fault-injector using unix ptrace interface. Tech. rep., Citeseer

51. Zhao, Y., Nasrullah, Z., Li, Z.: Pyod: a python toolbox for scalable outlier detection. J. Mach. Learn. Res. **20**(96), 1–7 (2019). http://jmlr.org/papers/v20/19-011.html

AI for Ocean Science and Engineering

Diffusion Probabilistic Models for Underwater Image Super-Resolution

Kai Wang and Guoqiang Zhong[✉]

College of Computer Science and Technology, Ocean University of China,
Qingdao, China
gqzhong@ouc.edu.cn

Abstract. In recent years, single image super-resolution (SISR) has been extensively employed in the realm of underwater machine vision. However, the unique challenges posed by the underwater environment, including various types of noise, blurring effects, and insufficient illumination, have rendered the recovery of detailed information from underwater images a complex task for most existing methodologies. In this paper, we address and propose solutions to these challenges inherent in the application of super-resolution techniques in underwater machine vision. We introduce a novel underwater SISR diffusion probability model, termed as DiffUSR. This marks the first instance of utilizing a diffusion probability model in the domain of underwater SISR. Our innovative model enhances the data likelihood by employing a unique variant of variational constraints. Notably, DiffUSR is capable of providing diverse and realistic super-resolution (SR) predictions by progressively transforming Gaussian noise into SR images based on low-resolution (LR) inputs via a Markov Chain process. This approach represents a significant advancement in the field of underwater image super-resolution.

Keywords: Underwater image super-resolution · Diffusion probabilistic models · Underwater image dataset · Deep learning

1 Introduction

Autonomous Underwater Vehicles (AUVs), guided by visual navigation, play a pivotal role in a broad spectrum of critical applications. These applications range from monitoring marine species and inspecting subsea cables and debris, to exploring the intricate topography of the seafloor. The significance of these tasks is profound, as they contribute substantially to our understanding of marine ecosystems, the upkeep of underwater infrastructure, and the unearthing of new geological features.

This work was partially supported by the National Key Research and Development Program of China under Grant No. 2018AAA0100400, the Natural Science Foundation of Shandong Province under Grants No. ZR2020MF131 and No. ZR2021ZD19, and the Science and Technology Program of Qingdao under Grant No. 21-1-4-ny-19-nsh.

Both AUVs and their counterparts, Remotely Operated Vehicles (ROVs), are extensively employed in these applications. They utilize synthesized images to model visual attention, a process that informs navigational decisions. This process is integral to the successful operation of these vehicles, enabling them to navigate the complex underwater environments with a high degree of precision and efficiency.

However, despite the deployment of advanced cameras and sophisticated imaging technologies, the quality of underwater imagery is often severely compromised. The unique conditions of the underwater environment, characterized by poor visibility, absorption, and scattering, pose significant challenges to the effective execution of these tasks. These factors can distort the captured images, making it difficult for the AUVs and ROVs to accurately interpret their surroundings. Poor visibility underwater is primarily caused by the presence of particulate matter, such as sediment and plankton, which can obscure the camera's view. Absorption refers to the loss of light energy as it travels through water, which can significantly reduce the clarity and range of underwater images. Scattering, the redirection of light by the water and particulate matter, can further degrade image quality by creating a 'haze' effect.

The prevailing methods for underwater Single Image Super-Resolution (SISR) predominantly rely on Convolutional Neural Networks (CNNs) [1,2] and Generative Adversarial Networks (GANs) [3,4]. However, the application of these models in the context of underwater imagery presents a unique set of challenges. Underwater images are subjected to a series of unique distortions. For instance, scattering induces irregular non-linear distortions, leading to images characterized by low contrast and blurriness [5]. As a result, standard SISR models, trained on a diverse range of images, often fail to generate realistic high-resolution underwater images.

Another significant challenge in this field is the scarcity of large-scale underwater datasets. The lack of such datasets hampers the ability to conduct extensive research, as it limits the opportunities to train and evaluate the performance of SISR models on underwater images. Without sufficient data, it is difficult to fine-tune these models to handle the unique challenges posed by underwater imaging.

In light of these challenges, our work aims to make significant contributions in several key areas. Our work makes contributions in the following aspects:

(a) We propose a novel SISR diffusion probability model (DiffUSR) for underwater images, designed to operate at scaling factors of $2\times$, $4\times$, and $8\times$. This model effectively mitigates issues such as excessive smoothing and information loss in underwater SISR.

(b) We present RedSea-2K, a comprehensive dataset specifically curated for the training of underwater image super-resolution models. This dataset provides 960 paired images for training and an additional 100 paris for testing.

(c) Additionally, we substantiate the efficacy of our proposed DiffUSR model through a series of qualitative and quantitative experiments, juxtaposing the outcomes with the performance metrics of several cutting-edge models.

2 Related Work

SISR has been a field of research in computer vision for nearly 20 years [6], but there are few methods for underwater images. Underwater SISR is crucial for improving image quality and holds great research value. Compared to the extensive research conducted on enhancing the resolution of natural images, the realm of SISR techniques for underwater images remains relatively underexplored.

2.1 SISR for Underwater Imagery

The scarcity of large-scale datasets that accurately capture the distinctive distortion distribution found in underwater images serves as the primary hindrance to further advancements in this field. Underwater datasets currently available in the research community primarily concentrate on object detection [7] and image enhancement [5] tasks. However, these datasets often have limitations in terms of resolution, typically restricted to 256 × 256 pixels. In addition to the aforementioned datasets, there are a few available for underwater image super-resolution, such as USR248 [3] and UFO120 [2]. However, existing training sets have limitations due to their maximum image size of 640 × 480 pixels. This constraint hinders the accurate extraction and modeling of fine details in underwater images. As a result, the performance and applicability of both established and innovative SISR models on underwater images have not been thoroughly examined.

In recent years, there have been some research efforts in the field of underwater SISR. These endeavors primarily aim to enhance the quality of underwater images by reconstructing them from low-resolution images that are often affected by noise and blurriness. [8–10] Despite their commendable performance in specific applications, there is still significant room for improvement in order to achieve SOTA results. Therefore, the purpose of this paper is to thoroughly examine these aspects and explore potential avenues for further advancement.

2.2 Diffusion Probabilistic Models

Diffusion probabilistic models belong to the category of generative models that utilize a Markov chain for transforming latent variables, assumed to follow simple distributions (such as Gaussian), into data that follows complex distributions. In this process, the transformation facilitates the diffusion of variables, enabling the generation of more intricate data patterns. The prevalent employment of diffusion models in the field largely stems from the groundbreaking 2020 research paper, DDPM: Denoising Diffusion Probabilistic Models [11]. Contrasting with GANs, DDPM primarily concentrates on effectively modeling noisy images and employing a reverse process termed "denoising" to generate the original, unblemished images. Conversely, GANs achieve image fidelity by utilizing discriminators to optimize the fitting of unaltered images, thus emphasizing a fundamental divergence between these two methodologies. In the past few years, a number of studies have emerged that utilize diffusion models in the domain of image restoration. These investigations encompass diverse applications, including but not limited to image super-resolution, image denoising, and image deblurring.

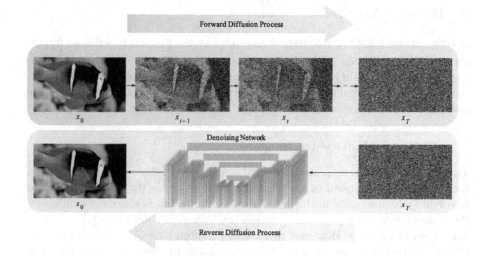

Fig. 1. The training of the Diffusion model can be divided into two parts: forward diffusion process add noise to the picture and reverse diffusion process remove noise from pictures.

3 Method

3.1 Diffusion Model

A diffusion model belongs to the category of generative models that utilize parameterized Markov chains trained through variational inference. This model operates by progressively generating data x_0 from a latent variable x_N, transitioning from a simple distribution to a complex distribution over a sequence of diffusion steps, denoted as $n \in \{1, 2, ..., N\}$. In this context, N represents the total number of diffusion steps. By leveraging the power of variational inference and Markov chains, diffusion models provide an effective framework for generating data from complex distributions while maintaining tractability through the use of simpler distributions.

The architecture of a diffusion model, as depicted in Fig. 1, encompasses two crucial processes: the forward diffusion process and the reverse diffusion process. In the diffusion process, the model gradually diffuses or spreads information throughout the input to generate increasingly coherent representations. This iterative procedure aids in capturing complex dependencies and refining the latent structure of the data. Conversely, the reverse process aims to reconstruct the original input from the diffused representation by iteratively restoring the finer details and reducing the noise introduced during diffusion. By leveraging these complementary processes, diffusion models effectively balance information flow and reconstruction fidelity, enabling robust and high-quality image generation, denoising, and restoration capabilities.

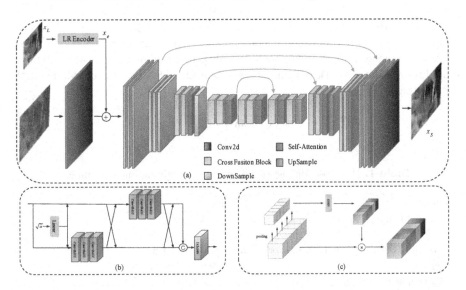

Fig. 2. (a) depicts the network architecture of our proposed DiffUSR model. (b) illustrates the structure of the Cross Fusion Block (CFB), where continuous noise is introduced by means of an affine transformation of the features. (c) is the structure of Simplified Channel Attention (SCA).

In the training process, diffusion model define a diffusion process that transforms an input image x_0 to Gaussian noise $x_N \sim \mathcal{N}(0,1)$ by N iterations. Each iteration of the diffusion process can be described as follows:

$$q\left(x_n \mid x_{n-1}\right) = \mathcal{N}\left(x_n; \sqrt{1-\beta_n}x_{n-1}, \beta_n \mathbf{I}\right), \tag{1}$$

where x_n is the noised image at time-step n, β_n is the predefned scale factor, and $\mathcal{N}$ represents the Gaussian distribution. The Eq. (1) can be further simplified as follows:

$$q\left(\mathbf{x}_n \mid \mathbf{x}_0\right) = \mathcal{N}\left(\mathbf{x}_n; \sqrt{\bar{\alpha}_n}\mathbf{x}_0, (1-\bar{\alpha}_n)\mathbf{I}\right), \tag{2}$$

where $\alpha_n = 1 - \beta_n$, $\bar{\alpha}_n = \prod_{i=0}^{n} \alpha_i$.

In the inference phase (reverse process), the diffusion model initiates by sampling from a Gaussian random noise map x_N and then incrementally denoise x_T until it achieves a high-quality output x_0:

$$p\left(\mathbf{x}_{n-1} \mid \mathbf{x}_n, \mathbf{x}_0\right) = \mathcal{N}\left(\mathbf{x}_{n-1}; \mu_n\left(\mathbf{x}_n, \mathbf{x}_0\right), \sigma_n^2 \mathbf{I}\right), \tag{3}$$

where mean $\mu_t\left(\mathbf{x}_n, \mathbf{x}_0\right) = \frac{1}{\sqrt{\alpha_n}}\left(\mathbf{x}_n - \epsilon\frac{1-\alpha_n}{\sqrt{1-\bar{\alpha}_n}}\right)$ and variance $\sigma_t^2 = \frac{1-\bar{\alpha}_{n-1}}{1-\bar{\alpha}_n}$ β_n. ϵ indicates the noise in x_t, which is the only uncertain variable in the reverse process. The diffusion model employs a denoising network $\epsilon_\theta(x_n, n)$ to estimate ϵ.

To facilitate the training of $\epsilon_\theta(x_n, n)$, this model requires an input of a clean image x_0, and then randomly selects a time step n and a noise $\epsilon \sim \mathcal{N}(0,1)$ to generate noisy images x_n according to Eq. (2). Following this, the diffusion model refines the network parameters, symbolized as θ, of ϵ_θ, in accordance with the approach [11]:

$$\nabla_\theta \, || \, \epsilon - \epsilon_\theta \left(\sqrt{\bar{\alpha}_n} \mathbf{x}_0 + \epsilon\sqrt{1 - \bar{\alpha}_n}, n \right) ||_2^2. \tag{4}$$

Table 1. Pseudo-code algorithm process of training and reasoning

Training	Inference
1. Input: LR image x_L, total diffusion step N	1. $x_N \sim \mathcal{N}(0,1)$
2. Initialize: conditional noise predictor ϵ_θ	2. for n=N, ..., 1 do
3. repeat	3. $\epsilon \sim \mathcal{N}(0,1)$ if n>1, else $\epsilon = 0$
4. Upsample x_L to $x_0 q(x_0)$	4. $x_n = \sqrt{\bar{\alpha}_n} x_0 + \epsilon\sqrt{1 - \bar{\alpha}_n}$
5. Sample $\epsilon \sim \mathcal{N}(0,1)$	$x_{n-1} = \frac{1}{\sqrt{\bar{\alpha}_n}} \left(x_n - \frac{1-\alpha_n}{\sqrt{1-\bar{\alpha}_n}} \epsilon_\theta(x_n, \sqrt{\bar{\alpha}_n}) \right) + \sigma_n \epsilon$
6. Take gradient descent step on	5. end for
$\nabla_\theta \, || \, \epsilon - \epsilon_\theta \left(\sqrt{\bar{\alpha}_n} \mathbf{x}_0 + \epsilon\sqrt{1 - \bar{\alpha}_n}, n \right) ||_2^2.$	6. return x_0

3.2   DiffUSR

The Diffusion Underwater image Super-Resolution (DiffUSR) model that we propose is depicted in Fig. 2 (a). This model's architecture is reminiscent of the U-Net structure, a well-established design known for its effectiveness in various image processing tasks. The DiffUSR network is composed of three primary components: the downsampling layer, the feature fusion layer, and the upsampling layers. The downsampling layer is responsible for reducing the spatial dimensions of the input image while concurrently increasing the depth of the feature maps. In each downsampling operation, except for the final step, the number of channels is doubled, allowing the model to capture more complex and abstract representations of the input data as it progresses through the network.

The feature fusion layer serves as a bridge between the downsampling and upsampling layers. It is at this juncture that the model learns to integrate and interpret the abstracted features extracted during the downsampling process. The upsampling layers, designed to mirror the structure of the downsampling layers in a symmetrical pattern, gradually restore the spatial dimensions of the feature maps. This is achieved while simultaneously reducing the depth of the feature maps, ultimately producing an output that matches the original input image's dimensions.

Drawing inspiration from LatticeNet [12], we introduce a Cross Fusion Block (CFB) into our model, as illustrated in Fig. 2 (b). The CFB decomposes the input signal into multi-order representations, allowing for a more nuanced and detailed interpretation of the input data. The cross-structured network design facilitates better integration of information between features, while also reducing computational complexity.

Furthermore, to enhance the model's ability to capture and represent noise in the input data, we introduce continuous noise into the cross-fusion network. This is achieved through an affine transformation of the features before each feature cross-fusion. This design choice allows our model to better handle and interpret noise in the input data, improving the overall quality and clarity of the super-resolved output. The Low-Resolution (LR) Encoder's purpose is to extract and encode the shallow-level information from the low-resolution image x_L into x_e. The encoded x_e is subsequently incorporated into each reverse step to provide guidance for generating the corresponding high-resolution space. In this work, we employ the Residual Channel Attention Block (RCAB) structure proposed by RCAN [13] as the LR Encoder. This structure utilizes residual groups (RG) and long skip connections (LSC) to enable the network's main components to focus on the more informative elements within the LR features. Specifically, we intentionally exclude the final convolution layer of the RCAB architecture. This decision is driven by our primary objective, which is to capture the intrinsic LR image information rather than solely aiming for specific SR results.

In the interlayer spaces, we adopt the Simplified Channel Attention (SCA) method proposed by NAFNet [14], as demonstrated in Fig. 2 (c). This attention mechanism upholds the two fundamental roles of channel attention: the aggregation of global information and the facilitation of channel information interaction. The algorithms for training and sampling are succinctly summarized in Table 1.

Fig. 3. A few instances sampled from the RedSea-2K trainset.

4 Experiments

4.1 Implementation Details

Our RedSea-2K dataset provides large-scale 960 pairs images for training 2×, 4×, and 8× underwater SISR models. In order to uncover the spatial distribution of the distinct distortions commonly found in underwater imagery, our RedSea-2K dataset comprises meticulously collected underwater images obtained from oceanic explorations and field experiments. After selecting HR images with a resolution of 2K at different positions with various visibility conditions, we gradually generate corresponding ×2, ×4, ×8 LR images through Bicubic (BI) [15] downsampling operations. Each paired set consists of 960 RGB images for training purposes, while an additional 100 test images are provided for benchmark evaluation. A few sample images from the dataset are provided in Fig. 3.

Our experiments were conducted on a server running Ubuntu 20.04, equipped with an Nvidia RTX 3090 graphics card, which was used for both the training and testing of the dataset. Throughout the training process, we configured the batch size to be 16. We employed an initial learning rate of $1e-4$ and selected Adam as the optimizer for training the model using the L1 loss function. Each dataset was subjected to a training regimen of 100 epochs.

4.2 Experimental Analysis

Table 2 presents the quantitative results derived from the RedSea-2K test set, evaluated under a variety of scaling factors. As the data in Table 2 clearly demonstrates, our proposed DiffUSR model outperforms in terms of upscaling factors of $\times 2$, $\times 4$, and $\times 8$ on the RedSea-2K test set. This superior performance suggests that DiffUSR is not only capable of effectively handling different levels of image magnification but also demonstrates its robustness and versatility in the field of underwater image super-resolution. This is a significant finding, as it indicates that our model can be effectively applied across a range of scenarios, making it a valuable tool for researchers and practitioners in this field.

In Fig. 4, we provide a visual comparison at a $\times 4$ scale from the RedSea test dataset. This figure offers a more detailed view of the performance of our model, allowing for a closer examination of its capabilities. It is evident from the close-up view that our proposed DiffUSR model excels in terms of high-frequency local textures. This is a critical aspect of image super-resolution, as it contributes to the overall quality and clarity of the upscaled images.

Furthermore, our model also outperforms in terms of color richness, clarity, and contrast consistency. These factors are particularly important in the context of underwater imaging, where the unique lighting conditions and water properties can often lead to images that are dull or lacking in contrast. The ability of our model to enhance these aspects of the images is a testament to its effectiveness and potential for application in real-world scenarios.

4.3 Ablation Study

In our quest to understand the impacts of the total diffusion step (T), the noise predictor channel size (c), and the efficacy of Cross Fusion Block(Crs.), we have embarked on a series of ablation studies, the results of which are presented in Table 3. A careful analysis of rows 1, 2, 3, and 4 reveals a positive correlation between the total diffusion steps and the quality of the synthesized images, with the latter improving as the former increases.

Further insights can be gleaned from rows 1, 5, and 6, which suggest that a larger model width, or noise predictor channel size, is conducive to better performance. However, it's important to note that an increase in both the total diffusion steps and model width tends to slow down the inference due to the

Table 2. Quantitative evaluation under scale factors ×2, ×4, and ×8. We report the average PSNR [16], SSIM [17], NIQE [18], MA [19], PI [20] on RedSea-2K testset.

Method	Scale	Params	PSNR	SSIM	NIQE	MA	PI
Bicubic [15]	×2	–	29.11	0.8132	5.7645	5.5132	5.1235
DSRCNN [21]	×2	368K	30.65	0.8791	5.8643	6.2837	4.9209
SRResNet [22]	×2	1,491K	30.88	0.8878	5.7386	6.3327	4.7394
SRDRM-GAN [3]	×2	617K	29.73	0.8495	5.5374	6.5092	4.5323
IMDN [23]	×2	694K	31.47	0.8934	5.6923	6.6162	4.5873
SRFlow [24]	×2	694K	33.05	0.9094	5.2243	6.7302	4.5437
ELAN [25]	×2	536K	32.84	0.9132	5.2746	6.7073	4.5124
ESRT [26]	×2	770K	33.37	0.9128	5.2621	6.7239	4.6382
DiffUSR	×2	576K	33.56	0.9134	5.2039	6.7273	4.5269
Bicubic [15]	×4	–	27.85	0.7283	8.7429	3.6203	7.5323
DSRCNN [21]	×4	368K	28.25	0.7623	8.5284	4.4129	7.0938
SRResNet [22]	×4	1,491K	29.07	0.7903	8.3295	4.4591	6.8247
SRDRM-GAN [3]	×4	617K	29.25	0.7862	8.1923	4.6092	6.6328
IMDN [23]	×4	694K	29.36	0.7989	8.2093	4.6823	6.209
SRFlow [24]	×4	694K	29.81	0.8106	8.0293	4.7059	6.4382
ELAN [25]	×4	536K	29.51	0.8035	7.9391	4.8039	6.5394
ESRT [26]	×4	770K	30.15	0.7985	8.1029	4.7736	6.4382
DiffUSR	×4	576K	30.64	0.8119	8.0232	4.7329	6.3794
Bicubic [15]	×8	–	24.24	0.6382	9.8926	2.3722	8.3924
DSRCNN [21]	×8	368K	25.98	0.6873	9.5101	2.5811	8.3533
SRResNet [22]	×8	1,491K	26.43	0.691	9.5162	2.6725	8.3072
SRDRM-GAN [3]	×8	617K	25.79	0.5724	9.3828	2.8082	8.142
IMDN [23]	×8	694K	25.97	0.6965	9.3683	2.9081	8.2052
SRFlow [24]	×8	694K	27.13	0.7163	9.4091	3.0536	7.8505
ELAN [25]	×8	536K	27.35	0.7231	9.3924	3.1293	7.8293
ESRT [26]	×8	770K	27.03	0.7175	9.2127	3.2646	7.7893
DiffUSR	×8	576K	27.31	0.7233	9.2044	3.2928	7.7222

heightened computational demands. As a result, after a careful consideration of the trade-offs, we have settled on T = 100 and c = 64 as the optimal parameters for our model.

A comparison of rows 1 and 7 provides compelling evidence of the effectiveness of the Cross Fusion Block. The data clearly shows that the implementation of the Cross Fusion Block not only significantly boosts the quality of the images but also accelerates the training process, thereby underscoring the value of the Cross Fusion Block in enhancing both the quality of the output and the efficiency of the model training.

Fig. 4. A comparative analysis of the reconstructed high-resolution images of RedSea-2K test-set is conducted, utilizing various methodologies with a scale factor of ×4.

Table 3. Ablations of DiffUSR for SR (8×) on RedSea-2K. T, c and Crs. denote the total diffusion step, channel size of the noise predictor, and the Cross Fusion Block.

T	c	$Crs.$	↑PSNR	↑SSIM	↑NIQE	↑LR-PSNR	↓Steps
100	64	✓	27.31	0.7233	9.2044	53.56	300k
25	64	✓	27.05	0.7024	9.1136	52.27	300k
200	64	✓	27.12	0.7055	9.1472	52.42	300k
1000	64	✓	27.16	0.7162	9.152	52.49	300k
100	32	✓	27.08	0.7129	9.1278	52.31	300k
100	128	✓	27.36	0.7242	9.2073	53.61	300k
100	64	✗	26.24	0.7012	9.1463	52.05	600k

5   Conclusion

In this paper, we propose a novel underwater SISR diffusion probability model (DiffUSR), designed to operate at scaling factors of 2×, 4×, and 8×. This model is capable of providing diverse and realistic SR predictions by progressively transforming Gaussian noise into SR images based on LR inputs, utilizing a Markov Chain process. This approach represents a significant advancement in the field of underwater image super-resolution, offering a promising solution to the challenges of image enhancement in underwater environments. Additionally, we present a comprehensive underwater image dataset, termed RedSea-2K. This dataset is specifically curated to facilitate the supervised training of underwater SISR models, thereby providing a robust platform for further research and development in this domain. In the future, we aspire to augment the performance of this model in the context of 8× SISR, and further enhance its computational efficiency to cater to diverse types of underwater visual domains.

References

1. Benmoussa, A., et al.: MSIDN: mitigation of sophisticated interest flooding-based DDoS attacks in named data networking. Futur. Gener. Comput. Syst. **107**, 293–306 (2020)
2. Islam, M.J., Luo, P., Sattar, J.: Simultaneous enhancement and super-resolution of underwater imagery for improved visual perception. arXiv preprint arXiv:2002.01155 (2020)
3. Islam, M.J., Enan, S.S., Luo, P., Sattar, J.: Underwater image super-resolution using deep residual multipliers. In: 2020 IEEE International Conference on Robotics and Automation (ICRA), pp. 900–906. IEEE (2020)
4. Li, L., et al.: Super-resolution reconstruction of underwater image based on image sequence generative adversarial network. Math. Probl. Eng. **2020**, 1–10 (2020)
5. Islam, M.J., Xia, Y., Sattar, J.: Fast underwater image enhancement for improved visual perception. IEEE Robot. Autom. Lett. **5**(2), 3227–3234 (2020)
6. Yang, W., Zhang, X., Tian, Y., Wang, W., Xue, J.H., Liao, Q.: Deep learning for single image super-resolution: a brief review. IEEE Trans. Multimedia **21**(12), 3106–3121 (2019)
7. Islam, M.J., Ho, M., Sattar, J.: Understanding human motion and gestures for underwater human-robot collaboration. J. Field Robot. **36**(5), 851–873 (2019)
8. Chen, Y., Yang, B., Xia, M., Li, W., Yang, K., Zhang, X.: Model-based super-resolution reconstruction techniques for underwater imaging. In: Photonics and Optoelectronics Meetings (POEM) 2011: Optoelectronic Sensing and Imaging, vol. 8332, pp. 119–128. SPIE (2012)
9. Fan, F., Yang, K., Fu, B., Xia, M., Zhang, W.: Application of blind deconvolution approach with image quality metric in underwater image restoration. In: 2010 International Conference on Image Analysis and Signal Processing, pp. 236–239. IEEE (2010)
10. Yu, Y., Liu, F.: System of remote-operated-vehicle-based underwater blurred image restoration. Opt. Eng. **46**(11), 116002–116002 (2007)
11. Ho, J., Jain, A., Abbeel, P.: Denoising diffusion probabilistic models. Adv. Neural. Inf. Process. Syst. **33**, 6840–6851 (2020)
12. Luo, X., Xie, Y., Zhang, Y., Qu, Y., Li, C., Fu, Y.: LatticeNet: towards lightweight image super-resolution with lattice block. In: Vedaldi, A., Bischof, H., Brox, T., Frahm, J.M. (eds.) Computer Vision-ECCV 2020: 16th European Conference, Glasgow, UK, 23–28 August 2020, Proceedings, Part XXII 16, pp. 272–289. Springer, Cham (2020). https://doi.org/10.1007/978-3-030-58542-6_17
13. Zhang, Y., Li, K., Li, K., Wang, L., Zhong, B., Fu, Y.: Image super-resolution using very deep residual channel attention networks. In: Ferrari, V., Hebert, M., Sminchisescu, C., Weiss, Y. (eds.) ECCV 2018. LNCS, vol. 11211, pp. 294–310. Springer, Cham (2018). https://doi.org/10.1007/978-3-030-01234-2_18
14. Chen, L., Chu, X., Zhang, X., Sun, J.: Simple baselines for image restoration. In: Avidan, S., Brostow, G., Cissé, M., Farinella, G.M., Hassner, T. (eds.) European Conference on Computer Vision. LNCS, vol. 13667, pp. 17–33. Springer, Cham (2022). https://doi.org/10.1007/978-3-031-20071-7_2
15. Keys, R.: Cubic convolution interpolation for digital image processing. IEEE Trans. Acoust. Speech Signal Process. **29**(6), 1153–1160 (1981)
16. Hore, A., Ziou, D.: Image quality metrics: PSNR vs. SSIM. In: 2010 20th International Conference on Pattern Recognition, pp. 2366–2369. IEEE (2010)

17. Wang, Z., Bovik, A.C., Sheikh, H.R., Simoncelli, E.P.: Image quality assessment: from error visibility to structural similarity. IEEE Trans. Image Process. **13**(4), 600–612 (2004)
18. Liu, L., Liu, B., Huang, H., Bovik, A.C.: No-reference image quality assessment based on spatial and spectral entropies. Signal Process. Image Commun. **29**(8), 856–863 (2014)
19. Ma, C., Yang, C.Y., Yang, X., Yang, M.H.: Learning a no-reference quality metric for single-image super-resolution. Comput. Vis. Image Underst. **158**, 1–16 (2017)
20. Blau, Y., Mechrez, R., Timofte, R., Michaeli, T., Zelnik-Manor, L.: The 2018 PIRM challenge on perceptual image super-resolution. In: Leal-Taixé, L., Roth, S. (eds.) ECCV 2018. LNCS, vol. 11133, pp. 334–355. Springer, Cham (2019). https://doi. org/10.1007/978-3-030-11021-5_21
21. Tang, Y., Wu, X., Bu, W.: Deeply-supervised recurrent convolutional neural network for saliency detection. In: Proceedings of the 24th ACM International Conference on Multimedia, pp. 397–401 (2016)
22. Ledig, C., et al.: Photo-realistic single image super-resolution using a generative adversarial network. In: Proceedings of the IEEE Conference on Computer Vision and Pattern Recognition, pp. 4681–4690 (2017)
23. Hui, Z., Gao, X., Yang, Y., Wang, X.: Lightweight image super-resolution with information multi-distillation network. In: Proceedings of the 27th ACM International Conference on Multimedia, pp. 2024–2032 (2019)
24. Lugmayr, A., Danelljan, M., Van Gool, L., Timofte, R.: SRFlow: learning the super-resolution space with normalizing flow. In: Vedaldi, A., Bischof, H., Brox, T., Frahm, J.M. (eds.) Computer Vision-ECCV 2020: 16th European Conference, Glasgow, UK, 23–28 August 2020, Proceedings, Part V 16, pp. 715–732. Springer, Cham (2020). https://doi.org/10.1007/978-3-030-58558-7_42
25. Zhang, X., Zeng, H., Guo, S., Zhang, L.: Efficient long-range attention network for image super-resolution. In: Avidan, S., Brostow, G., Cissé, M., Farinella, G.M., Hassner, T. (eds.) European Conference on Computer Vision, vol. 13677, pp. 649–667. Springer, Cham (2022). https://doi.org/10.1007/978-3-031-19790-1_39
26. Lu, Z., Li, J., Liu, H., Huang, C., Zhang, L., Zeng, T.: Transformer for single image super-resolution. In: Proceedings of the IEEE/CVF Conference on Computer Vision and Pattern Recognition, pp. 457–466 (2022)

Classification Method for Ship-Radiated Noise Based on Joint Feature Extraction

Libin Du, Mingyang Liu, Zhichao Lv[✉], Zhengkai Wang, Lei Wang, and Gang Wang

College of Ocean Science and Engineering, Shandong University of Science and Technology, Qingdao, China
lvzhichao@hrbeu.edu.cn

Abstract. In order to address the problem of poor recognition performance from single signal features in ship identification and to enhance the accuracy of Convolutional Neural Networks (CNNs) in underwater acoustic target recognition, this paper proposes a method of joint feature extraction for ship target identification, combining energy features extracted from wavelet decomposition and frequency domain features extracted from Mel filters. Subsequently, two types of CNNs are constructed to train the joint features, evaluating the recognition performance of the joint features for ship targets. Through result analysis, it is found that the joint features can effectively identify ship targets. When compared to the recognition performance of the single Mel frequency domain features, the recognition accuracy of the joint features is significantly higher, providing a useful reference for underwater acoustic target recognition.

Keywords: Radiated Noise · Feature Extraction · Wavelet Energy Features · Mel Spectrum Features · Convolutional Neural Networks

1 Introduction

The recognition of ship-radiated noise is a hot research topic in the field of underwater acoustics engineering, which carries significant implications for national naval defense construction, ship navigation safety monitoring, and maritime traffic control.

The radiated noise from ship targets primarily results from the mechanical movement of the vessel, the rotation of the propellers, and cavitation effects. Therefore, the radiated noise from a ship contains abundant target information. The features extracted from the processing of ship-radiated noise serve as an important reference for ship target identification.

The main steps involved in identifying ship targets using radiated noise include filtering, feature extraction, and target recognition. Filtering primarily involves removing ocean environmental noise to improve the signal-to-noise ratio of ship noise. Literature [1] combines wavelet analysis with minimum mean

C. Cruz et al. (Eds.): IC 2023, CCIS 2036, pp. 75–90, 2024.
https://doi.org/10.1007/978-981-97-0065-3_6

adaptive filtering technology to enhance the signal-to-noise ratio of ship-radiated noise. Literature [2] applies the idea of norm regularization to the least squares algorithm, proposing norm regularization sparse least squares algorithm. Feature extraction technology uses signal processing methods to extract feature vectors from radiated noise for target recognition. Currently, the main methods for analyzing underwater acoustic signals include wavelet analysis, spectrum analysis, and non-linear theoretical analysis. Literature [3] extracts the envelope spectrum of ship-radiated noise signals through the characteristics of Morlet wavelets. It utilizes the modulation information components in the spectral features, using $1(1/2)$ spectral analysis to obtain the relevant features of the ship propeller's shaft frequency. Literature [4] applies wavelet transformation to process ship signals, extracting the coefficients of the signal and noise in the wavelet transformation domain. Literature [5] extracts the wavelet packet energy features of the signal and combines them with Mel cepstrum coefficients to form joint features for classification. Literature [6] extracts the wavelet packet energy of the signal and uses principal component analysis to reduce the dimensionality of the signal, thus reducing the dimension of the feature signal. Literature [7] extracts the Mel cepstrum coefficient features and spectral density features of cyclostationary analysis from ship-radiated noise. It integrates these features at the feature and decision levels, and the recognition results show that the fusion algorithm at the decision level can significantly improve the target recognition rate. Literature [8] studies the application of chaos theory in underwater acoustic signals and analyzes the maximum Lyapunov exponent of ship-radiated noise. When this index is finite and positive, ship-radiated noise shows significant chaotic characteristics, which indicates that the ship-radiated noise signal has obvious nonlinear characteristics.

Traditional ship target recognition techniques primarily rely on sonar soldiers and other technical professionals for manual identification [9]. With the development and progress of computers, recognition techniques based on statistical features and machine learning recognition techniques based on neural networks have emerged. Literature [10] analyzes and extracts the modulation noise envelope detection spectrum of ship-radiated noise, and optimizes the method of support vector machine using the distribution of neighboring samples. The recognition results show that this method is suitable for recognizing the modulation noise envelope detection spectrum of ship-radiated noise. Literature [11] extracts the Mel cepstrum coefficients of radiated noise and builds a Long Short-Term Memory (LSTM) network, which identifies the Mel cepstrum coefficients of target radiated noise through this neural network. Literature [12] proposes a target recognition network that combines a one-dimensional convolutional neural network and an LSTM network. The recognition accuracy is improved by cascading the two networks.

This paper uses wavelet analysis technology and Mel spectrum analysis technology, constructs a target recognizer using Convolutional Neural Networks (CNN), and proposes a research method for ship-radiated noise recognition that combines wavelet energy features and Mel time-frequency features.

The main research contents of this paper are as follows:

(1) Wavelet Energy Feature Extraction. The radiated noise signal of the ship is decomposed using wavelet decomposition, and the signals of different frequency bands are reconstructed. The energy of the reconstructed signal is calculated and normalized to obtain the energy features of the signal in different frequency bands.
(2) Mel Spectrum Feature Extraction. Perform a Short-Time Fourier Transform (STFT) on the ship signal to obtain signal frequency domain information, convert the frequency domain information to the Mel frequency domain, extract the Mel frequency domain features of the ship-radiated noise signal, and plot the Mel spectrum feature of the signal.
(3) Deep Learning-based Target Recognition. Based on the energy features of the signal and the Mel spectrum features, a joint feature model of the feature layer is constructed by combining the above two types of ship signal feature vectors. The Convolutional Neural Network (CNN) in deep learning is used for ship recognition. The method is validated using real measurement data of radiated noise from seven kinds of ships.

2 Theoretical Work

2.1 Wavelet Transform

The radiated noise from ships is a form of non-stationary signal, which rarely has a fixed time interval. Traditional Fourier Transform struggles to effectively process this signal, as it can only capture the frequency domain information of the signal. Consequently, it finds it challenging to depict the most essential time-frequency local characteristics of non-stationary signals. The wavelet transform, on the other hand, possesses superior non-stationary signal analysis capabilities. This has led to its extensive application in the field of feature extraction from ship-radiated noise [13–15].

The wavelet transform is based on a wavelet basis function. Let's define the wavelet basis function as $\Psi(t)$, we have:

$$\Psi_{a,b}(t) = \frac{1}{\sqrt{a}}\Psi(\frac{t-b}{a})$$ (1)

In the formula, both a and b are constants, and $a > 0$. These are obtained by the translation and scaling transformations of the base function. If a and b keep changing, a set of functions $\Psi_{a,b}(t)$ can be obtained. Given a square-integrable signal $x(t)$, where $x(t) \in L^2(R)$, let the wavelet transform of $x(t)$ be:

$$W_{T_x}(a,b) = \frac{1}{\sqrt{a}}\int x(t)\Psi^*\left(\frac{t-b}{a}\right)dt$$
$$= \int x(t)\Psi_{a,b}{}^*(t)dt$$ (2)
$$= \langle x(t),\ \Psi_{a,b}(t)\rangle$$

In this formula, a, b, and t are all continuous variables. The formula is referred to as the continuous wavelet transform. The wavelet transform of the signal function $x(t)$ is a function of a and b. Here, a is called the scale factor, b is the time-shift factor, $\Psi(t)$ is the basic wavelet, and $WT_x(a,b)$ is the wavelet basis function. This function $WT_x(a,b)$ is a set of functions obtained from the basic wavelet through a sequence of translation and scaling transformations. To ensure that the energy of the wavelet transform of $x(t)$ remains the same for different values of a, the following condition needs to be met:

$$\int |\Psi_{a,b}(t)|^2 dt = \frac{1}{a} \int |\Psi(\frac{t-b}{a})|^2 dt \tag{3}$$

The process mentioned above is referred to as the continuous wavelet transform of signal $x(t)$ in the time domain. For the discretization of the wavelet transform, the scale factor a is usually discretized according to a power series, i.e., $a = a_0^j$. Under the same scale, the shift factor b is discretized uniformly, i.e., $b = kb_0$, where a_0, b_0 are real numbers greater than 0, and j, k are integers. The introduction of scale and shift discretization transforms the wavelet basis function and the wavelet transform as follows:

$$\Psi_{a_0^j, kb_0} = a_0^{\frac{j}{2}} \Psi(a_0^{-j} t - kb_0) \tag{4}$$

$$WT_x(a_0^j, kb_0) = a_0^{\frac{j}{2}} \int x(t) \Psi^*(a_0^{-j} t - kb_0) dt \tag{5}$$

The method of wavelet energy feature extraction in this article is developed based on the above wavelet analysis principles. The radiated noise signal of the ship is decomposed by wavelet, and the corresponding wavelet features are extracted.

2.2   Mel Spectrum Features

The human ear has different auditory sensitivities to sound waves of different frequencies. The Mel frequency is closely related to the auditory characteristics of the human ear. The analysis method of Mel Frequency Cepstral Coefficients (MFCC), which is formed by combining Mel frequency with cepstral analysis, plays an important role in the field of underwater acoustic signals [16–18]. Figure 1 shows the relationship between Mel frequency and signal frequency. The transformation relationship between Mel frequency and signal frequency is as follows:

$$m = 2595 \log(1 + \frac{f}{700}) = 1127 \ln(1 + \frac{f}{700}) \tag{6}$$

The transformation from frequency to Mel frequency can be achieved through a set of triangular filters. Define a filter group with M filters, where the center frequency is $f(m)$, $m = 1, 2, ..., $ M. The intervals between each $f(m)$ decrease as

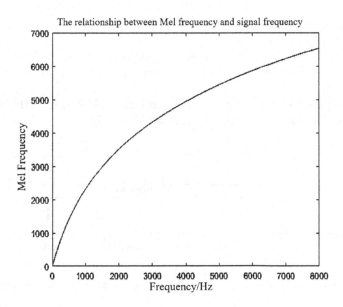

The relationship between Mel frequency and signal frequency

Fig. 1. The relationship between Mel frequency and signal frequency

the value of m decreases and widen as the value of m increases. The frequency response of the triangular filter is defined as:

$$H_m(k) = \begin{cases} 0, & k < f(m-1) \\ \dfrac{2(k - f(m-1))}{(f(m+1) - f(m-1))(f(m) - f(m-1))}, & f(m-1) \le k \le f(m) \\ \dfrac{2(f(m+1) - k)}{(f(m+1) - f(m-1))(f(m+1) - f(m))}, & f(m) \le k \le f(m+1) \\ 0, & k > f(m+1) \end{cases} \tag{7}$$

In the equation, $\sum\limits_{m=0}^{M-1} H_m(k) = 1$.

The steps of extracting Mel spectrum features are shown in Fig. 2 and can be divided into four steps:

(1) Signal pre-emphasis. The original signal is pre-emphasized to enhance the proportion of the high-frequency part of the signal while keeping the low-frequency part of the signal unchanged. Pre-emphasis generally uses a first-order high-pass filter to process the signal.

(2) Signal framing and windowing. Signal framing is to divide the continuous signal into several short-term signal blocks, each block is called a frame, for local analysis and processing, reducing the amount of computation. Signal windowing is based on signal framing, applying windowing processing to

each frame of signal, to eliminate abrupt changes at the signal boundaries and can also reduce errors.

(3) Short-time Fourier transform. Perform a short-time Fourier transform on the framed and windowed signal to obtain the frequency domain information of the signal.

(4) Mel filtering. Use a set of Mel filters to perform Mel filtering on the frequency domain signal to obtain the Mel features of the signal.

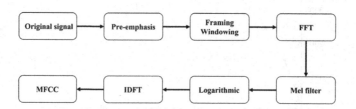

Fig. 2. The extraction process of Mel Frequency Cepstral Coefficients (MFCC)

2.3   Convolutional Neural Networks

Convolutional Neural Networks (CNN) are a type of feed-forward neural network, initially applied in the field of image recognition. With the development of this neural network, it has also started to be used in fields such as speech recognition and natural language recognition [19,20]. Using CNN can effectively solve the problem of slow training efficiency with large data. Generally speaking, convolutional neural networks are composed of one or more convolutional layers and top fully connected layers, and also include pooling layers corresponding to the convolutional layers. Taking a two-dimensional convolutional neural network as an example, Fig. 3 shows the general structure of this neural network. The convolutional layer is the core of the convolutional neural network. The idea is to achieve data dimension reduction and compression by convolving the input data with the convolutional kernel. CNN obtains different feature matrices of the data after multiple convolutional layers and pooling layers, and finally obtains an output feature matrix through the fully connected layer. This feature matrix contains different feature information of the input signal and is an important reference for signal recognition.

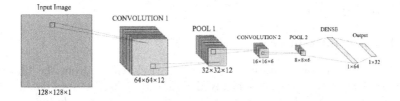

Fig. 3. General Structure of Convolutional Neural Networks

The convolutional layer is the core of the convolutional neural network, and convolution is the core content of the convolutional layer. Although the data volume can be reduced through the convolutional layer, the data volume after convolution is still very large for a general computer, which can easily lead to overfitting. Therefore, it is necessary to perform a downsampling on the data after convolution to reduce the training parameters. This process is achieved through the pooling layer. A convolutional layer and a pooling layer together form the basic structure of the convolutional neural network. The specific workflow of the convolutional layer and the pooling layer is shown in Fig. 4.

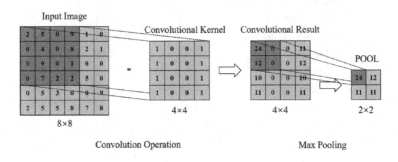

Fig. 4. Workflow of Convolutional and Pooling Layers

Figure 4 defines a 4×4 convolutional kernel. It performs convolution operations with the corresponding content of the input data to obtain the matrix after convolution. This is the basic content of the convolutional layer. Then, the pooling layer is used to reduce the data dimension. The dimension of the pooling layer needs to be smaller than that of the convolutional layer. The pooling layer generally has three methods: maximum pooling, mean pooling, and global maximum pooling.

After several convolutional layers and pooling layers, the fully connected layer forms the reduced-dimensional data features into a feature vector of the convolutional neural network. The target is classified through this feature vector.

3   Experiments

3.1   Dataset Introduction

The ship-radiated noise dataset used in this article comes from the ShipsEar database. This database is specifically designed to collect ship-radiated noise. The database contains twelve kinds of radiated noise signals, including radiated noise from 11 different ship types and 1 group of ocean environment background noise, totaling 90 groups of radiated noise data. The files are stored in .wav audio format. Table 1 summarizes the data file types in this database, and Fig. 5 shows the signal characteristics of a certain ship in this database.

Table 1. ShipsEar database data

Ship type	Number of data	Duration/s
Dredger	5	262
Fishboat	4	510
Motorboat	13	1008
Mussel boat	5	726
Natural ambient noise	12	1134
Ocean liner	7	938
Passenger ship	30	4256
Pilot ship	2	138
RORO	5	1512
Sailboat	4	404
Trawler	1	162
Tugboat	2	206
In Total	90	11256

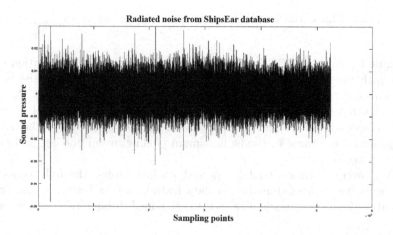

Fig. 5. Radiated noise from ShipsEar database

3.2   Joint Feature Extraction

The dataset consists of actual measured data. During the signal collection process, interference from background noise is present, and some data have issues with high noise levels and blank signals. Therefore, before extracting features from the signals, it is necessary to manually remove the blank segment signals and eliminate some signals where the background noise is too intense. Due to the insufficient amount of data collected for certain types of ships, this study discarded the data for Dredger, Pilot ship, Sailboat, Trawler, and Tugboat types. Only the data for Fish boat, Motorboat, Mussel boat, Natural ambient noise, Ocean liner, Passenger, and RORO were retained and processed.

In order to expand the training set, this study segmented the original data into 2-second clips, resulting in a total of 4400 data groups. Each group of data underwent corresponding feature extraction to derive the relevant feature vectors. For the segmented data, two types of features were extracted.

Wavelet Energy Feature Extraction. In this study, the db4 wavelet is used as a basis function to carry out a five-level wavelet decomposition on the data. The decomposition results are shown in Fig. 6. The db4 wavelet is chosen as the basis function because it has four vanishing moments, allowing it to perfectly reconstruct any cubic polynomial. Meanwhile, the five-level wavelet decomposition provides a good balance between time and frequency resolution.

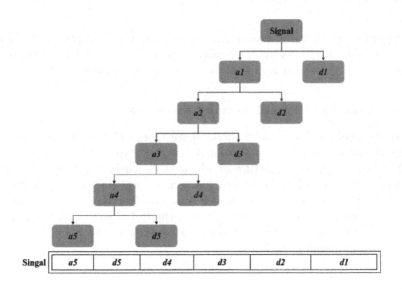

Fig. 6. Schematic Diagram of Wavelet Decomposition

Signals of different decomposition levels are reconstructed in turn to obtain two components of different levels, approximation and detail, denoted as $C_{n,a}$ and $C_{n,d}$ respectively. Here, n represents the number of levels of wavelet decomposition, and a and d respectively represent the approximation and detail components. The energy of each layer's approximation and detail components is calculated as follows:

$$E_{n,c} = \sum_{i=1}^{N} |x_{ni}|^2 \tag{8}$$

In the formula, n represents the number of levels of wavelet decomposition, c stands for the approximation and detail components at that level, and i represents the N discrete points of that component, with x_{ni} denoting the amplitude of the discrete points of the reconstructed signal.

Finally, the energies of the approximation and detail components are normalized as follows:

$$E_a = [E_{1a}, E_{2a}, E_{3a}, E_{4a}, E_{5a}]/E_{sum,a} \tag{9}$$

In the formula, $E_{sum,a}$ represents the total energy of the approximation components. Similarly, the normalized feature matrix E_d for the energy of the detail components can be obtained. Finally, the feature matrices of the approximation and detail components are combined to obtain the final wavelet energy feature matrix $E = [E_a, E_d]$ for the signal.

Mel Spectrogram. In this study, a Hamming window of length 512 is used to window the signal, with a window overlap of 256 and a discrete Fourier transform length of 1024. The signal is filtered using 64 Mel bandpass filters.

3.3   Convolutional Neural Network Structure

This study combines the wavelet energy features and Mel spectrogram features of signals and designs two convolutional neural network structures. One uses the AlexNet model, the structure of which is shown in Fig. 7. It is an eight-layer network structure. If the pooling layers and Local Response Normalization layers (LRN) are not counted, the model consists of five convolutional layers and three fully connected layers. The model uses the ReLU activation function and Dropout as a regularization term to prevent overfitting, enhancing the model's robustness and learning efficiency.

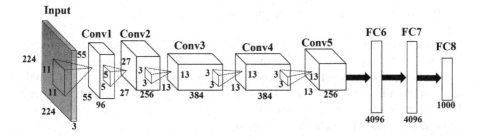

Fig. 7. AlexNet Model

The other convolutional neural network used in this study is a network structure composed of four convolutional layers and pooling layers. The convolutional kernel is set as 3 × 3, the stride is set as 1, the pooling layer is defined as max pooling, the pooling window is designed as 2 × 2, and the stride is set as 2. The structure of the model is shown in Fig. 8. The specific parameters of the network are shown in Table 2.

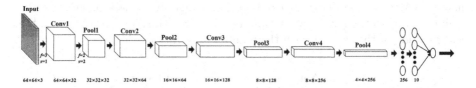

Fig. 8. 4-layer Convolutional Network Structure

Table 2. Network training parameter

Parameter name	Parameter settings
loss	categorical_crossentropy
optimizer	SGD
epochs	40
metrics	accuracy
batch_size	Auto

4 Results

Based on the dataset described in Sect. 3, we fuse the two feature vectors of all audio segments, randomly divide the neural network's training set and test set, and respectively carry out the training and testing of the network. In this study, we allocate 90% of the data for model training and 10% of the data for model testing.

We set the number of iterations to 40, with each iteration automatically calculating the input size for training, and the number of training times also automatically determined by the computer hardware. We use both the single Mel spectrum feature of the signal and the combined feature as input, and train the data with two types of convolutional neural networks. The accuracy of the training and test sets is shown in Table 3.

Table 3. Accuracy of the Training and Test Sets

Input	AlexNet (Train Set/Test Set)	4-layers CNN (Train Set/Test Set)
Mel Spectral Features	90.13%/84.79%	95.95%/7.22%
Joint Features	91.87%/86.17%	96.31%/14.06%

Figure 9 shows the change in accuracy of the training set with the number of iterations, while Fig. 10 shows the change in accuracy of the test set with the number of iterations. Figure 11 exhibits the change in the loss function of the training set with the number of iterations, and Fig. 12 displays the change in the loss function of the test set with the number of iterations.

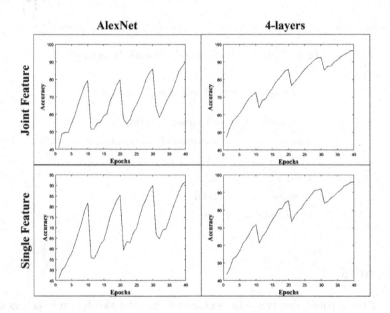

Fig. 9. Changes in the Accuracy of the Training Set with the Number of Iterations

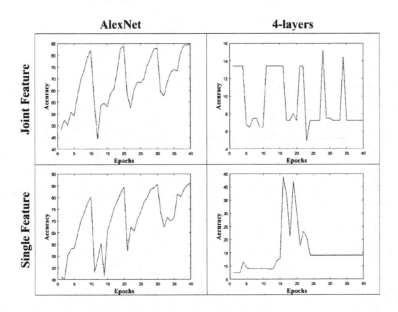

Fig. 10. Changes in the Accuracy of the Test Set with the Number of Iterations

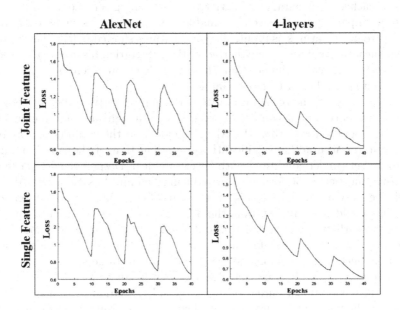

Fig. 11. Changes in the Loss Function of the Training Set with the Number of Iterations

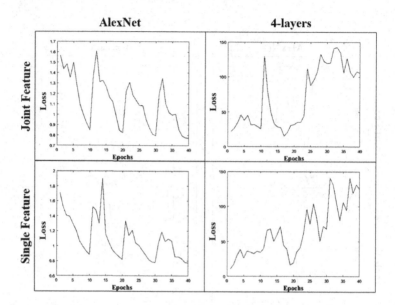

Fig. 12. Changes in the Loss Function of the Test Set with the Number of Iterations

In the AlexNet model, the joint features achieved an accuracy of 91.87% on the training set and 86.17% on the test set. In the 4-layer model, the joint features reached an accuracy of 96.31% on the training set and 14.06% on the test set. Compared with the accuracy and loss function results when using single Mel spectrogram features as input, the joint features showed better classification performance on the dataset. Furthermore, the recognition accuracy of the joint features was significantly higher than that of single feature scenarios when there were a limited number of training samples.

Upon comparing the recognition rate performances of two different convolutional neural networks, it can be observed that although the 4-layer model has a higher recognition rate than the AlexNet model on the training set samples, its recognition accuracy is significantly lower than that of the AlexNet model on the test set (under conditions with fewer samples). Despite the 4-layer model demonstrating better performance in reducing training loss than the AlexNet model, the AlexNet model surpasses it in terms of classification prediction accuracy. This could also suggest that the 4-layer model may be overly complex, leading to overfitting on the training data and a lack of generalization ability on the test data. Given the aforementioned factors, the AlexNet model holds a more pronounced advantage in the recognition of ship targets.

In summary, it can be seen that the joint features have a certain optimizing effect on the recognition accuracy of ship targets. In situations with a low number of samples, the fusion of features combined with the AlexNet network can achieve better recognition performance.

5 Conclusion

To train a ship target recognition network with higher recognition accuracy and to overcome the shortcomings of poor recognition accuracy with single type feature vector inputs, this study combines the wavelet transform domain and Mel frequency domain features of ship-radiated noise. We have constructed a target feature extraction method that combines wavelet energy features and Mel spectrogram features at the feature level, and used two convolutional neural networks to recognize the combined features.

Through experimental validation on the ShipEar dataset, the convolutional neural network recognition method proposed in this study, which combines wavelet energy features and Mel spectrogram features, outperforms the neural network that uses a single feature vector input. Furthermore, the recognition performance of the AlexNet network is superior to that of the 4-layer convolutional neural network. The approach of training neural networks through different feature fusion methods also provides a new research direction for underwater acoustic target recognition.

The advantage of using ship-radiated noise for recognition is that data collection is highly covert, making it particularly suitable for the field of underwater acoustic countermeasures. Since the underwater acoustic channel is a very complex type of channel, it contains a large amount of background noise and is affected by reverberation. Therefore, the target radiation noise that is collected often contains significant interference, which affects the accuracy of the signal neural network training to a large extent.

The shortcomings of this study lie in the fact that the training data used had a large amount of background noise. After excluding data with a low signal-to-noise ratio, the remaining data was relatively scarce. In addition, the training set used in this study was relatively uniform, making it difficult to verify the recognition performance of this method in actual marine environments. Finally, while the wavelet transform can display the local characteristics of the signal, it only decomposes the low-frequency coefficient part and cannot further decompose the high-frequency components.

References

1. Sun, J.: Research on line spectrum enhancement of ship radiated noise based on modern signal theory. Kunming University of Science and Technology (2013)
2. Qin, Z.: Sparse adaptive filtering technique and its application in underwater acoustic channel equalization. Southeast University (2020)
3. Zhao, M., Song, Y., Zheng, W.: Modulation feature extraction of ship-radiated noise based on wavelet demodulation-1(1/2) D spectrum. Ship Sci. Technol. **41**(15), 122–126 (2019)
4. Wu, G.: Research on feature extraction method of ship noise signal based on wavelet transform. Ship Sci. Technol. **40**(20), 22–24 (2018)
5. Xu, Q., Wang, B.: Ship radiated noise recognition based on wavelet packet analysis and deep learning. Mar. Eng. **43**(05), 29–34+43 (2021)

6. Wu, C., Wang, B., Xu, Q.: Ship radiated noise recognition technology based on wavelet packet decomposition and PCA-Attention-LSTM. Acoust. Tech. **41**(02), 264–273 (2022)

7. Wang, Y., Hou, P., Wu, D.: Ship target recognition method based on Feature Fusion. Ship Sci. Technol. **44**(01), 146–149 (2022)

8. Yang, L., Zheng, S.: Feature extraction of ship-radiated noise based on chaos theory. J. Naval Eng. Univ. **26**(04), 50–54+62 (2014)

9. Maks, J.N., Wang, Y.: Application of expert system in underwater acoustics. Acoust. Electron. Eng. **03**, 39–45 (2015)

10. Dai, W., Zhang, Z., Liu, Q.: Neighbor sample distribution weighted SVM and its application in ship propeller blade number recognition. In: 2015 Academic Conference of Hydroacoustic Branch of China Acoustic Society, pp. 360–363 (2010)

11. Zhang, S., Tian, D.: Intelligent classification method of Mel frequency cepstrum coefficient for underwater acoustic targets. Appl. Acoust. **38**(02), 267–272 (2019)

12. Ren, C., Wang, L., Han, X.: Underwater acoustic target recognition method based on joint neural network. Ship Sci. Technol. **44**(01), 136–141 (2022)

13. Guo, T., Zhang, T., Lim, E., López-Benítez, M., Ma, F., Yu, L.: A review of wavelet analysis and its applications: challenges and opportunities. IEEE Access **10**, 58869–58903 (2022)

14. Kim, K.I., Pak, M.I., Chon, B.P., Ri, C.H.: A method for underwater acoustic signal classification using convolutional neural network combined with discrete wavelet transform. Appl. Acoust. **19**(04), 2050092 (2021)

15. Quan, T., Yang, X., Jingjing, W.: DOA estimation of underwater acoustic array signal based on wavelet transform with double branch convolutional neural network. In: Proceedings of the 15th International Conference on Underwater Networks & Systems, pp. 1–2 (2021)

16. Wang, W., Li, S., Yang, J., Liu, Z., Zhou, W.: Feature extraction of underwater target in auditory sensation area based on MFCC. In: 2016 IEEE/OES China Ocean Acoustics (COA), pp. 1–6. IEEE (2016)

17. Song, G., Guo, X., Wang, W., Ren, Q., Li, J., Ma, L.: A machine learning-based underwater noise classification method. Appl. Acoust. **184**, 108333 (2021)

18. Tong, Y., Zhang, X., Ge, Y.: Classification and recognition of underwater target based on MFCC feature extraction. In: 2020 IEEE International Conference on Signal Processing, Communications and Computing (ICSPCC), pp. 1–4. IEEE (2020)

19. Doan, V.S., Huynh-The, T., Kim, D.S.: Underwater acoustic target classification based on dense convolutional neural network. IEEE Geosci. Remote Sens. Lett. **19**, 1–5 (2020)

20. Liu, F., Shen, T., Luo, Z., Zhao, D., Guo, S.: Underwater target recognition using convolutional recurrent neural networks with 3-D Mel-spectrogram and data augmentation. Appl. Acoust. **178**, 107989 (2021)

AI in Finance

Forecasting the Price of Bitcoin Using an Explainable CNN-LSTM Model

SiXian Chen, Zonghu Liao[✉], and Jingbo Zhang

China University of Petroleum, Beijing, China
chensixian@student.cup.edu.cn, zong@cup.edu.cn

Abstract. Artificial Intelligence (AI) significantly improves time series forecasting in the financial market, yet it is challenging to establish reliable real-world finance applications due to a lack of transparency and explainability. This paper prototypes an explainable CNN-LSTM model that combines the advantages of CNN and LSTM (Long and Short Term) to train and forecast the price of Bitcoin using a group of 11 determinants. By avoiding information loss and information superposition, it combines long-term context information and short-term feature information to obtain comprehensive and accurate feature representation. Experiments show that CNN-LSTM generally has higher accuracy than a single LSTM network when processing and predicting Bitcoin sequence data, as measured by a mean absolute percentage error (MAPE) of 2.39% and an accuracy of 89.54%. Additionally, the CNN-LSTM model explains that trading volume and prices (Low, High, Open) contribute to the price dynamics, while oil and Dow Jones Index (DJI) influence the price behavior at a low level. We argue that understanding these underlying explanatory determinants may increase the reliability of AI's prediction in the cryptocurrency and general finance market.

Keywords: Bitcoin · Determinants · Price Forecast · Explainable AI · Accuracy

1 Introduction

In recent years, the remarkable expansion of the cryptocurrency market, led by Bitcoin, has attracted the interest of investors, analysts, and researchers. The inherent volatility in Bitcoin prices poses a unique challenge for forecasting, which has led to the exploration of advanced machine-learning techniques [1]. This study focuses on developing and applying an explainable CNN-LSTM model for forecasting Bitcoin prices, in line with the changing landscape of predictive analytics in the cryptocurrency field.

The application of deep learning techniques, such as Convolutional Neural Networks (CNNs) and Long Short-Term Memory networks (LSTMs) [2], has gained prominence in time series forecasting because of their ability to capture complex patterns within nonlinear and dynamic data.

C. Cruz et al. (Eds.): IC 2023, CCIS 2036, pp. 93–101, 2024.
https://doi.org/10.1007/978-981-97-0065-3_7

Furthermore, the significance of interpretability in machine learning models cannot be overstated, particularly in the realm of financial forecasting. The SHAP (Shapley Additive Explanations) algorithm has emerged as a powerful tool for providing transparent insights into the decision-making processes of complex models [3]. Through SHAP, our goal is to elucidate the contributing factors behind the predictions of our CNN-LSTM model, in order to foster a deeper understanding of the dynamics that drive Bitcoin price movements.

As we navigate this evolving landscape, this research not only contributes to the expanding knowledge in cryptocurrency forecasting but also aims to address the requirement for transparency and interpretability in predictive models. In the subsequent sections, we will elaborate on our methodology, data selection, and model architecture, providing a comprehensive overview of our approach and its implications in the context of recent advancements in the field. Through this exploration, we aim to provide valuable insights for market participants and researchers who are looking to navigate the complexities of Bitcoin price prediction.

2   Methodology

2.1   Convolution Neural Network (CNN)

Convolutional neural networks are an important branch of deep learning neural networks. It was first proposed by Yann LeCun [4] and others in the late 1980 s and early 1990 s. CNN is primarily inspired by biological visual systems, and its design is influenced by the mechanism of feature extraction in the visual cortex neurons. With the advancement of computing power and the availability of vast datasets, Convolutional Neural Networks (CNN) have achieved significant progress in computer vision tasks, including image recognition, object detection, and image generation [5]. As a result, CNN has emerged as a fundamental technology in the field of deep learning [6].

The convolutional neural network can be considered as a feature extractor with outstanding performance and a high level of automation. Its basic structure consists of two layers, one of which is the feature extraction layer. The input of each neuron is connected to the local receptive field of the previous layer, and the local features are extracted. Once the local feature is extracted, its positional relationship with other features is also determined. The second layer is the feature mapping layer. Each computing layer of the network is composed of multiple feature maps, and each feature map represents a plane. All neurons on the plane have equal weights. The feature mapping structure utilizes a sigmoid function as the activation function of the convolutional network, resulting in a small impact on the function kernel and making the feature map shift invariant. In addition, since neurons on a mapping surface share weights, the number of free parameters in the network is reduced. Each convolutional layer in the convolutional neural network is followed by a pooling layer for local averaging and feature extraction. This unique two-stage feature extraction structure reduces the feature resolution.

Traditional time series prediction methods often struggle with handling non-linear and non-stationary time series data. However, CNN, as a powerful deep learning technology, possesses excellent feature extraction capabilities. By utilizing the sliding window convolution kernel, CNNs can automatically learn local features in time series data. They can also gradually extract more abstract and advanced features between multiple convolutional layers, making them ideal for handling time series prediction tasks.

2.2 Long Short-Term Memory (LSTM)

LSTM (Long Short-Term Memory) is a special type of recurrent neural network (RNN) model [7] that addresses the issues of gradient explosion and gradient disappearance in long sequence training [8]. Compared to the traditional RNN model, LSTM incorporates memory units in the hidden layer and regulates the input and output of historical information through a gating structure. This allows for better control of memory information in time series data.

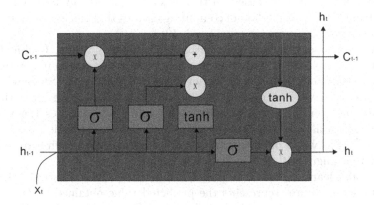

Fig. 1. LSTM network structure diagram.

As shown in Fig. 1, the network structure of LSTM consists of three gating units, namely the input gate (It), the forgetting gate (Ft), and the output gate (Ot). The calculation formula is as follows:

$$I_t = \sigma \left(X_t W_{xi} + H_{t-1} W_{hi} + b_i \right) \tag{1}$$

$$F_t = \sigma \left(X_t W_{xf} + H_{t-1} W_{hg} + b_f \right) \tag{2}$$

$$O_t = \sigma \left(X_t W_{xo} + H_{t-1} W_{ho} + b_o \right) \tag{3}$$

where: σ is the sigmoid activation function; W_{xi} is the weight matrix from the input x to the input gate; X_t is the input at time t; W_{hi} is the weight matrix from the hidden layer state to the input gate; b_i is the linear offset of the input gate; W_{xi} is the weight matrix from the input x to the input gate.

The memory unit state value C_t at time t consists of two parts, namely the memory unit state value C_{t-1} at the previous moment and the input gate input information to be updated $\widetilde{C}_t$. These two parts are controlled by the input gate and the forget gate respectively, thereby obtaining the current memory. The calculation formulas of unit state C_t are:

$$\widetilde{C}_t = \tanh\left(X_t W_{xc} + H_{t-1} W_{hc} + b_c\right) \tag{4}$$

$$C_t = F_t \odot C_{t-1} + I_t \odot \widetilde{C}_t \tag{5}$$

2.3   Shapley Additive Explanations (SHAP)

SHAP (Shapley Additive Explanations) is an explanatory method for explaining the prediction results of machine learning models. It was proposed by Lundberg and Lee in 2017 [9] to offer global and local explanations for black-box models, such as deep learning models. The SHAP method is based on the Shapley value in cooperative game theory. It sorts the value of each "player" in all possible "cooperations" according to their respective contributions to the total expenditure and obtains income based on this, ensuring consistency between income and contributions. This method of computing ensures that the interaction effects between features are properly considered [10].

SHAP is a model explanation tool based on Shapley values. It aims to use algorithms to explain the underlying model, enabling people to understand its working method and the basis for the decision-making process. Shapley's value is the core of SHAP. The principle is derived from "cooperative game theory," similar to the concept of "more work, more gain." It calculates the value of each "player" in all possible situations based on their contribution to the total expenditure. The values of "cooperation" are sorted, and the benefits are obtained in a way that aligns with the contributions made.

In machine learning, the term "player" refers to the characteristics of the sample, "total expenditure" represents the predicted value obtained by the sample after the model operation, and "profit" is the difference between the predicted value of this sample and the mean predicted value of all samples. By calculating the Shapley value of each feature, you can understand its contribution to the prediction results in the model's decision-making process. This allows you to learn the basis for the model's predictions.

Due to the nature of post hoc explanation, the explanation process of SHAP is independent of the model, allowing it to run concurrently with the prediction model. Therefore, most machine learning models can use SHAP for explanatory analysis. SHAP explains the model's prediction results for the sample by calculating the contribution of each feature in the sample. The Shapely value is represented as a linear model with feature attribution additivity, which converts the model's prediction value into the sum of the coefficient values of each feature [11]. It not only clarifies the importance of each feature but also explains the direction of influence that each feature has on model decision-making. Therefore, SHAP can not only be used as an interpretation tool for machine learning models but also as an effective feature selection method [12].

3   Experiment

The model training data is obtained from the investment data platform. The dataset includes a total of 11 attributes, such as market data, mainstream currency data, and macroeconomic indicators, from 2015 to 2020 as the research objects. Among them, the market data includes the opening price, highest price, lowest price, closing price, and trading volume of BTC. The mainstream currency data includes the closing price of XRP and LTC. The macroeconomic indicators include crude oil price, gold price, and Dow Jones Index. As there are missing values in the data set, the mean value is used to fill them. 90% of the data in the dataset is used to train the model, while the remaining 10% is used to test the model's prediction ability (Fig. 2).

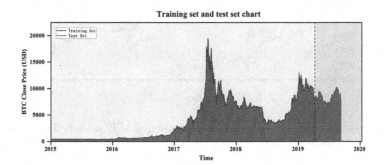

Fig. 2. Bitcoin Price Dataset.

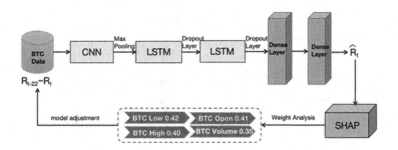

Fig. 3. CNN-LSTM Bitcoin prediction model.

The prediction network structure model mainly includes an input layer, a hidden layer, and an output layer. The input layer contains the relevant data from the training set R. As shown in Fig. 3, the hidden layer mainly consists of one layer of the CNN network, two layers of the LSTM network, and a Dense layer. To enhance the network's robustness and nonlinear prediction ability, a Dropout layer is included in the network model design. The output layer primarily generates the prediction results for the closing price of Bitcoin. It utilizes the Shap method to analyze the interpretability of the prediction model, thereby enhancing the model's ability to forecast Bitcoin prices.

As shown in Fig. 4, the trained CNN-LSTM network model is utilized to conduct research on bitcoin price prediction, using 11 determinants related to bitcoin prices. The MAPE value is 3.65% and the accuracy is 72.54%. It can be seen from the figure that the CNN-LSTM network model accurately predicts the overall bitcoin price, which aligns well with the actual price. But there is a problem. The prediction error in the second half of the test set is relatively large, which may be attributed to the overall market growth. The model fails to adjust to fluctuations in the Bitcoin market.

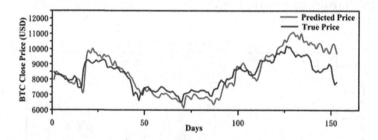

Fig. 4. Predict bitcoin price using 11 determinants, MAPE value is 3.65%, Accuracy is 72.54% (Tolerance=5%).

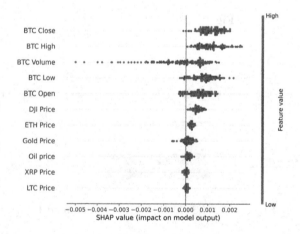

Fig. 5. The Shap model explains the prediction results.

For the model mentioned above, we utilized the SHAP method to perform a comprehensive interpretability analysis of the prediction data. The final calculation results of the analysis are shown in Fig. 5. The Bitcoin transaction volume has a significant impact on predicting Bitcoin's closing price, and there is a positive correlation between Bitcoin's lowest and highest prices and the results. As for factors such as the price of crude oil, ETH, XRP, and LTC, most data points are distributed within the range of a SHAP value of 0. This indicates that they have a minimal impact on most prediction results.

Based on the interpretability analysis results of the SHAP method, we conducted an in-depth investigation into several key parameters that have a significant impact on the predicted price of Bitcoin. These parameters include the opening price, closing price, lowest price, and highest price. A total of four attributes were used to train the model. The prediction results are shown in Fig. 6 below. Compared to the original model, we have enhanced its accuracy in predicting the price of Bitcoin.

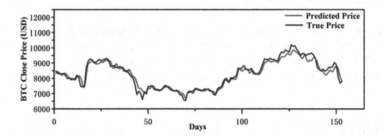

Fig. 6. Predict bitcoin price using 5 determinants, MAPE value is 2.39%, Accuracy is 89.54% (Tolerance=5%).

The performance of the enhanced model has significantly improved, demonstrating impressive statistical indicators. The mean absolute percentage error (MAPE) value is 2.39%, and the accuracy is 89.54%. Compared to the previous model, the accuracy rate has increased by 23%. This indicates that the improved model is better equipped to handle the diverse fluctuations in Bitcoin prices. Such results not only demonstrate the effectiveness of the SHAP method in selecting specific parameters, but also emphasize the superior performance of the improved model in enhancing prediction accuracy. These findings have significant implications for enhancing our understanding of Bitcoin price fluctuations and for making more accurate predictions about Bitcoin price trends in financial decision-making and investments.

4 Conclusion

In this study, we utilized the tensorflow and mindspore machine learning library and the Keras open-source artificial neural network library to construct a novel CNN-LSTM model. By training the model on attributes related to Bitcoin prices and conducting interpretability analysis using the SHAP algorithm, we obtained insights into the performance of the model and the importance of its features during the testing phase. Based on the results of the interpretability analysis, we selected several attributes that were most helpful in training the model and further optimized it.

The constructed CNN-LSTM model not only performs well but also has a strong prediction ability for Bitcoin price changes. This discovery not only has practical significance for financial market participants but also provides strong support for deepening the understanding of Bitcoin price fluctuations.

In future research, we will continue to focus on the rapid development of the deep learning field and explore more advanced deep learning models to further enhance the accuracy of Bitcoin price predictions. This method of continuous iteration and optimization will enable us to better adapt to dynamic changes in the market and establish a solid foundation for research and application in the field of Bitcoin price prediction.

Acknowledgments. We thank the reviewers, Jingbo Zhang and Weiyan Liu, for the useful comments. We thank Huawei for providing a Mindspore framework for AI analysis and CUP-SHFE Finance Lab for providing a Bloomberg facility. The support fund was sponsored by the CAAI-Huawei MindSpore Open Fund (No. CAAIXSJLJJ-2022-001A).

References

1. Grinberg, R.: Bitcoin: an innovative alternative digital currency. Hastings Sci. Technol. Law J. **4**, 159–208 (2011)
2. Fawaz, H.I., Forestier, G., Weber, J., Idoumghar, L., Muller, P.-A.: Deep learning for time series classification: a review. Data Min. Knowl. Discov. **33**, 917–963 (2018)
3. Guidotti, R., Monreale, A., Ruggieri, S., Turini, F., Pedreschi, D., Giannotti, F.: A survey of methods for explaining black box models. ACM Comput. Surv. **93**, 1–42 (2018)
4. LeCun, Y., Boser, B., Denker, J.S., Henderson, D., Howard, R.E., Hubbard, W., Jackel, L.D.: Backpropagation applied to handwritten zip code recognition. Neural Comput. **1**(4), 541–551 (1989)
5. He, K., Zhang, X., Ren, S., Sun, J.: Deep residual learning for image recognition. In: Proceedings of the IEEE Conference on Computer Vision and Pattern Recognition, pp. 770–778 (2016)
6. Simonyan, K., Zisserman, A.: Very deep convolutional networks for large-scale image recognition (2014). arXiv preprint arXiv:1409.1556
7. Hochreiter, S., Schmidhuber, J.: Long short-term memory. Neural Comput. **9**(8), 1735–1780 (1997)
8. Greff, K., Srivastava, R.K., Koutník, J., Steunebrink, B.R., Schmidhuber, J.: LSTM: a search space odyssey. IEEE Trans. Neural Netw. Learn. Syst. **28**(10), 2222–2232 (2015)
9. Lundberg, S.M., Lee, S.I.: A unified approach to interpreting model predictions. In: Proceedings of the 31st International Conference on Neural Information Processing Systems (NIPS 2017), pp. 4765–4774 (2017)
10. Ancona, M., Ceolini, E., Öztireli, C., Gross, M.: Explaining deep neural networks with a polynomial time algorithm for Shapley value approximation. In: Proceedings of the IEEE/CVF Conference on Computer Vision and Pattern Recognition (CVPR 2019), pp. 8557–8566 (2019)

11. Lundberg, S.M., Lee, S.-I.: A unified approach to interpreting model predictions. Adv. Neural Inf. Process. Syst. **30**, 4768–4777 (2017)
12. Marcilio, W.E., Eler, D.M.: From explanations to feature selection: assessing SHAP values as feature selection mechanism C. In: 2020 33rd SIBGRAPI Conference on Graphics Patterns and Images (SIBGRAPD), pp. 340–347 (2020)

Augmenting Bankruptcy Prediction Using Reported Behavior of Corporate Restructuring

Xinlin Wang$^{(\boxtimes)}$ 🄳 and Mats Brorsson 🄳

Interdisciplinary Centre for Security, Reliability and Trust, University of
Luxembourg, 1855 Kirchberg, Luxembourg
{xinlin.wang,mats.brorsson}@uni.lu

Abstract. Credit risk assessment of a company is commonly conducted
by utilizing financial ratios that are derived from its financial statements.
However, this approach may not fully encompass other significant aspects
of a company. We propose the utilization of a hybrid dataset that combines financial statements with information about corporate restructuring behavior in order to construct diverse machine learning models to
predict bankruptcy. Utilizing a hybrid data set provides a more comprehensive and holistic perspective on a company's financial position
and the dynamics of its business operations. The experiments were carried out using publicly available records of all the files submitted by
small and medium-sized enterprises to Luxembourg Business Registers.
We conduct a comparative analysis of bankruptcy prediction using six
machine learning models. Furthermore, we validate the effectiveness of
the hybrid dataset. In addition to the conventional testing set, we deliberately chose the timeframe encompassing the years of the Covid-19 pandemic as an additional testing set in order to evaluate the robustness of
the models. The experimental results demonstrate that the hybrid data
set can improve the performance of the model by 4%–13% compared to
a single source data set. We also identify suitable models for predicting
bankruptcy.

Keywords: Machine learning · Bankruptcy prediction · Credit risk

1 Introduction

Small and medium-sized enterprises (SMEs) are of paramount importance in
diverse economies. According to the World Bank [1], SMEs constitute approximately 90% of all companies and play a substantial role in generating over 50% of
global employment. SMEs operate predominantly within localized communities,
providing employment opportunities to local residents and acting as drivers of
economic advancement by fostering competition, innovation, and increased pro-
ductivity. Moreover, SMEs demonstrate a strong sense of social responsibility
and commitment to sustainability, frequently prioritizing community engagement and environmental preservation [2]. Therefore, helping SMEs to run in
good health is of great importance to society and the economy.

Predicting bankruptcy for SMEs can help enterprises to proactively identify potential risks, adapt their business strategies, and improve their overall competitiveness and stability in a timely manner. Furthermore, bankruptcy prediction can facilitate the assignment of credit ratings for SMEs, providing credit endorsements and enabling them to access better financial services. This, in turn, can promote their growth and development [3]. Finally, bankruptcy prediction can help governments and social organizations identify potential financial crises and implement proactive measures to mitigate the adverse economic and social consequences.

The origins of bankruptcy prediction models can be traced back to the late 1960s, when the Logit model was proposed by Beaver in 1966 [4] and the Z-score model was proposed by Altman in 1968 [5]. Both models were formulated using financial ratios and played a significant role during their respective periods, establishing a fundamental framework for subsequent investigations into the prediction of bankruptcy. Financial ratios are accounting-based ratios used to assess the financial health of a company, typically derived from its financial statements [6]. With the development of the financial industry and the field of data science, numerous studies have been conducted on the prediction of studies primarily rely on accounting-based ratios and employ various models to predict [7–12]. Although several studies have also incorporated various types of data, including market-based variables [13–15] and macroeconomic indicators [16], studies on input data (or features) had been a largely under explored domain compared to studies on models.

In this study, our goal is to improve the accuracy of bankruptcy prediction models by including data on reported corporate restructuring behavior, in addition to using accounting-based ratios as input variables. We used a publicly available dataset from Luxembourg Business Registers (LBR)[1] to implement experiments and validate our hypothesis. Registered companies in Luxembourg are required to submit their basic information, business operation files, and financial statements to LBR. We create a hybrid dataset consisting of accounting-based ratios and features related to restructuring behavior. We compare the bankruptcy prediction results of six machine learning models: logistic regression (LR), random forest (RF), lightGBM (LGB), multilayer perceptron (MLP), convolutional neural network (CNN), and long short-term memory (LSTM). We validate the effectiveness of a hybrid dataset and identify suitable models for predicting bankruptcy. We specifically compare the time periods before and after the pandemic as the testing sets to assess the robustness of the models during the special economic period.

This paper makes several contributions to the field of bankruptcy prediction, including the following:

– We present the first large-sample bankruptcy prediction using corporate restructuring behavior, which, to the best of our knowledge, has not been explored before;

---

[1] https://www.lbr.lu/.

- We conduct a comparative study of six well-known bankruptcy prediction models using real-world data from Luxembourg Business Registers;
- We evaluate the performance of the models in response to Covid-19 pandemic period and analyze the drift of prediction models.

2 Related Works

Researchers often prioritize finding ways to improve the effectiveness of the model while overlooking the importance of studying input data. In [17]'s review, the authors summarized over 60 studies that apply accounting-based ratios, also known as financial ratios, to various models. In Table 1, we have provided a list of recent studies that focus on input data and selected representative works that are based on financial ratios. We present these studies by year of publication, categorizing them according to data category, data type, prediction models, evaluation approaches, sample size and publication year. In recent decades, there has been a notable rise in studies that examine different input data. The development of computer technology and data science has made it easier to collect, store, process, and model data.

In recent years, studies have shown the benefits of incorporating diverse input data into bankruptcy prediction models. These studies have expanded beyond traditional data, such as financial ratios and market-based variables, to explore various types of input data. This work [24] confirms that financial ratios are predictive indicators of firm failure. The study also suggests that non-financial variables, such as localization and economic conditions, are drivers of SMEs failure. The study [16] combines financial ratios and macroeconomic data to analyze their impact on firms, providing evidence for the reliability of macroeconomic data. Another study [22] focuses on using SMEs' transaction data for the prediction of bankruptcy, without relying on accounting data. The results show that this approach outperformed the benchmark method. Some research pairs also include studies on different types of data. The authors [21] use shared directors and managers to establish a connection between two companies and developed a model using relational data to identify the companies with the highest risk. In contrast, this study [15] focuses on using a deep learning model to extract textual information as a complementary variable to accounting and market data to improve prediction accuracy. This paper focuses on corporate restructuring behaviors, such as changes in registered address, management, and corporate regulations. There are two pieces of work [7] and [9] use similar indicators related to corporate restructuring, but they are static and cannot reflect corporate behavior. The present study addresses this gap by focusing on data that reflect changes in corporate behavior.

As demonstrated in Table 1, many studies have been dedicated to improving prediction accuracy using various models. Bankruptcy prediction models must be applied practically in the financial industry, necessitating both model accuracy and explainability. The work [19] compares the accuracy and explainability of different data mining methods for predicting bankruptcy. The study compares

Table 1. Studies on Bankruptcy Prediction

Study	Data Category	Data Type	Prediction Models	Evaluation Approaches	Sample Size	Publish Year
[7]	Financial ratios, basic firm information, reported and compliance, operational risk	Numerical data	Altman's Z-score, generic model	AUC, roc curve	3,462,619	2008
[13]	Financial ratios, market-based variables	Numerical data	Black and Scholes models, Altman's Z-score	ROC curve, information content tests	15,384	2008
[14]	Financial ratios, market-based variables	Numerical data	MLP, CART, LR, RF, SVM, ensemble, boosting	Accuracy, sensitivity, specificity	16816	2009
[8]	Financial ratios	Numerical data	MLP, boosting, bagging	Accuracy ratio, AUC	1458	2009
[18]	Financial ratios, corporate governance indicators	Numerical data	Altman's Z-score, SVM	Type I error, Type II error, average accuracy, brier score	108	2010
[19]	Financial ratios	Numerical data	DT, LR, MLP, RBFN, SVM	Correct classification rate	1321	2012
[20]	Accounting, market and macroeconomic data	Numerical data	LR, Altman's Z-score, MLP	AUC, Gini rank coefficient, Kolmogorov-Smirnov	23,218	2013
[9]	Financial ratios, corporate governance indicators	Numerical data	SVM, KNN, NB,CART, MLP	ROC curve	478	2016
[10]	Financial ratios	Image data	CNN	Identification rates, ROC curve	7520	2019
[21]	Financial ratios, relational data	Numerical data, graph data	SVM,GNN	AUC	60,000	2017
[11]	Financial ratios	Numerical data	LR, ANN, SVM, PLS-DA, SVM-PLS	Confusion matrix, accuracy, sensitivity, specificity, AUC	212	2017
[16]	Financial ratios, macroeconomic indicators, industrial factors	Numerical data	MDS	/	165	2019
[15]	Accounting-based ratio, market-based variables, textual discolures	Numerical data, text data	Word embedding, CNN, DNN	Accuracy ratio, AUC	11,827	2019
[12]	Financial ratios, basic firm information	Numerical data, categorical data	LR, RF, XGBoost, LightGBM, ANN	AUC	977,940	2019
[22]	Basic firm informaion, SME network-based variables, transactional data	Numerical data, graph data, categorical data	LDA, LR, SVM, DT, RF, XGB, NN	AUC	340,531	2021
[23]	Textual sentiment	Text	SVM, Bayes, KNN, DT, CNN, LSTM	AUC	10,034	2022

algorithms such as neural networks, support vector machines, and decision trees, and concludes that decision trees are both more accurate and easier to interpret compared to neural networks and support vector machines. In [25] it was demonstrated that LightGBM achieved the highest performance, with fast and cost-effective training, and the model's results could be interpreted using SHAP value analysis. In contrast, authors [15] argue that simple deep learning models outperform other data mining models. This paper conduct a comparative analysis study by selecting multiple models from Table 1 which are more universal for data modeling and comparison. The authors [26] use the classical bankruptcy prediction models from the study [27] and the study [28] to validate the model performance for different time periods, however, applying these models to a time period other than the one in which they were developed can significantly reduce their accuracy. Furthermore, the study [29] shows that forecasting models perform significantly worse during crisis periods compared to non-crisis periods. In this paper, we compare the model performance in two periods: pre-Covid19 and post-Covid19, to verify whether the model performance changes due to the pandemic.

In summary, this paper aims to enhance the performance of bankruptcy prediction model by utilizing a hybrid dataset that combines corporate restructuring behavior data with accounting-based ratios. To determine the robustness of the model during the Covid-19 period, separate testing sets will be used for observation.

3   Methodology

The main focus of our study is to examine the effectiveness of reported corporate restructuring behavior in predicting bankruptcy. We also aim to analyze the robustness of various bankruptcy models during the Covid-19 pandemic. In this section, we will first present the overall framework for investigating these problems. Then, we will focus on the details of the input data and explain the experimental design.

3.1   Conceptual Framework

Figure 1 illustrates the six stages of a framework designed to conduct a comparative study of bankruptcy models using different input data. The data used in the study consists of financial statements and reporting documents. However, since the cash flow statement and profit & loss statement were not included, the financial statements only consisted of the balance sheet. The reporting documents that companies submit to disclose their operational behaviors are usually classified as textual files.

Different methods are used to extract three types of features from the raw data. The first is financial ratios. Since we do not have a cash flow statement or profit & loss statement, we will create as many financial ratios as possible. The second type also includes accounting-based features. These variables are

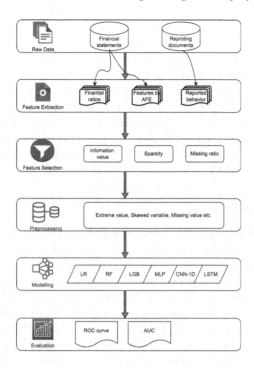

Fig. 1. Conceptual framework of experimental design

constructed using an automatic feature engineering method that we developed in our previous work. This method is capable of creating highly effective features even with limited data [30]. The final type of features is behavior-related features. We design these variables based on corporate restructuring behavior, such as changes in registered addresses, manager resignations, and mergers and acquisitions.

The next stage of the framework involves selecting features from the current variables to eliminate unfavorable and redundant variables caused by sparsity, missing, and repetition. The information value (IV) is an indicator used to measure the predictive power of an independent feature [31]. A higher information value indicates that the feature has greater predictive power. The formula for calculating information value is as follows [32]:

$$IV = \sum_{i=1}^{n} (\frac{G_i}{G} - \frac{B_i}{B}) * \ln \frac{G_i/G}{B_i/B} \tag{1}$$

We select features with an IV value greater than 0.02 and a missing rate less than 0.7.

The data are pre-processed to address missing values, infinite values, and skewed variables, making it more suitable for modeling. We also exclude abnormal samples, such as companies that have submitted financial reports prior to the reference year. We replace infinite values with the highest finite value. In the fifth stage, we assess the effectiveness of behavior-related features by comparing the prediction performance of hybrid datasets that include both behavior-related features and accounting-based ratios with datasets that only contain accounting-based ratios. We trained six popular models, including logistic regression (LR), random forest (RF), LightGBM (LGB), multiple perceptron (MLP), convolutional neural network (CNN) and long-short-term memory (LSTM) to compare their prediction results. We use the receiver operating characteristic curve (ROC curve) and Area under the ROC curve (AUC) as indicators of evaluating the performance of models, which are commonly used and discussed in Sect. 2.

3.2 Variables and Data

Since over 90% of the firms in the Luxembourg industry distribution are finance-related, this paper excludes these firms and focuses only on SMEs as our target samples to avoid an imbalanced distribution of samples. The state of a company is not static and can change over time, either by being established or going bankrupt. This means that the company may enter or exit the sample set. We utilize a sliding time window approach to sample from the raw data. The sliding-time window is a technique used to extract data from a time series dataset by defining a fixed period of time (window) and moving it forward by a certain interval (step size). This technique allows for continuous monitoring of system states [33].

As of June 2022, there are 74,611 companies in Luxembourg. The average lifespan of companies is approximately 3.5 years. Therefore, we have selected a timeframe of up to 3 years for predicting bankruptcy. We create datasets with three different windows (1-year, 2-year, and 3-year) to predict one step forward (one year). According to the timeline (Fig. 3), the three datasets consist of 1-year data from t_{-1} to t_0, 2-year data from t_{-2} to t_0, and 3-year data from t_{-3} to t_0. The sample size of these three datasets, including solvent and bankrupted companies, was summarized in Table 2. However, there is another category of companies with an unknown status. Some companies have not uploaded annual reports or declared bankruptcy, which contributes to the variation in data from year to year. As depicted in the Fig. 2, the bankruptcy rate of SMEs in Luxembourg has decreased over the past decade. It may indicate that the business conditions of SMEs are improving or that the overall economic environment has improved, resulting in greater stability for SMEs. Additionally, other factors such as policy support or industry changes may also influence the bankruptcy rate of SMEs. It is surprising to find that the bankruptcy rate of SMEs has actually increased during the Covid-19 pandemic, suggesting that fewer SMEs are going bankrupt compared to previous periods. We hypothesize that this could be attributed to government financial assistance during the special period. Some companies may be technically bankrupt but have not yet filed for bankruptcy due to delays in filing.

As mentioned earlier, we derive three types of features from raw data: two accounting-based variables and one behavior-based variable. The statistics and

Table 2. Summary of three datasets

Year	1-year		2-year		3-year	
	Solvent	Bankrupt	Solvent	Bankrupt	Solvent	Bankrupt
2012	21738	621	/	/	/	/
2013	23804	687	17087	461	/	/
2014	25686	669	18790	451	16512	361
2015	27331	663	20301	436	18188	361
2016	28781	653	21475	461	19477	384
2017	30748	661	22789	449	20755	378
2018	32718	606	24419	392	22061	322
2019	34557	431	25793	319	23504	267
2020	36309	179	27034	138	24596	121
2021	22387	34	17195	28	15571	24

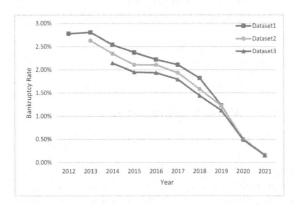

Fig. 2. Bankruptcy rate of three datasets from 2012 to 2021

descriptions of these features can be found in Table 3. SMEs are not required to prepare and disclose cash flow statements and income statements. Therefore, we can only calculate 18 financial indicators based on the available data [34–37]. We developed an algorithm for automatic feature engineering [30] to derive as many useful features as possible from financial statements to address issues caused by the absence of certain financial statements or data quality problems. This algorithm maximizes data mining to generate high-quality features that enhance prediction accuracy. Behavior-based variables are derived from information reported by SMEs regarding corporate restructuring, including both statistical and trend variables.

3.3   Experimental Setup

Dataset Description. We divide the datasets into two parts: the training set and the testing set, as outlined in Table 4. To maintain consistency between the training set and the testing set, we divided the data from 2012 to 2018 into a 70% training set and a 30% testing set. Additionally, we created two additional testing sets: one using solvent and bankrupt SMEs from 2019 as a pre-Covid testing set, and another using solvent and bankrupt SMEs from 2020 and 2021 as a post-Covid testing set. To train our models, we utilized the 5-fold cross-validation method and did not set aside a separate validation set. Table 4 have a bankruptcy rate below 3%, making them highly imbalanced. The negative datasets are typically large in size, so we used the under-sampling method during data preprocessing to balance the rate to 25%.

$$t_{-3} \qquad t_{-2} \qquad t_{-1} \qquad t_0$$

Fig. 3. Definition of time period

Table 3. Description of variables in this study

Variable	Description
Financial ratios (FR)	Current ratio, debt to equity, working capital to total assets, total liabilities to total assets, equity to total assets, quick ratio, current assets to total assets, cash to total assets, cash to current liabilities, long term debt to equity, total assets growth rate, quick assets to total assets, current assets to current liabilities, (cash or marketable securities) to total assets, total debt to total assets, equity to fixed assets, current assets to total liabilities, short-term liabilities to total assets
Automatic feature engineering (AFE)	Automatically generate features from financial statements, which can adapt to any kind of numerical data
Reported corporate restructuring behavior-related features (RB)	Modification of name or corporate name, registered office, social object, administrator/manager, daily management delegate, associate, person in charge of checking the accounts, Social capital/social funds, managing director/steering committee, duration, legal form, social exercise, permanent representative of the branch, merger/demerger, depositary, transfer of business assets, assets or business sectors, address, trading name, activities, manager, seat, reason, name, chairman/director, personne autorisée à gérer, administrer et signer, person with the power to commit the company, ministerial approval

Table 4. Summary of datasets splitting

		1-year	2-year	3-year
Training	Solvent (Negative)	110805	87467	67920
	Bankrupt (Positive)	2625	1797	1244
	Bankruptcy Rate	2.31%	2.01%	1.80%
Testing	Solvent (Negative)	47458	37403	29081
	Bankrupt (Positive)	1155	853	562
	Bankruptcy Rate	2.38%	2.23%	1.90%
Pre-Covid	Solvent (Negative)	28730	25793	23504
	Bankrupt (Positive)	368	319	267
	Bankruptcy Rate	1.26%	1.22%	1.12%
Post-Covid	Solvent (Negative)	48846	44229	40167
	Bankrupt (Positive)	181	166	145
	Bankruptcy Rate	0.37%	0.37%	0.36%

Models. In this paper, we have chosen six bankruptcy prediction models, which include statistical, machine learning, and deep learning models, by synthesizing the statistics from previous studies in Part II. We comprehensively evaluate the behavior-based features by comparing the performance of representative models. Table 5 displays the environmental information used for model training.

Logistic regression predicts the likelihood of a binary outcome using one or more predictor variables. Logistic regression models have advantages in bankruptcy prediction due to their simplicity, fast computation, and better results when dealing with smaller datasets. In this study, we adopt *LogisticRegression* from *sklearn* package and use *GridSearch* to determine the optimal parameters within a specific range.

Table 5. Information of training machine

Device name	Tesla V100-SXM2-32 GB
Linux version	Red Hat 8.5.0–10
Python version	3.8.6
Pytorch version	1.10.1+cu111
Cuda version	11.1
Cudnn version	8005
Sklearn version	1.2.1
Number of GPU	2
Number of CPU	16

Random forest creates a forest of decision trees, with each tree being trained on a random subset of the data and a random subset of predictor variables. Random forest models outperform single decision tree models and other classification models in terms of predictive performance and robustness, and can effectively handle high-dimensional and complex datasets. In this study, we utilize *RandomForestClassifier* from *sklearn* package and employ *GridSearch* to determine the optimal parameters within a specific range.

LightGBM prioritizes speed and efficiency, specifically for managing large datasets. The method utilizes a gradient-based approach to construct decision trees and incorporates various optimization techniques to accelerate the training process. In this study, we adopt *LGBMClassifier* from *lightgbm* package and use *GridSearch* to decide the best parameters from a specific range.

Multilayer perceptron is commonly used for classification and regression tasks. It has the ability to learn complex non-linear relationships between inputs and outputs, making it a powerful tool for various applications. In this study, we incorporate embedding layers for sparse reported behavior features, as depicted in Fig. 4. We train the model by Pytorch. We choose *BCEWithLogitsLoss* as loss function, *Adam* as optimizer, and *auc* as metric function. We set batch size to 64, epoch to 50 and learning rate to 0.00001.

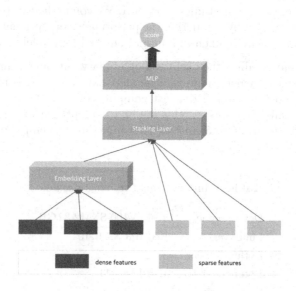

Fig. 4. Structure of MLP

Convolutional Neural Network. In this study, we only have tabular data, so we use a one-dimensional CNN (CNN-1D) for prediction. CNN-1D is more effective at capturing local features in the data and has a strong ability to adapt. The convolutional layer extracts features from input data, the pooling layer reduces the number of features and improves model robustness, and the fully connected layer maps the features to the output space for classification. We choose *BCE-WithLogitsLoss* as loss function, *Adam* as optimizer, and *auc* as metric function. We set batch size to 1024, epoch equals to 50 and learning rate to 0.00005.

Long short-term memory aims to address the vanishing gradient problem commonly encountered in traditional recurrent neural networks. The model is capable of retaining long-term dependencies in the input data, making it suitable for various sequence prediction tasks. We reshape the data to fit the time step and features for LSTM in order to predict bankruptcy several years in advance. In this study, we choose *BCEWithLogitsLoss* as loss function, *Adam* as optimizer, and *auc* as metric function. We set batch size to 64, epochs to 50 and learning rate to 0.00001.

Performance Evaluation. In selecting the performance measures for the model, we refer to and synthesize previous studies in Sect.2 and select two metrics, AUC and ROC curve, to assess the effectiveness of the model.

Area under the Receiver Operating Characteristic Curve (AUC) is a performance metric that assesses a classification model's ability to differentiate between positive and negative samples. AUC is not affected by sample imbalance or threshold selection, making it a more comprehensive measure of classifier performance compared to accuracy. The interpretation is straightforward as it summarizes the model's performance with a single scalar value. The formula for calculating AUC is:

$$\text{AUC} = \int_0^1 \text{TPR}(FPR^{-1}(t)) \, dt \tag{2}$$

Receiver Operating Characteristic Curve (ROC curve) is a graphical representation of the True Positive Rate (TPR) plotted against the False Positive Rate (FPR) at various classification thresholds. TPR represents the proportion of positive samples correctly classified as positive. On the other hand, FPR represents the proportion of negative samples incorrectly classified as positive. ROC curve is a useful tool for visualizing the trade-off between TPR and FPR at various classification thresholds. The curve is created by plotting the TPR against FPR for every possible classification threshold. It offers a visual representation of the model's performance and helps in selecting the right classification threshold, considering the desired balance between TPR and FPR. The formula for calculating TPR and FPR is:

$$\text{FPR} = \frac{\text{False Positives}}{\text{False Positives} + \text{True Negatives}} \tag{3}$$

and

$$\text{FPR} = \frac{\text{False Positives}}{\text{False Positives} + \text{True Negatives}} \tag{4}$$

And we plot ROC curve by

$$\text{ROC curve: TPR vs FPR} \tag{5}$$

4   Results and Discussion

4.1   Features Evaluation

Information value (IV) is a widely used metric for selecting features in binary classification models. Assesses the ability of a feature to predict the target variable by analyzing its relationship. In essence, IV quantifies the amount of information that a feature provides about the target variable. It is commonly used to rank the importance of different features in a predictive model. We calculate the IV for AFE features, financial ratios, and behavior-related features. The results are displayed in Fig. 5. When performing feature selection using IV, features with high IV scores are generally considered more important and informative than those with low IV scores. By eliminating features with low IV scores, we can potentially simplify the model, enhance its performance, and identify the most significant predictors for a specific problem. Additionally, IV provides a standardized and interpretable measure of feature importance that can be easily communicated to stakeholders and decision-makers.

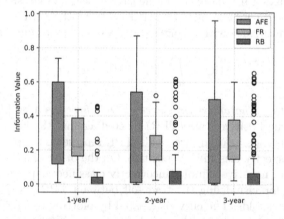

Fig. 5. IV for features created from AFE, FR and RB

We observe that the number of AFE features is the highest, and most of these features have relatively high IV values. Financial ratios, while fewer in number compared to AFE features, have higher IV values and are less varied. On the other hand, behavior-related features exhibit a wide range of IV values, with some having very high values and the majority clustered towards the lower end of the y-axis. This suggests that these features have little impact on predicting bankruptcy. Behavioral correlation features are often sparse matrices, with the majority of eigenvalues being 0. To mitigate the drawbacks of high coefficient matrices, we will employ feature filtering and summing techniques to maximize the utilization of the data.

4.2 Ablation Experimental Results

We implement the ablation experiments to evaluate if the behaviour-related features can improve the model performance. We create four datasets: AFE, AFE+RB, FR and FR+RB to compare the model performance of with RB features and without RB features. The experiments were carried out on 6 models and 3 time periods. We select 2 out of the 18 results as the representative results and include all the other experimental results in the appendix for reference. Figure 6 summarizes the performance of different features on lightGBM and LSTM by comparing their ROC curves. We use a green line to represent AFE features, a yellow line to represent FR features, a red line to represent AFE and RB features, and a brown line to represent FR and RB features. From this figure, it can be clearly seen that models trained on hybrid datasets of financial and behavior-related features outperform datasets that only include financial features.

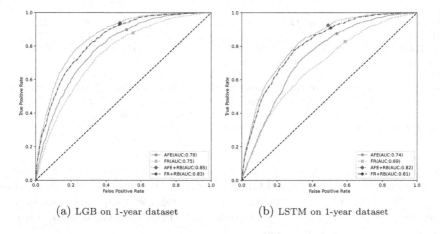

(a) LGB on 1-year dataset   (b) LSTM on 1-year dataset

Fig. 6. ROC curve of lightGBM and LSTM on 1-year datasets

The results of LR (Fig. 7a, Fig. 7b, Fig. 7c), LGB(Fig. 7g, Fig. 6b, Fig. 7h) and LSTM (Fig. 6b, Fig. 7n, Fig. 7p) very clearly show the advantages of hybrid datasets for bankruptcy prediction. Although the results of RF(Fig. 7d, Fig. 7d, Fig. 7f) and the results of MLP (Fig. 7i, Fig. 7j, Fig. 7k), we can still find the advantage of hybrid datasets, but not very obvious. The results of CNN-1D(Fig. 7l, Fig. 7m, Fig. 7n) are inconclusive, as the performance of financial-related features is comparable to random guessing. Additionally, the performance of models improves with longer training data periods. This suggests that using a larger data set can capture more accurate trends and patterns that indicate

Table 6. AUC of models on testing, pre-Covid and post-Covid sets

	1-year			2-year			3-year		
	Testing	Pre-Covid	Post-Covid	Testing	Pre-Covid	Post-Covid	Testing	Pre-Covid	Post-Covid
LR									
AFE	0.7522	0.7539	**0.7669**	0.7494	0.7486	**0.7632**	**0.7702**	0.7599	0.7548
FR	0.7231	**0.7493**	0.7425	0.7384	**0.7621**	0.7605	0.7706	**0.7755**	0.7660
AFE+RB	**0.8234**	0.7433	0.7129	**0.8556**	0.7636	0.7124	**0.8767**	0.7755	0.7226
FR+RB	**0.8118**	0.7375	0.6948	**0.8642**	0.7758	0.7037	**0.8918**	0.7885	0.7178
RF									
AFE	0.7894	**0.8018**	0.7786	0.7860	**0.8077**	0.7775	0.8024	0.8052	**0.8078**
FR	0.7564	**0.7687**	0.7337	0.7676	**0.7838**	0.7454	0.7836	0.7810	**0.7873**
AFE+RB	0.7733	0.7739	**0.8195**	0.7733	0.7658	**0.8423**	0.7876	0.7711	**0.8589**
FR+RB	0.7294	0.7316	**0.7941**	0.7450	0.7309	**0.8164**	0.7524	0.7286	**0.8418**
LGB									
AFE	0.7887	0.7980	**0.8092**	0.7925	0.7976	**0.8156**	0.8133	0.8108	**0.8147**
FR	0.7490	0.7629	**0.7741**	0.7634	0.7752	**0.7912**	**0.7990**	0.7865	0.7903
AFE+RB	**0.8542**	0.7732	0.7705	**0.8783**	0.8065	0.7623	**0.8930**	0.8168	0.7621
FR+RB	**0.8312**	0.7421	0.7226	**0.8706**	0.7759	0.7299	**0.8903**	0.7919	0.7313
MLP									
AFE	0.7408	0.7383	**0.7536**	0.7428	0.7447	**0.7752**	**0.7489**	0.7240	0.7263
FR	0.7282	**0.7482**	0.7299	0.7204	**0.7585**	0.7537	0.7301	**0.7462**	0.7329
AFE+RB	**0.8109**	0.6799	0.6775	**0.8014**	0.6841	0.7032	**0.7743**	0.6710	0.6643
FR+RB	**0.7980**	0.6904	0.6740	**0.8145**	0.6946	0.6775	**0.8218**	0.7238	0.6902
CNN-1D									
AFE	0.7277	0.7288	**0.7335**	0.5607	0.5857	**0.6705**	0.4318	0.4629	**0.5317**
FR	0.6153	0.6495	**0.7055**	0.4922	0.4954	**0.5524**	0.7165	0.7196	**0.7333**
AFE+RB	**0.7442**	0.7090	0.7188	**0.7306**	0.6361	0.6229	**0.7243**	0.6298	0.5995
FR+RB	**0.7508**	0.6456	0.6504	**0.7139**	0.615	0.6141	**0.7217**	0.6339	0.6114
LSTM									
AFE	0.7404	0.7468	**0.7579**	0.7066	0.7036	**0.7406**	0.7193	0.7248	**0.7363**
FR	0.6879	0.7247	**0.7497**	0.7064	0.7369	**0.7628**	0.7419	0.7554	**0.7563**
AFE+RB	**0.8245**	0.7342	0.7145	**0.8158**	0.7094	0.6924	**0.8046**	0.7236	0.6951
FR+RB	**0.8087**	0.7159	0.6920	**0.8211**	0.7461	0.7146	**0.8187**	0.7470	0.7114

potential bankruptcy. Furthermore, it is worth noting that machine learning models such as LR, RF, and LGB outperform deep learning models such as MLP, CNN-1D, and LSTM. Overall, hybrid datasets offer significant advantages over single-source datasets for predicting bankruptcy. LightGBM model outperforms all other models in 3 time periods.

4.3   Performance About Covid Period

As described in Sect. 3, the bankruptcy rate decreases significantly since 2019. It only has a 1% bankruptcy rate in 2019 and less than a 5% bankruptcy rate for 2020 and 2021. There are several reasons for the drop in the bankruptcy rate. First, the implementation of fiscal stimulus policies. Many countries adopted large-scale fiscal stimulus policies to ease the economic pressure caused by the epidemic, such as providing loans, tax cuts, and direct funding to businesses. Implementing these policies may help companies maintain cash flow and reduce the risk of bankruptcy. Second, debt moratorium and grace period. Many companies obtained debt moratorium and grace period arrangements during the pandemic, which allowed them to delay debt repayment, thereby easing short-term financial stress and reducing the risk of bankruptcy.

However, this downward trend in bankruptcy rates may only be temporary, as these policies and arrangements may be unsustainable and companies are facing various uncertainties and challenges. For now, we can not see any evidence directly from the data but just observe that the distribution of both pre-Covid set and post-Covid set drift a lot from the training set. The experimental results verify this observation Table 6.

More than 50% results show that the model performances of pre-Covid sets and post-Covid sets are better than those of testing sets, which is contrary to common sense. Furthermore, we find the hybrid datasets perform less favorable for both pre-Covid and post-Covid time period. We assume that this is because the reporting behavior of companies changed during the pandemic period, which means companies may not submit or report their restructuring behavior in time due to the pandemic. This inconsistency on reported behavior data will confuse the model thus make the prediction performance not as good as testing set.

5   Conclusion

In conclusion, this study introduces the historical background of bankruptcy prediction models, which have traditionally used accounting-based ratios as input variables. It also presents a new approach to improving these models by incorporating data on reported corporate restructuring behavior. The study compares six models and identifies the most suitable one for predicting bankruptcy. The study validates the effectiveness of the hybrid dataset and analyzes the potential drift of the model during the pandemic. The experimental results demonstrate that utilizing a hybrid dataset can enhance the performance of bankruptcy prediction models by 4%–13% compared to using a single source dataset. We also assess the performance of these models during the pandemic and analyze their drift. This study offers valuable insights into bankruptcy prediction models and highlights areas for future research. The findings of this study can assist SMEs in identifying risks, adapting their business strategies, and enhancing their competitiveness and stability in a timely manner. Furthermore, the proposed bankruptcy prediction model can be used to assign credit ratings for SMEs and provide credit endorsements, thereby facilitating their growth. Finally, this study can assist governments and social organizations in identifying potential financial crises and implementing proactive measures to mitigate adverse economic and social impacts.

Acknowledgements. We thank the Luxembourg National Research Fund (FNR) under Grant 15403349 and Yoba S.A. for supporting this work.

A   Appendix A

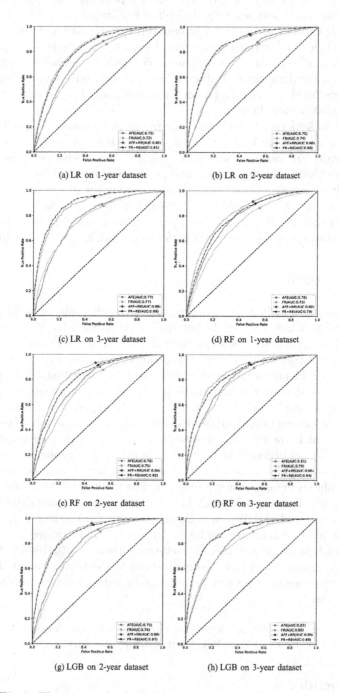

(a) LR on 1-year dataset

(b) LR on 2-year dataset

(c) LR on 3-year dataset

(d) RF on 1-year dataset

(e) RF on 2-year dataset

(f) RF on 3-year dataset

(g) LGB on 2-year dataset

(h) LGB on 3-year dataset

Fig. 7. The rest results for ROC curve of 6 models on 3 datasets

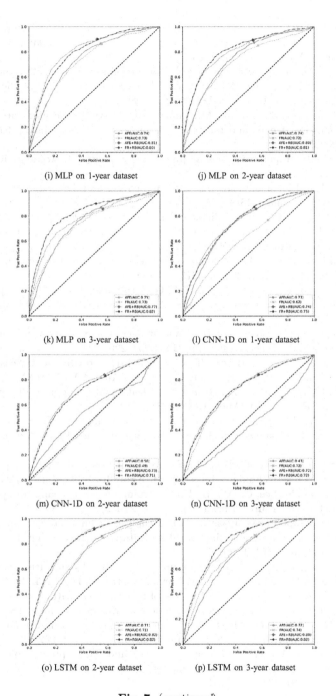

(i) MLP on 1-year dataset

(j) MLP on 2-year dataset

(k) MLP on 3-year dataset

(l) CNN-1D on 1-year dataset

(m) CNN-1D on 2-year dataset

(n) CNN-1D on 3-year dataset

(o) LSTM on 2-year dataset

(p) LSTM on 3-year dataset

Fig. 7. (*continued*)

References

1. Bank, W.: Improving smes' access to finance and finding innovative solutions to unlock sources of capital (2023). https://www.worldbank.org/en/topic/smefinance. Accessed 10 Apr 2023
2. Varga, J.: Defining the economic role and benefits of micro small and medium-sized enterprises in the 21st century with a systematic review of the literature. Acta Polytechnica Hungarica **18**(11), 209–228 (2021)
3. Rao, P., Kumar, S., Chavan, M., Lim, W.M.: A systematic literature review on SME financing: trends and future directions. J. Small Bus. Manag. **61**(3), 1247–1277 (2023)
4. Beaver, W.H.: Financial ratios as predictors of failure. J. Account. Res. 71–111 (1966)
5. Altman, E.I.: Financial ratios, discriminant analysis and the prediction of corporate bankruptcy. J. Finan. **23**(4), 589–609 (1968)
6. Damodaran, A.: Corporate Finance. Wiley, Hoboken (1996)
7. Altman, E.I., Sabato, G., Wilson, N.: The value of non-financial information in sme risk management. SSRN 1320612 (2008)
8. Kim, M.-J., Kang, D.-K.: Ensemble with neural networks for bankruptcy prediction. Expert Syst. Appl. **37**(4), 3373–3379 (2010)
9. Liang, D., Lu, C.-C., Tsai, C.-F., Shih, G.-A.: Financial ratios and corporate governance indicators in bankruptcy prediction: a comprehensive study. Eur. J. Oper. Res. **252**(2), 561–572 (2016)
10. Hosaka, T.: Bankruptcy prediction using imaged financial ratios and convolutional neural networks. Expert Syst. Appl. **117**, 287–299 (2019)
11. Mselmi, N., Lahiani, A., Hamza, T.: Financial distress prediction: the case of French small and medium-sized firms. Int. Rev. Finan. Anal. **50**, 67–80 (2017)
12. Son, H., Hyun, C., Phan, D., Hwang, H.J.: Data analytic approach for bankruptcy prediction. Expert Syst. Appl. **138**, 112816 (2019)
13. Agarwal, V., Taffler, R.: Comparing the performance of market-based and accounting-based bankruptcy prediction models. J. Bank. Finan. **32**(8), 1541–1551 (2008)
14. Chandra, D.K., Ravi, V., Bose, I.: Failure prediction of dotcom companies using hybrid intelligent techniques. Expert Syst. Appl. **36**(3), 4830–4837 (2009)
15. Mai, F., Tian, S., Lee, C., Ma, L.: Deep learning models for bankruptcy prediction using textual disclosures. Eur. J. Oper. Res. **274**(2), 743–758 (2019)
16. Khoja, L., Chipulu, M., Jayasekera, R.: Analysis of financial distress cross countries: using macroeconomic, industrial indicators and accounting data. Int. Rev. Finan. Anal. **66**, 101379 (2019)
17. Kumar, P.R., Ravi, V.: Bankruptcy prediction in banks and firms via statistical and intelligent techniques-a review. Eur. J. Oper. Res. **180**(1), 1–28 (2007)
18. Lin, F., Liang, D., Chu, W.-S.: The role of non-financial features related to corporate governance in business crisis prediction. J. Mar. Sci. Technol. **18**(4), 4 (2010)
19. Olson, D.L., Delen, D., Meng, Y.: Comparative analysis of data mining methods for bankruptcy prediction. Decis. Supp. Syst. **52**(2), 464–473 (2012)
20. Tinoco, M.H., Wilson, N.: Financial distress and bankruptcy prediction among listed companies using accounting, market and macroeconomic variables. Int. Rev. Finan. Anal. **30**, 394–419 (2013)
21. Tobback, E., Bellotti, T., Moeyersoms, J., Stankova, M., Martens, D.: Bankruptcy prediction for SMEs using relational data. Decis. Supp. Syst. **102**, 69–81 (2017)

22. Kou, G., et al.: Bankruptcy prediction for SMEs using transactional data and two-stage multiobjective feature selection. Decis. Supp. Syst. **140**, 113429 (2021)
23. Huang, B., Yao, X., Luo, Y., Li, J.: Improving financial distress prediction using textual sentiment of annual reports. Ann. Oper. Res. **330**, 1–28 (2022)
24. Schalck, C., Yankol-Schalck, M.: Predicting French sme failures: new evidence from machine learning techniques. Appl. Econ. **53**(51), 5948–5963 (2021)
25. Lextrait, B.: Scaling up SMEs' credit scoring scope with lightGBM. Appl. Econ. **55**, 1–19 (2022)
26. Grice, J.S., Dugan, M.T.: The limitations of bankruptcy prediction models: some cautions for the researcher. Rev. Quant. Finan. Acc. **17**, 151–166 (2001)
27. Zmijewski, M.E.: Methodological issues related to the estimation of financial distress prediction models. J. Account. Res. 59–82 (1984)
28. Ohlson, J.A.: Financial ratios and the probabilistic prediction of bankruptcy. J. Account. Res. 109–131 (1980)
29. Papík, M., Papíková, L.: Impacts of crisis on SME bankruptcy prediction models' performance. Expert Syst. Appl. **214**, 119072 (2023)
30. Wang, X., Kräussl, Z., Zurad, M., Brorsson, M.: Effective automatic feature engineering on financial statements for bankruptcy prediction (2022)
31. Howard, R.A.: Information value theory. IEEE Trans. Syst. Sci. Cybern. **2**(1), 22–26 (1966)
32. Siddiqi, N.: Credit Risk Scorecards: Developing and Implementing Intelligent Credit Scoring, vol. 3. John Wiley & Sons, Hoboken (2012)
33. Zhang, L., Lin, J., Karim, R.: Sliding window-based fault detection from high-dimensional data streams. IEEE Trans. Syst. Man Cybern. Syst. **47**(2), 289–303 (2016)
34. Boguslauskas, V., Mileris, R., Adlytė, R.: The selection of financial ratios as independent variables for credit risk assessment. Econ. Manag. **16**(4), 1032–1040 (2011)
35. Yu, Q., Miche, Y., Séverin, E., Lendasse, A.: Bankruptcy prediction using extreme learning machine and financial expertise. Neurocomputing **128**, 296–302 (2014)
36. Zhu, Y., Zhou, L., Xie, C., Wang, G.-J., Nguyen, T.V.: Forecasting SMEs' credit risk in supply chain finance with an enhanced hybrid ensemble machine learning approach. Int. J. Prod. Econ. **211**, 22–33 (2019)
37. Zięba, M., Tomczak, S.K., Tomczak, J.M.: Ensemble boosted trees with synthetic features generation in application to bankruptcy prediction. Expert Syst. Appl. **58**, 93–101 (2016)

AI for Education

A New Dataset and Method for Creativity Assessment Using the Alternate Uses Task

Luning Sun[1], Hongyi Gu[2], Rebecca Myers[1], and Zheng Yuan[2,3]($\boxtimes$)

[1] University of Cambridge, Cambridge, UK
{ls523,rm804}@cam.ac.uk
[2] NetMind.AI, London, UK
hongyi.gu@netmind.ai
[3] King's College London, London, UK
zheng.yuan@kcl.ac.uk

Abstract. Creativity ratings by humans for the alternate uses task (AUT) tend to be subjective and inefficient. To automate the scoring process of the AUT, previous literature suggested using semantic distance from non-contextual models. In this paper, we extend this line of research by including contextual semantic models and more importantly, exploring the feasibility of predicting creativity ratings with supervised discriminative machine learning models. Based on a newly collected dataset, our results show that supervised models can successfully classify between creative and non-creative responses even with unbalanced data, and can generalise well to out-of-domain unseen prompts.

Keywords: Creativity · Alternate uses task · Automated scoring

1 Introduction

Creativity, defined as the production of novel and useful products [24], is one of the most important skills for student and young people development [3], and a valuable employee outcome associated with organisational sustainability and innovation [13]. A core element of creativity is divergent thinking in problem solving [15,20]. One of the most widely used divergent thinking tests is the alternate uses task (AUT) [14,31], which asks respondents to list as many uses for common items (e.g. newspaper) as possible, and usually within a time limit. The responses are then rated on dimensions such as *fluency, originality, flexibility*, and *elaboration* [1]. Similar to many other creativity tests, it requires human raters to score the responses manually, rendering the results subjective, unreliable, and undermining their validity [17]. Consequently, education and training in creativity are severely constrained by the lack of an objective and efficient measurement of creativity [30].

To automate the scoring process of the AUT, researchers have capitalised on recent developments in natural language processing (NLP) and proposed that semantic distance could be calculated to predict human creativity ratings. For

C. Cruz et al. (Eds.): IC 2023, CCIS 2036, pp. 125–138, 2024.
https://doi.org/10.1007/978-981-97-0065-3_9

instance, [12] found that GloVe [26], among a number of publicly available word embeddings models, produced the most reliable and valid *originality* scores on the AUT. [5] constructed a latent semantic distance factor based on five non-contextual semantic spaces, and found strong correlations between the semantic distances and the respondent-level (i.e. person-level) human ratings of creativity in the AUT responses.

Unlike previous work, we propose to address the AUT scoring as a supervised discriminative machine learning problem and particularly as a binary classification problem: classifying between creative and non-creative responses. In addition to examining the relationship between semantic distance variables and human ratings of creativity in the AUT responses, we explore supervised machine learning models for the prediction of creativity ratings. Our results show that the proposed method generalises well to unseen tasks and prompts. We also compare the performance of our proposed models to that of OpenAI's ChatGPT,[1] and discuss its potential application in creativity assessment.

This paper makes the following contributions. First, we introduce a new dataset of AUT responses, the Cambridge AUT Dataset,[2] and make it publicly available to facilitate future research on creativity assessment. Second, to our knowledge, we present the first comparison between the application of contextual and non-contextual semantic spaces in the context of creativity assessment. Finally, as far as we know, this is the first attempt to apply a supervised learning model to the scoring of AUT responses, which demonstrates performance improvement across a set of different prompts.

2   The Cambridge AUT Dataset

2.1   Data Collection

The AUT data used in this study was collected as part of a larger project on creativity assessment [25] that received ethics approval from both the Faculty of Education, University of Cambridge and Cambridge Judge Business School. Two common objects were implemented as prompts for the AUT, namely *bowl* and *paperclip*. For each prompt, participants were given 90 s to come up with as many different uses as possible (see Sect. A).

A total of 1,297 participants (Gender: 693 female, 567 male, 14 other, 23 missing; Age: mean 26.26 years, SD 9.68 years, 13 missing; Ethnicity: 883 White, 54 Asian, 54 Black, 110 mixed, 124 other, 72 missing), who were recruited through Cambridge University mailing lists, social media, and a testing website,[3] took part in the task online between April 2020 and January 2021.[4] 1,027 of them provided non-empty answers for *bowl* (each with an average of 7.40 uses; SD:

---

[1] https://chat.openai.com/.

[2] https://github.com/ghydsgaaa/Cambridge-AUT-dataset.

[3] https://discovermyprofile.com/.

[4] Participants were not paid but given the opportunity to opt into a draw to win one of ten £10 Amazon vouchers.

Table 1. Response examples of different average ratings for each prompt.

Average rating	Prompt: bowl	Prompt: paperclip
1.0	fish holder	drawing
2.0	doing an inhalation	make a logo
3.0	space ship	pasta mold
4.0	sending mail through river	holding nose while swimming

3.49) and 1,020 for *paperclip* (each with an average of 6.23 uses; SD: 2.94). For each object, all uses (referred to as *responses* below) were pooled together and only the English ones were subject to annotation.

2.2 Annotation

We applied the subjective scoring method based on the Consensual Assessment Technique [2,10]. A group of psychology students were trained on how to evaluate the responses on their *originality*, using a Likert scale from 0 to 4, where 0 indicates a not valid or not relevant use, 1 a common use without any originality, 2 an uncommon use with limited originality, and 3 and 4 original uses with moderate and extreme creativity, respectively.

Three raters were initially recruited to annotate the AUT responses. Each of them was tasked with a random sample of the responses. The assignment of the responses among the raters ensured that each unique response would be rated by at least two raters. Due to time constraints, one of the raters had to quit midway and the remaining annotation was completed by a fourth rater (their ratings were combined in the dataset).

After removing duplicate responses, a total of 3,380 responses for *bowl* and 3,650 for *paperclip* were annotated. Both objects received the same average rating (1.27, SDs: 0.49 for *bowl* and 0.45 for *paperclip*). 95 responses for *bowl* and 86 for *paperclip* received average ratings of below 1, which means that at least one of the raters rated the responses as invalid uses, hence being removed from the subsequent analyses. Response examples of different average ratings for each prompt are presented in Table 1.

Notably, the dataset is severely unbalanced, with more than half responses rated 1 and only a few responses rated 3 and above - see Table 2. This is expected, as creative responses are less frequent by nature. Nonetheless, less frequent responses may not necessarily be creative. The creativity ratings in this work focus on the absolute originality in the responses rather than their relevant frequency. It is also worth noting that the inter-rater agreement is not particularly high (correlations range from 0.39 to 0.58 - see Table 3) compared to other assessment tasks such as essay scoring [4]. This is likely due to the nature of human ratings in creativity assessment, which are based on their own subjective perception of creativity [10,23].

Table 2. Number of responses per average rating in the Cambridge AUT dataset. Responses with an average rating below 1 (i.e. a not valid or not relevant use) are excluded from the analyses.

Average rating	#responses combined	#responses (bowl)	#responses (paperclip)
<1.0	181	95	86
1.0	4,167	2,096	2,071
1.5	1,717	638	1,079
2.0	666	392	274
2.5	188	104	84
3.0	92	48	44
3.5	15	6	9
4.0	4	1	3
Total	7,030	3,380	3,650

3 Semantic Models

Following previous work [5,12], we analyse the AUT responses collected in our dataset and test whether combining multiple models of semantic distance into a single latent variable can approximate human creativity ratings.

3.1 Semantic Distance

Pre-trained semantic models are used to compute the semantic distance (i.e. cosine distance) between the prompt and the response. We employ four contextual models: Universal Sentence Encoder [9],[5] Sentence-Transformers [27],[6] DistilRoBERTa [28],[7] and GPT-3 [8];[8] and three non-contextual models: GloVe [26],[9] Word2vec [21],[10] and fastText [7].[11]

For non-contextual models, we first extract embeddings for each word in the response, and then take the multiplicative composition as suggested by [5,22]. For contextual models, we extract the sentence embeddings directly.

3.2 Confirmatory Factor Analysis

Table 3 presents zero-order correlations among human ratings and semantic distance variables. Confirmatory factor analysis (CFA) is performed to investigate

---

[5] https://tfhub.dev/google/universal-sentence-encoder/4.
[6] https://huggingface.co/sentence-transformers/all-MiniLM-L6-v2.
[7] https://huggingface.co/distilroberta-base.
[8] https://beta.openai.com/docs/models/gpt-3.
[9] glove-wiki-gigaword-300.
[10] word2vec-google-news-300.
[11] fasttext-wiki-news-subwords-300.

Table 3. Correlations among human ratings and semantic distance variables: poly-choric correlations between human raters, polyserial correlations between human raters and semantic distances, and pearson correlations between semantic distances. r1-3: rater1-3; USE: Universal Sentence Encoder; ST: Sentence-Transformers.

	r1	r2	r3	USE	ST	RoBERTa	GPT-3	GloVe	Word2vec	fastText
r1	1.00	-	-	-	-	-	-	-	-	-
r2	0.58	1.00	-	-	-	-	-	-	-	-
r3	0.39	0.44	1.00	-	-	-	-	-	-	-
USE	0.14	0.16	0.16	1.00	-	-	-	-	-	-
ST	0.19	0.17	0.31	0.62	1.00	-	-	-	-	-
RoBERTa	0.12	0.16	0.22	0.57	0.76	1.00	-	-	-	-
GPT-3	0.08	0.20	0.11	0.51	0.50	0.55	1.00	-	-	-
GloVe	0.05	0.02	0.06	0.12	0.05	-0.08	-0.14	1.00	-	-
Word2vec	0.03	0.04	0.05	0.22	0.17	0.09	0.21	0.40	1.00	-
fastText	−0.03	−0.01	0.03	0.15	0.11	0.07	0.14	0.35	0.24	1.00

the latent correlation between human ratings and a semantic distance factor underlying different semantic models.[12]

We specify two models to examine the relationship between the response-level human ratings and the semantic distance factors built upon contextual ($\mathbf{Model_{contextual}}$) and non-contextual semantic models ($\mathbf{Model_{non\text{-}contextual}}$), respectively. Both contextual and non-contextual models yield good model fit to the data.[13] The contextual model reveals a higher correlation between the latent semantic distance factor and the human ratings than the non-contextual model ($r = 0.065$, $p < .001$ for the non-contextual model - see Fig. 1, Sect. B; and $r = 0.293$, $p < .001$ for the contextual model - see Fig. 2, Sect. B).

Nevertheless, these latent correlations between the response-level human ratings and the semantic distance factors are still considerably low, in comparison to those correlations reported in previous studies based on the respondent-level data [5,12], suggesting that these semantic distance variables cannot be used reliably as an unsupervised model to predict human creativity ratings.

4   Binary Classification Models

Since the semantic distance variables reported above fail to adequately predict the human creativity ratings, in this section we turn to supervised machine learning methods. In light of the availability of a labeled dataset, we conduct experiments, where we fine-tune pre-trained language models to improve their prediction accuracy. Since the dataset is severely unbalanced (see Table 2), we

---

[12] CFA is a statistical technique used to verify the factor structure of a set of observed variables and test if the relationship between observed variables and their underlying latent constructs exist.

[13] Detailed CFA results are presented in Table 6, Sect. B.

Table 4. Micro-average F1 scores on the AUT test sets. The highest scores for each prompt are in bold.

Tested on	$Model_{bowl}$	$Model_{paperclip}$	$Model_{bowl+paperclip}$	ChatGPT	Baseline
Bowl	**0.79**	0.73	0.76	0.65	0.70
Paperclip	0.61	0.65	**0.67**	0.56	0.60
Combined	0.69	0.68	**0.72**	0.60	0.65

cast the task as a binary classification between creative (average rating > 1, i.e. at least one of the raters assigned 2 or above) and non-creative (average rating $= 1$) responses. Take prompt *bowl* as example, "mixing stuff" is considered a non-creative response with average rating 1 and "knee caps" is considered a creative response with average rating 3. We further split the dataset into a training set (90%) and a test set (10%).

4.1   Fine-Tuned Models

Fine-tuning pre-trained language models via supervised learning is key to achieving state-of-the-art performance in many NLP tasks. Adopting this approach, we experiment with three transformer-based pre-trained language models: BERT [11], RoBERTa [19], and GPT-3 [8].

To fine-tune BERT and RoBERTa, we use them as the underlying language model and add a linear layer on the top, which allows for binary classification. We construct the input by concatenating the prompt w and the response $R = r_1, r_2, ..., r_n$:

$$[CLS]; w; [SEP]; r_1, r_2, ..., r_n; [SEP] \qquad (1)$$

where the $[CLS]$ representation is then fed into the output layer for classification. During training, the model is optimised in an end-to-end manner. We fine-tune *bert-base-uncased*[14] and *roberta-base*[15] on the AUT data, with a batch size of 32 and a learning rate of $3 \times e^{-05}$ for 5 epochs.

For GPT-3, we fine-tune the GPT-3 babbage model using the OpenAI's API.[16]

In our experiments, 5-fold cross validation is performed and detailed results are presented in Table 7, Table 8 and Table 9, Sect. C. The fine-tuned BERT models are chosen for later experiments due to their superior micro-average F1 scores.

4.2   Results

Prediction results of our fine-tuned BERT models on the test sets for each prompt as well as both prompts combined are reported in Table 4. Three binary

---

[14] https://huggingface.co/bert-base-uncased.

[15] https://huggingface.co/roberta-base.

[16] https://openai.com/blog/openai-api.

Table 5. Micro-average F1 scores on the dataset from [6]. The highest scores for each prompt are in bold.

Tested on	Model$_{bowl}$	Model$_{paperclip}$	Model$_{bowl+paperclip}$	ChatGPT	Baseline
Box	0.64	**0.72**	0.69	0.54	0.58
Rope	**0.62**	0.61	**0.62**	0.51	0.51

classification models trained on different data are compared: **Model$_{bowl}$** is trained on responses for *bowl* only; **Model$_{paperclip}$** is trained on responses for *paperclip* only; and **Model$_{bowl+paperclip}$** is trained on the data for both prompts.

Using the majority class as **Baseline**, we observe an increase in the F1 scores on the prompt-specific level (i.e. in-domain) and the same for the cross-prompt predictions (i.e. out-of-domain). The best model for prompt *bowl* is the prompt-specific model **Model$_{bowl}$**, achieving a micro-average F1 score of 0.79. Notably, **Model$_{bowl+paperclip}$** yields the best performance when tested on prompt *paperclip*, outperforming its prompt-specific model **Model$_{paperclip}$** (0.67 vs. 0.65). These results suggest that given more data (even from out-of-domain prompts), the model is able to improve the overall performance on different prompts, hence showing a potential to serve as prompt-independent filters for creative responses in the AUT.

4.3   A Case Study with New AUT Prompts

In order to explore the generalisability of our models, we apply our classification models to the AUT responses collected in a previous study with different prompts than those here, namely *rope* and *box* [6]. Since a different annotation scheme was used - a scale from 1 (not at all creative) to 5 (very creative), we split their data into two classes: non-creative (responses with an average human rating of 1), and creative (those with an average human rating of 2 or above).

In Table 5 we report the prediction results of our models on the responses to prompts *box* and *rope*.[17] In general, all our models outperform the majority class baseline, indicating a prompt independence and a cross-dataset applicability. The result suggests that using training data from only a few prompts (even just one or two), it is possible to develop supervised machine learning models that can work as a generic, automated scoring tool for the AUT with any unseen prompt.

4.4   Comparison with ChatGPT Predictions

Inspired by recent progress on using generative, pre-trained large language models as evaluators in tasks like machine translation [18], code generation [32] and grammatical error correction [29], we explore how these models can be applied in creativity assessment. We apply ChatGPT (gpt-3.5-turbo at temperature 0)

---

[17] Per-class precision, recall and F1 scores are reported in Table 10, Sect. D.

to the same task on both our dataset and that from [6],[18] and report results in Table 4 and Table 5. We can see that **ChatGPT** underperforms the majority **Baseline** on both datasets, revealing its limitation in evaluating abstract concepts like creativity.

Detailed per-class analysis reveals that **ChatGPT** achieves high precision, yet considerably low recall for non-creative responses on both datasets, while an opposite pattern is observed for creative responses.[19] As its performance seems complementary to that of our fine-tuned models, we see a potential of integrating both methods, which may result in further performance gains in creativity assessment.[20]

5 Conclusions

In this paper, we performed confirmatory factor analysis to investigate the latent correlations between the semantic distance factors and the human ratings of creativity in a newly collected AUT dataset, the Cambridge AUT Dataset. On the response level, we observed significant but lower correlations than those on the respondent level as reported in previous studies. It was also noted that contextual semantic models appear to show greater resemblance to the human ratings than non-contextual models. One step further, we experimented with several fine-tuned models, which showed encouraging performance improvement in classifying between creative and non-creative responses under both in-domain and out-of-domain settings. When applied to an external dataset with new prompts, the models trained on our dataset exhibited reasonably well predictions, showing promising generalisability.

With the above findings, we see a possibility of developing an automated scoring tool for the AUT using supervised machine learning models. To extend this line of research, we plan to examine different model architecture and gather more data with different prompts, in order to better understand the generalisability of the supervised models in the general creativity assessment.

6 Limitations

We notice relatively low agreement among the annotators. One possible explanation is that the annotators come from different countries (e.g. the UK, India, and China) with different native languages and cultural backgrounds. Past literature [16] found cross-cultural differences in both the idea generation and the idea evaluation phases of the divergent thinking task. It is likely that the annotators do not share entirely the same conceptual framework for creative ideas

---

[18] The prompt we used for experiments with ChatGPT is provided in Sect. E.

[19] Per-class precision, recall and F1 scores are reported in Table 11 and Table 12, Sect. F.

[20] One viable solution is employing a voting ensemble technique, which involves assigning weights to results of both models and striking a balance between precision and recall. Alternatively, we could prompt ChatGPT to generate quantified results and establish a threshold for comparing its outputs with those of the fine-tuned models.

around the prompts, resulting in inconsistent ratings. Future work is warranted to confirm this.

Due to data imbalance and sparsity, this paper addresses the AUT scoring as a binary classification between creative and non-creative responses. The proposed approach may therefore fail to evaluate creativity at detailed levels of granularity. It would be ideal to collect more responses with higher ratings so as to develop an automated creativity assessment system with greater precision. Moreover, to address the concern of overfitting in our experiments, we used 5-fold cross validation and applied our models to unseen data, which showed comparable results.

The results with regard to ChatGPT is based on preliminary experiments. A more thorough investigation using different parameters, prompts, and models is warranted. We are excited to see how large language models like ChatGPT may help with creativity assessment in the future.

Acknowledgement. We would like to thank all participants who took part in the AUT and all raters who annotated the responses. LS acknowledges financial support from Invesco through their philanthropic donation to Cambridge Judge Business School.

A   The Instructions Used for the AUT

General instruction: For the next four questions, there will be a time limit. For each task, please read the instructions and enter each possible answer separately by pressing the enter key after each one. If you run out of answers you may move on by pressing the next button, otherwise your question will automatically change after the allocated time.

Each task requires you to come up with as many different answers as possible. Try to be creative as there is no right or wrong answer.

Prompt 1: List as many different uses of a bowl as you can think of.

Prompt 2: Think of many different uses of a paperclip.

B   Detailed CFA Results

Table 6. Latent correlations between human creativity ratings and semantic distance factors (**Model$_{non-contextual}$** and **Model$_{contextual}$**) on the Cambridge AUT dataset.

Tested on	Model$_{non-contextual}$	Model$_{contextual}$
Bowl	0.127	0.278
Paperclip	-	0.296
Combined	0.065	0.293

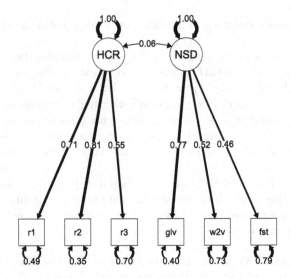

Fig. 1. CFA diagram of **Model$_{non-contextual}$** on the Cambridge AUT dataset. r1-3: rater1-3; glv: GloVe; w2v: Word2vec; fst: fastText; HCR: human creativity rating factor, NSD: non-contextual semantic distance factor.

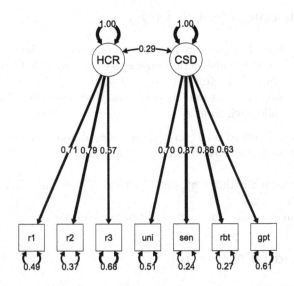

Fig. 2. CFA diagram of **Model$_{contextual}$** on the Cambridge AUT datseta. r1-3: rater1-3; uni: Universal Sentence Encoder; sen: Sentence-Transformers; rbt: RoBERTa; gpt: GPT-3; HCR: human creativity rating factor, CSD: contextual semantic distance factor.

C   Cross Validation Results

Table 7. Fine-tuned BERT cross validation results on the Cambridge AUT training sets. P: precision; R: recall.

Model	Non-creative			Creative			Micro-average
	P	R	F1	P	R	F1	F1
BERTbowl	0.86	0.88	0.87	0.72	0.68	0.70	0.82
BERTpaperclip	0.85	0.69	0.76	0.44	0.66	0.53	0.69
BERTbowl+paperclip	0.91	0.82	0.86	0.53	0.72	0.61	0.80

Table 8. Fine-tuned RoBERTa cross validation results on the Cambridge AUT training sets. P: precision; R: recall.

Model	Non-creative			Creative			Micro-average
	P	R	F1	P	R	F1	F1
RoBERTabowl	0.83	0.86	0.85	0.66	0.60	0.63	0.79
RoBERTapaperclip	0.70	0.81	0.76	0.63	0.49	0.55	0.68
RoBERTabowl+paperclip	0.80	0.74	0.77	0.58	0.67	0.62	0.71

Table 9. Fine-tuned GPT-3 babbage cross validation results on the Cambridge AUT training sets. P: precision; R: recall.

Model	Non-creative			Creative			Micro-average
	P	R	F1	P	R	F1	F1
GPT-3bowl	0.87	0.87	0.87	0.69	0.71	0.70	0.82
GPT-3paperclip	0.74	0.76	0.75	0.62	0.61	0.62	0.70
GPT-3bowl+paperclip	0.80	0.79	0.80	0.64	0.63	0.63	0.71

D   Model Performance on the Dataset from [6]

Table 10. Prediction performance on the dataset from [6]. P: precision; R: recall.

Tested on	Model	Non-creative			Creative			Micro-average
		P	R	F1	P	R	F1	F1
Box	**Model**bowl	0.57	0.65	0.61	0.70	0.63	0.66	0.64
	Modelpaperclip	0.60	0.78	0.68	0.84	0.69	0.75	0.72
	Modelbowl+paperclip	0.75	0.66	0.70	0.64	0.73	0.68	0.69
Rope	**Model**bowl	0.67	0.62	0.65	0.57	0.62	0.60	0.62
	Modelpaperclip	0.41	0.71	0.52	0.83	0.57	0.67	0.61
	Modelbowl+paperclip	0.60	0.64	0.62	0.64	0.60	0.62	0.62

E   ChatGPT Prompt

You are a judge in the alternate uses task, where respondents are asked to list different uses for a common object. You will be presented with the object and a response that illustrates one of its uses. Please judge if the response is creative or non-creative. Inappropriate, invalid, irrelevant responses, and responses with common uses are considered non-creative, whereas appropriate, valid, novel and unusual uses are considered creative.
The object is: {prompt}
The response is: {response}
Please give your answer in "creative" or "non-creative".
Your answer:

F   ChatGPT Classification Results

Table 11. ChatGPT results on the Cambridge AUT dataset. P: precision; R: recall.

Tested on	Non-creative			Creative			Micro-average
	P	R	F1	P	R	F1	F1
Bowl	0.85	0.60	0.70	0.45	0.75	0.56	0.65
paperclip	0.82	0.35	0.48	0.48	0.88	0.62	0.56
Combined	0.84	0.48	0.61	0.46	0.83	0.60	0.60

Table 12. ChatGPT results on the dataset from [6]. P: precision; R: recall.

Tested on	Non-creative			Creative			Micro-average
	P	R	F1	P	R	F1	F1
Box	0.77	0.28	0.42	0.48	0.88	0.62	0.54
Rope	0.69	0.35	0.47	0.42	0.76	0.54	0.51

References

1. Amabile, T.M.: Social psychology of creativity: a consensual assessment technique. J. Pers. Soc. Psychol. **43**(5), 997–1013 (1982)
2. Amabile, T.M.: The social psychology of creativity: a componential conceptualization. J. Pers. Soc. Psychol. **45**(2), 357–376 (1983)
3. Ananiadou, K., Claro, M.: 21st century skills and competences for new millennium learners in OECD countries. In: OECD Education Working Papers (41) (2009). https://www.oecd-ilibrary.org/content/paper/218525261154
4. Andersen, Ø.E., Yuan, Z., Watson, R., Cheung, K.Y.F.: Benefits of alternative evaluation methods for automated essay scoring. In: Proceedings of the 14th International Conference on Educational Data Mining (EDM 2021), Paris, France (2021)

5. Beaty, R.E., Johnson, D.R.: Automating creativity assessment with SemDis: an open platform for computing semantic distance. Behav. Res. Methods **53**(2), 757–780 (2021)
6. Beaty, R.E., et al.: Robust prediction of individual creative ability from brain functional connectivity. Proc. Natl. Acad. Sci. **115**(5), 1087–1092 (2018)
7. Bojanowski, P., Grave, E., Joulin, A., Mikolov, T.: Enriching word vectors with subword information. Trans. Assoc. Comput. Linguist. **5**, 135–146 (2017)
8. Brown, T., et al.: Language models are few-shot learners. In: Advances in Neural Information Processing Systems, vol. 33, pp. 1877–1901 (2020)
9. Cer, D., et al.: Universal sentence encoder. arXiv preprint arXiv:1803.11175 (2018)
10. Cseh, G.M., Jeffries, K.K.: A scattered CAT: a critical evaluation of the consensual assessment technique for creativity research. Psychol. Aesthet. Creat. Arts **13**(2), 159–166 (2019)
11. Devlin, J., Chang, M.W., Lee, K., Toutanova, K.: BERT: pre-training of deep bidirectional transformers for language understanding. arXiv preprint arXiv:1810.04805 (2018)
12. Dumas, D., Organisciak, P., Doherty, M.: Measuring divergent thinking originality with human raters and text-mining models: a psychometric comparison of methods. Psychol. Aesthet. Creat. Arts **15**(4), 645–663 (2021)
13. George, J.M., Zhou, J.: Dual tuning in a supportive context: joint contributions of positive mood, negative mood, and supervisory behaviors to employee creativity. Acad. Manag. J. **50**(3), 605–622 (2007). https://doi.org/10.5465/AMJ.2007.25525934
14. Guilford, J.P.: The Nature of Human Intelligence. McGraw-Hill, New York, NY (1967)
15. Guilford, J.P.: Creative Talents: Their Nature, Uses and Development. Bearly Limited, Buffalo, NY (1986)
16. Ivancovsky, T., Shamay-Tsoory, S., Lee, J., Morio, H., Kurman, J.: A dual process model of generation and evaluation: a theoretical framework to examine cross-cultural differences in the creative process. Personal. Individ. Differ. **139**, 60–68 (2019)
17. Kim, K.H.: Can we trust creativity tests? A review of the Torrance tests of creative thinking (TTCT). Creat. Res. J. **18**(1), 3–14 (2006). https://doi.org/10.1207/s15326934crj1801_2
18. Kocmi, T., Federmann, C.: Large language models are state-of-the-art evaluators of translation quality. arXiv preprint arXiv:2302.14520 (2023)
19. Liu, Y., et al.: RoBERTa: a robustly optimized BERT pretraining approach. arXiv preprint arXiv:1907.11692 (2019)
20. McCrae, R.R.: Creativity, divergent thinking, and openness to experience. J. Pers. Soc. Psychol. **52**(6), 1258–1265 (1987)
21. Mikolov, T., Sutskever, I., Chen, K., Corrado, G.S., Dean, J.: Distributed representations of words and phrases and their compositionality. In: Advances in Neural Information Processing Systems, vol. 26 (2013)
22. Mitchell, J., Lapata, M.: Composition in distributional models of semantics. Cogn. Sci. **34**(8), 1388–1429 (2010)
23. Mouchiroud, C., Lubart, T.: Children's original thinking: an empirical examination of alternative measures derived from divergent thinking tasks. J. Genet. Psychol. **162**(4), 382–401 (2001)
24. Mumford, M.D.: Where have we been, where are we going? Taking stock in creativity research. Creat. Res. J. **15**(2–3), 107–120 (2003)

25. Myers, R.J.: Measuring creative potential in higher education: the development and validation of a new psychometric test (2020). Unpublished Master's dissertation, University of Cambridge

26. Pennington, J., Socher, R., Manning, C.D.: Glove: global vectors for word representation. In: Proceedings of the 2014 Conference on Empirical Methods in Natural Language Processing (EMNLP), pp. 1532–1543 (2014)

27. Reimers, N., Gurevych, I.: Sentence-BERT: sentence embeddings using Siamese BERT-networks. arXiv preprint arXiv:1908.10084 (2019)

28. Sanh, V., Debut, L., Chaumond, J., Wolf, T.: DistilBERT, a distilled version of BERT: smaller, faster, cheaper and lighter. arXiv preprint arXiv:1910.01108 (2019)

29. Sottana, A., Liang, B., Zou, K., Yuan, Z.: Evaluation metrics in the era of GPT-4: reliably evaluating large language models on sequence to sequence tasks. In: Proceedings of the 2023 Conference on Empirical Methods in Natural Language Processing (EMNLP) (2023)

30. Susnea, I., Pecheanu, E., Costache, S.: Challenges of an e-learning platform for teaching creativity. In: Proceedings of the 11th International Scientific Conference eLearning and Software for Education. Bucharest, Romania, April 2015

31. Torrance, E.P.: Torrance Tests of Creative Thinking - Norms Technical Manual Research Edition - Verbal Tests, Forms A and B - Figural Tests, Forms A and B. Personnel Press, Princeton, NJ (1966)

32. Zhuo, T.Y.: Large language models are state-of-the-art evaluators of code generation (2023)

AI for Materials Science
and Engineering

Convolutional Graph Neural Networks for Predicting Enthalpy of Formation in Intermetallic Compounds Using Continuous Filter Convolutional Layers

Zongxiao Jin[1], Yu Su[1($\boxtimes$)], Jun Li[1], Huiwen Yang[1], Jiale Li[1], Huaqing Fu[2], Zhouxiang Si[2], and Xiaopei Liu[2]

[1] School of Materials Science and Engineering,
Shanghai University of Engineering Science, Shanghai, China
`suyu@sues.edu.cn`
[2] Shanghai Eraum Alloy Materials Co., Ltd., Shanghai, China

Abstract. Accurately predicting the enthalpy of formation for intermetallic compounds plays a crucial role in materials design and optimization. This article proposes a novel deep learning approach for predicting formation enthalpy. This research develops a graph neural network combined with continuous filter convolutional layers to simulate quantum interactions between atoms. This enables direct learning from atom types and coordinates without simplifying into grid representations. This model demonstrates superior performance on the public JARVIS-DFT dataset compared to traditional machine learning methods. The introduction of continuous filter convolutional layers enhances the ability of graph convolutional neural networks to effectively learn atomic spatial features. This provides a new way to construct graph data structures from crystallographic information for materials science. Additionally, this work highlights the potential value of using graph convolutional neural networks to predict enthalpy of formation for intermetallic compounds.

Keywords: Graph neural network · Intermetallic compounds · Formation enthalpy prediction · Crystal structure modeling

1 Introduction

Intermetallic compounds, resulting from diverse element bonding, have wide applications. Titanium-aluminum [1] and nickel-aluminum compounds [2] serve aerospace due to high-temperature resistance. Silicon-germanium compounds [3] are essential in microelectronics for their thermoelectric properties. Low-temperature superconductors like Nb_3Sn, Nb_3Al, and $NbTi$ are used in particle accelerators and medical imaging [4]. Additionally, compounds like Co_7Mo_6, Fe_7Mo_6 [5], and Fe_3Al [6] find roles in energy catalysts and magnetic storage due to specific properties. Accurate prediction of formation enthalpy is also crucial in materials design, as lower formation enthalpies, such as $-15.87\,\mathrm{KJ/mol}$

C. Cruz et al. (Eds.): IC 2023, CCIS 2036, pp. 141–152, 2024.
https://doi.org/10.1007/978-981-97-0065-3_10

for titanium-copper ($TiCu$), often indicate higher stability, making it widely used in biometal alloys due to its superior high-temperature strength and oxidation resistance [8]. Consequently, accurately predicting the formation enthalpy of intermetallic compounds is of significant importance in the design and optimization process of materials.

Formation enthalpy predictions, once reliant on complex experiments and costly Schrödinger equation calculations [7], have now become data-driven through machine learning. Zhang et al. [9] achieved remarkable accuracy using Gaussian process regression with an MAE as low as 0.044 eV/atom. However, classic machine learning models like random forest, Gaussian process regression, and deep neural networks [10] still face limitations, including issues with high-dimensional and sparse data. Random forest, for example, is constrained by the quality and depth of decision trees, potentially leading to reduced predictive accuracy. Additionally, Zhang et al.'s Gaussian process regression model, while offering uncertainty estimates, experiences increased computational complexity with larger datasets, making it less effective in such cases.

Krajewski et al. [11] combined deep neural networks with structure-related features [12], achieving Mean Absolute Errors (MAEs) of 28, 40, and 42 meV/atom on large datasets. However, like other deep learning models, it can overfit when dealing with sparse or noisy datasets. These models also exhibit black-box characteristics, yielding poor interpretability of predictions. To optimize predictions of formation enthalpy of intermetallic compounds under big data environments and improve physical interpretability, we need to consider the expression of physical features and appropriate model selection in model design. Xie et al. [13] proposed a Crystal Graph Convolutional Neural Network (CGCNN) that learns material properties directly from atomic connectivity in crystals, exemplified in perovskite. Park et al. [14] improved upon CGCNN with their ICGCNN model, learning material properties from crystal graphs, achieving superior prediction performance. Chen et al. [15] further developed the MEGNet model, employing two strategies to circumvent material science data limitations. Kamal et al.'s [16] ALIGNN model surpasses most GNN models in handling bond angle information. Overall, for atom-level modeling of intermetallic compounds, GNNs typically outperform standard Deep Neural Networks and conventional machine learning methods.

Building upon the inspiration from three-dimensional protein modeling [17], our study introduces a novel method to predict the formation enthalpy of intermetallic compounds using Graph Convolutional Neural Networks (GCNNs). This approach employs continuous filter convolutional layers to construct a graph data structure for crystal structures. Unlike traditional convolutional layers fixed on grids, the continuous filter convolutional layer breaks the confinement of local data correlation to grids. It accepts arbitrary atomic coordinates as input, eliminating the need for quantizing atoms to fixed grids, which would lead to information loss. This allows the capture of interatomic correlations through convolution operations, extracting feature representations from the atomic coordinate space.

2   Methodologies

2.1   Graph Construction

To apply graph neural networks in data processing, it is essential to convert atomic data of intermetallic compounds into a graph structure. In the crystal structure of these compounds, atoms are connected by metallic bonds, resembling a graph structure with nodes and edges. Therefore, each atom represents a node, and an atomic bond forms between two atoms when their distance is below a specific threshold, analogous to an edge in a graph. In crystallography, the existence of a connection between atoms is initially assessed by comparing the sum of their van der Waals radii to their interatomic distance. Typically, interaction occurs when the distance between atom centers is within 0.8 to 1.2 times their van der Waals radii sum.

Considering that previous studies in this field have predominantly extracted crystal structure data from strings containing structural information without providing a detailed graph construction process, our research proposes the utilization of an extensible structural information dictionary to uniformly describe such data. This dictionary encompasses crystal lattice vectors ($latticeMat$), atomic point coordinates ($coords$), atomic elements ($elements$), crystal unit cell dimensions (abc), and crystal lattice angles ($angles$). $lattice - mat$ is represented as a $3 * 3$ matrix describing lattice vectors. Coords consists of a list of atomic points, each described by three values corresponding to Cartesian coordinates (x, y, z). Elements correspond to the elements associated with each coordinate point in coords. abc denotes the dimensions of the crystal unit cell, while angles describe the crystal's lattice angles.

In addition, our research introduces an algorithm for constructing a graph from the structural information dictionary. This algorithm adaptively computes atomic connectivity relationships, transforming the data within the structural information dictionary into a graph structure suitable for input into graph neural networks. As shown in Algorithm 1.

In the transformation of graph data structures, we primarily utilized the networkx library. In this study, we further tensorized the graphs from networkx, extending the from_networkx method in Pytorch Geometric (PyG) [18], enabling it to be seamlessly integrated with this popular graph deep learning framework, Pytorch Geometric (PyG). To facilitate convenient calls during the training process, these methods have been encapsulated and integrated into a dedicated Data class.

2.2   Continuous Filter Convolutional Layer in Graph Convolutional Network

Utilizing Graph Convolutional Neural Networks to directly model the atomic coordinates of intermetallic compounds with irregular distributions is challenging. Standard convolutional layers defined only on regular grids cannot be directly applied, as a straightforward rasterization would result in the loss of

---

Algorithm 1: Adaptive Thresholding Algorithm

Input: *atom_dict*
Output: graph G
1 Initialize an empty graph G;
2 **for** each index and (coordinate, element) pair in *atom_dict* **do**
3   Add a node to G with the index as its ID and properties: coordinate and element;
4 **for** $i = 0$ **to** length of *atom_dict*'s coordinates **do**
5   **for** $j = i + 1$ **to** length of *atom_dict*'s coordinates **do**
6   Compute distance between *atom_dict*["coords"][i] and *atom_dict*["coords"][j];
7   Get element of atom i as $element_i$;
8   Get element of atom j as $element_j$;
9   Calculate threshold as the sum of atomic radii of $element_i$ and $element_j$ divided by 100;
10   **if** *distance is less than threshold* **then**
11   Add an edge between node i and node j with weight as distance;

12 **return** G;

---

structural information. To address the issue of irregularly distributed data in Graph Convolutional Neural Networks, we employed continuous filter convolutional layer. The continuous filter convolutional layer takes the coordinate differences (dx, dy, dz) between two atoms as input and outputs the filter weights w between them. In our intermetallic compound model, by inputting the coordinate information of each atom, the continuous filter convolutional layer is capable of learning the filter weight matrix that represents the interactions and structural information between atoms. By multiplying the feature vectors of individual atoms with the filter weight matrix, structural information is allowed to propagate through the network. As a result, the features of each atom (node) encompass not only its intrinsic information but also the spatial information from the three-dimensional atomic coordinates. After several iterations, the structural information propagates further within the graph, and the features of each atom (node) will encompass the spatial information of the entire three-dimensional atomic coordinates. The entire network utilizes the shifted softplus function as its activation function. The shifted softplus function is illustrated in Equation (1), where $SSP(0) = 0$. This ensures continuity and differentiability while enhancing the convergence properties of the network.

$$SSP(X) = \ln(0.5e^x + 0.5) \tag{1}$$

What distinguishes our work from traditional convolution layers is the use of continuous function filters. Continuous convolution operations are performed using precise atomic coordinates as input, rather than coordinates that have been simply gridded. The workflow is illustrated in Fig. 1.

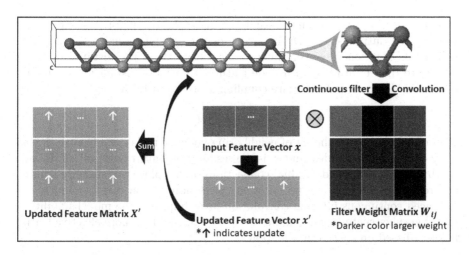

Fig. 1. Work flow of Continuous filter convolutional layer

2.3   Graph Neural Network Architecture

The graph neural network proposed in this study is specifically designed for predicting the formation enthalpy of intermetallic compounds, with the network architecture illustrated in Fig. 2. Crystal structure data is represented in the network as graphs, employing tensors such as node features, edge indices, position matrices, and edge weight matrices in specific layers, akin to how pixels in an image are represented in a network. The graph neural network primarily consists of continuous filter convolutional layers, feature update block, and the overall network architecture. In this section, we will delve into these three components in detail.

Continuous Filter Convolutional Layers. The continuous filter convolutional layer is a message-passing mechanism that combine continuous filtering and local convolution operations to learn relationships between nodes. Unlike SchNet [17], to address the property prediction problem of crystal structures in intermetallic compounds, we have adapted the continuous filter convolutional layers based on SchNet. Specifically, the continuous filter convolutional layers apply a Gaussian kernel function on each edge to capture distance information, and then encodes the distances as continuous filters applied to node features. The computational formula is shown in Eq. 2.

$$W_{ij} = exp(-\frac{\|p_i - p_j\|^2}{2\sigma^2}) \qquad (2)$$

where p_i and p_j are the position vectors of nodes i and j, $\|p_i - p_j\|$ denotes the computation of the Euclidean distance, and σ is the standard deviation of the Gaussian kernel. This layer comprises two linear layers and a non-linear normalization flow. The first linear layer maps the input node features to a hidden space, and the second linear layer maps the results of message passing back to the output space. During the message passing process, the Gaussian kernel function generates continuous weights based on the distance of edges, which are then multiplied with node features to achieve distance-based feature coupling. The formula for feature coupling is shown in Eq. 3.

$$x_i = \sum_{j \in N(i)} W_{ij} \cdot x_j \tag{3}$$

where $N(i)$ represents the set of neighboring nodes of node i, x_j denotes the features of neighboring nodes, and x_i is the result after feature aggregation. Finally, batch normalization and residual connections are employed to stabilize training and retain the original feature information. Building upon this foundation, our enhanced continuous filtering convolution layer improves the model's ability to model graph structure and distance information, enabling more effective feature learning based on spatial information.

Feature Update Block. The interaction block utilizes continuous filter convolutional layers (CFConv) to update node features. CFConv computes the messages received by each node as a Gaussian diffusion of its distance from neighboring nodes. A small multi-layer perceptron (MLP) is employed to learn the weight matrix W, where the cosine similarity C, related to the distances between nodes, is used as a coefficient to multiply with W, thus learning the weights of the associated edges. After several rounds of message passing through CFConv layers, node features are effectively updated, encoding local structural information.

Network Architecture. The network model begins with iterative node feature updates and local neighbor information aggregation via the Feature Update Block. We employ the effective GCNConv [19] for graph convolution layers, a widely used method in graph neural networks. GCNConv updates node features by aggregating neighboring node information, enabling relationship modeling within the graph. It provides unique graph representations for each crystal structure, enhancing the model's capacity for handling irregular graph data with iterative feature updates. After each graph convolution layer, we introduce ReLU activation for non-linearity and perform global max-pooling to reduce data dimensions before outputting results via fully connected layers.

Figure 2 illustrates the overall structure of the network, where (a) depicts the architecture of the continuous filter convolutional layers, (b) describes the entire framework flow for predicting the formation enthalpy of intermetallic compounds in this study, and (c) provides a detailed representation of the dimension changes of various tensors between different modules.

3 Experiment

3.1 Dataset

In this study, we utilized the Javis-dft-3d dataset, accessible at JARVISDFT. This comprehensive dataset encompasses 75,993 samples of 3D materials, computed employing the optB88vdW and TBmBJ methods. From the myriad of

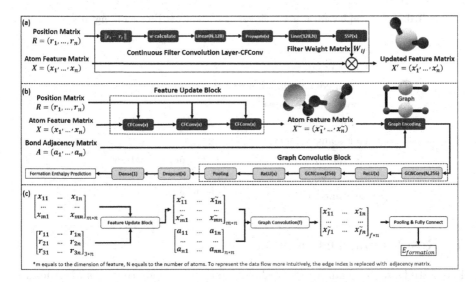

Fig. 2. Graph Neural Network Architecture of Prediction Formation Enthalpy: (a) Continuous Filter Convolutional Layer, (b) Overall Architecture of the Network, (c) Data Dimension Transformation in the Network

columns available, our focus was specifically directed towards the extraction of data pertaining to the 3D atomic structures and the formation energy, which is synonymous with enthalpy. Furthermore, to assemble the atomic features, we adopted the Magpie database embedded within the Matminer [20] and Pymatgen [21] software package.

3.2   Implementation Details

To predict the formation enthalpy of intermetallic compounds in this study, we employed a simple fully connected layer as the output and used the mean square error (MSE) function as the loss function for model training, along with the selection of Adam as the optimizer. In the initial training phase, we initially observed the convergence of the loss function with a limited number of data samples to perform parameter adjustments. Ultimately, a learning rate of 10^{-4} was adopted. All experiments were implemented in PyTorch [22] and its extension library PyTorch Geometric [18]. The network model was trained on a single NVIDIA TESLA T4 32G GPU. The batch size was set to 64, and for the initial parameter tuning, the number of epochs was set to 100. Subsequently, for fine-tuning the parameters, the number of epochs was set to 10. The Jarvis-dft-3d dataset was divided into training (0.7), validation (0.2), and test (0.1) sets, with the test set never used during the training process. Detailed model parameters are shown in Table 1.

Table 1. Graph neural network configuration used for training model.

Parameter name	Value
CFConv layers	3
Graph convolutional layers	2
Node input features	27
Hidden features	128
Gaussians number	50
Normalization	Batch normalization
Batch size	64
Learning rate	10^{-4}

4   Results and Discussion

In our study, we used adaptive threshold algorithm to construct crystal structure graphs for intermetallic compounds, improving the accuracy of spatial lattice representations. Taking TiCuSiAs as an example, our adaptive threshold algorithm outperforms ALIGNN's default method, which includes numerous extraneous edges. Our dynamic threshold adjustment, based on atomic distance distributions, filters out irrelevant connections, preserving structural information. This approach is versatile, even in complex systems. Our crystal graphs closely match actual structures, as shown in Fig. 3. In conclusion, our adaptive threshold algorithm enhances the quality of crystal graphs, offering a more reliable foundation for performance calculations and regression analyses compared to traditional fixed thresholds.

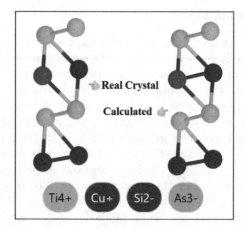

Fig. 3. Comparison of real crystal structure and calculated crystal structure of TiCuSiAs

We compared our formation enthalpy prediction results with PotNet [23], Matformer [24], ALIGNN [16], SchNet [17], and CGCNN [13]. As in previous studies, we used MAE (L1 Loss) to quantify predictions. In this research, our neural network model achieved an impressive prediction accuracy of 0.0337, marking a significant improvement over traditional and machine learning models, consistent with previous research (Table 2). The initial training of the model, as shown in Fig. 4(a), demonstrated a gradual convergence of the loss function, indicating the effectiveness of our designed graph neural network in learning from the dataset. Subsequently, we conducted performance tests on the network model. We modified the number of layers of the continuous filter convolutional layer CFConv [19] within the feature update module and assessed the impact of residual connections. Additionally, we compared the performance of the Point-NetConv [25], capable of handling input data node position matrices, with the traditional graph convolutional layer GCNConv. Our validation model, as presented in Table 3, employed an early stopping strategy. This strategy terminated training if the average loss on the test dataset reached the magnitude of $n * 10^{-2}$, and it allowed for up to four early stopping occurrences; otherwise, the model would run through all 10 epochs. According to models 1, 2, 5, 6 in Table 3 and Fig. 4(b), it is evident that directly feeding the unprocessed graph into the graph convolutional layer, PointNetConv exhibits predictive capability on the validation set, while GCNConv fails to converge the loss function. When the input graphs have undergone node feature updates, using PointNetConv causes the model to lose predictive ability, while GCNConv achieves lower loss. According

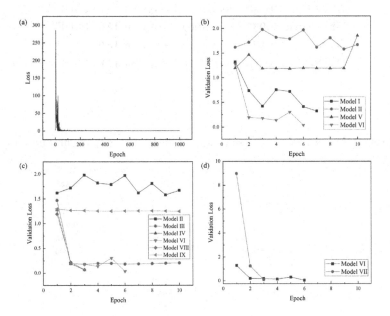

Fig. 4. Variation of the Loss on the validation set of different models with the increase of the number of iterations

to models 2, 3, 4, 6, 8, 9 in Table 3 and Fig. 4(c), with an increase in the number of layers in the continuous filter convolutional layer, the model's accuracy gradually improves, enhancing its predictive capability for formation enthalpy data of intermetallic compounds. The continuous filter convolutional layers introduce spatial information into node features, aiding the model in learning spatial information. However, as the number of layers in the continuous filtering convolution layer continues to increase, the model's predictive accuracy starts to decline, indicating the existence of an optimal threshold for the number of layers. Exceeding this threshold results in contamination of the original features by adding more spatial information. Referring to models 6, 7in Table 3 and Fig. 4(d), Model 7, which lacks residual connections, initially exhibits higher validation loss in the first epoch of training. However, by the second epoch, the validation loss rapidly decreases, approaching the performance level of Model 6 with residual connections. This highlights that in our study, the application of residual connections in the neural network model enhances its performance, enabling it to learn feature representations at a deeper level while effectively preventing overfitting and improving prediction accuracy.

Table 2. Comparison of different models to formation energy prediction MAEs.

Model	MAE	Dataset
PotNet	0.0294	Formation Energy on JARVIS-DFT
Matformer	0.0325	Formation Energy on JARVIS-DFT
ALIGNN	0.0331	Formation Energy on JARVIS-DFT
This work	**0.037**	Formation Energy on JARVIS-DFT
SchNet	0.045	Formation Energy on JARVIS-DFT
CGCNN	0.063	Formation Energy on JARVIS-DFT

Table 3. Performance test result on validation set of graph neural network.

Model NO	CFConv Layers	Graph Convolutional Layer	Residual Connection	MASE on validation set
1	0	PointNetConv	True	0.323
2	0	GCNConv	True	1.580
3	1	GNConv	True	0.178
4	2	GCNConv	True	0.066
5	3	PointNetConv	True	1.187
6	3	GCNConv	True	0.041
7	3	GCNConv	False	0.083
8	4	GCNConv	True	0.071
9	5	GCNConv	True	1.246

5 Conclusion

This work developed a graph convolutional neural network with continuous filter convolutional layers to predict the formation enthalpy of intermetallic compounds. This model shows promise in predicting formation enthalpy and introduces new perspectives for crystal structure modeling in materials science. Future research can explore deep graph neural networks and data augmentation techniques for further improvements. This study provides a valuable tool for AI applications in materials science.

Acknowledgement. This study is supported by the Natural Science Foundation of Shanghai (20ZR1422200), Class III Peak Discipline of Shanghai-Materials Science and Engineering (High-Energy Beam Intelligent Processing and Green Manufacturing) and Teaching Construction Project of Shanghai University of Engineering Science (NO: 1202305011). The authors declare that they have no competing financial interests.

References

1. Teschke, M., Moritz, J., Telgheder, L., Marquardt, A., Leyens, C., Walther, F.: Characterization of the high-temperature behavior of PBF-EB/M manufactured γ titanium aluminides. Prog Addit Manuf. **7**, 471–480 (2022)
2. Wang, X.-Y., Li, M., Wen, Z.-X.: The effect of the cooling rates on the microstructure and high-temperature mechanical properties of a nickel-based single crystal superalloy. Materials **13**, 4256 (2020)
3. Fonseca, L., et al.: Transitioning from Si to SiGe nanowires as thermoelectric material in silicon-based microgenerators. Nanomaterials **11**, 517 (2021)
4. Banno, N.: Low-temperature superconductors: Nb3Sn, Nb3Al, and NbTi. Superconductivity **6**, 100047 (2023)
5. Song, R., et al.: Ultrafine nanoporous intermetallic catalysts by high-temperature liquid metal dealloying for electrochemical hydrogen production. Nat. Commun. **13**, 5157 (2022)
6. Naveen Kumar, R., Koppolu, U.M.K., Rajasabai, S.P.: Magnetic behavior of ordered DO3-type Fe3Al Heusler alloy. Mater. Today Proc. **65**, 157–162 (2022)
7. Yang, J., Huang, J., Ye, Z., Fan, D., Chen, S., Zhao, Y.: First-principles calculations on structural energetics of Cu-Ti binary system intermetallic compounds in Ag-Cu-Ti and Cu-Ni-Ti active filler metals. Ceram. Int. **43**, 7751–7761 (2017)
8. Zhang, E., Wang, X., Chen, M., Hou, B.: Effect of the existing form of Cu element on the mechanical properties, bio-corrosion and antibacterial properties of Ti-Cu alloys for biomedical application. Mater. Sci. Eng. C **69**, 1210–1221 (2016)
9. Zhaohan, Z., Li, M., Flores, K., Mishra, R.: Machine learning formation enthalpies of intermetallics. J. Appl. Phys. **128**, 105103 (2020)
10. Ward, L.: A general-purpose machine learning framework for predicting. NPJ Computat. Mater. (2016)
11. Krajewski, A.M., Siegel, J.W., Xu, J., Liu, Z.-K.: Extensible structure-informed prediction of formation energy with improved accuracy and usability employing neural networks. Comput. Mater. Sci. **208**, 111254 (2022)
12. Seko, A., Hayashi, H., Nakayama, K., Takahashi, A., Tanaka, I.: Representation of compounds for machine-learning prediction of physical properties. Phys. Rev. B **95**, 144110 (2017)

13. Xie, T., Grossman, J.C.: Crystal graph convolutional neural networks for an accurate and interpretable prediction of material properties. Phys. Rev. Lett. **120**, 145301 (2018)
14. Park, C.W., Wolverton, C.: Developing an improved crystal graph convolutional neural network framework for accelerated materials discovery. Phys. Rev. Mater. **4**, 063801 (2020)
15. Chen, C., Ye, W., Zuo, Y., Zheng, C., Ong, S.P.: Graph networks as a universal machine learning framework for molecules and crystals. Chem. Mater. **31**, 3564–3572 (2019)
16. Choudhary, K., DeCost, B.: Atomistic line graph neural network for improved materials property predictions. NPJ Comput. Mater. **7**, 185 (2021)
17. Schütt, K.T., Kindermans, P.-J., Sauceda, H.E., Chmiela, S., Tkatchenko, A., Müller, K.-R.: SchNet: a continuous-filter convolutional neural network for modeling quantum interactions. In: Advances in Neural Information Processing Systems, vol. 30, pp. 992–1002 (2017)
18. Fey, M., Lenssen, J.E.: Fast graph representation learning with PyTorch geometric (2019). http://arxiv.org/abs/1903.02428
19. Kipf, T.N., Welling, M.: Semi-supervised classification with graph convolutional networks (2017). http://arxiv.org/abs/1609.02907
20. Ward, L., et al.: Matminer: an open source toolkit for materials data mining. Comput. Mater. Sci. **152**, 60–69 (2018). https://doi.org/10.1016/j.commatsci.2018.05.018
21. Ong, S.P., et al.: Python Materials Genomics (pymatgen): a robust, open-source python library for materials analysis. Comput. Mater. Sci. **68**, 314–319 (2013). https://doi.org/10.1016/j.commatsci.2012.10.028
22. Paszke, A., et al.: PyTorch: an imperative style, high-performance deep learning library (2019)
23. Lin, Y., Yan, K., Luo, Y., Liu, Y., Qian, X., Ji, S.: Efficient approximations of complete interatomic potentials for crystal property prediction (2023). http://arxiv.org/abs/2306.10045
24. Yan, K., Liu, Y., Lin, Y., Ji, S.: Periodic graph transformers for crystal material property prediction (2022). http://arxiv.org/abs/2209.11807
25. Qi, C.R., Yi, L., Su, H., Guibas, L.J.: PointNet++: deep hierarchical feature learning on point sets in a metric space (2017). http://arxiv.org/abs/1706.02413

Predicting Li Transport Activation Energy with Graph Convolutional Neural Network

Siqi Shi[1,3], Hailong Lin[1], Linhan Wu[2], Zhengwei Yang[2], Maxim Avdeev[4,5], and Yue Liu[2(✉)]

[1] State Key Laboratory of Advanced Special Steel, School of Materials Science and Engineering, Shanghai University, Shanghai 200444, China
[2] School of Computer Engineering and Science, Shanghai University, Shanghai 200444, China
yueliu@shu.edu.cn
[3] Materials Genome Institute, Shanghai University, Shanghai 200444, China
[4] Australian Nuclear Science and Technology Organisation, Sydney 2232, Australia
[5] School of Chemistry, The University of Sydney, Sydney 2006, Australia

Abstract. Exploring activation energy in ionic transport is one of the critical pathways to discovering high-performance inorganic solid electrolytes (ISEs). Although traditional machine learning methods have achieved relatively accurate activation energy predictions, they suffer from issues such as complex descriptor construction and poor generalization. Graph neural network (GNN) has gained widespread usage in accurate material property prediction due to their ability to uncover structure-property relationships latent in materials data in an end-to-end way. However, current graph representation methods and corresponding GNN models have not been widely applied to the ion transport properties of materials. Here, we introduce the interstitial network graph representation method, and design a GNN model to predict activation energy in Li-containing compounds. As a result, the dynamic ion migration process is characterized, enabling the GNN to automatically capture the inherent mechanisms of ion transport. Performance tests demonstrate that interstitial network representation method achieves high prediction accuracy, with a 10% improvement in prediction accuracy compared with the crystal structure representation method. The developed model can be used to screen and design ISEs and in general providing new ideas for applying machine learning in materials science.

Keywords: Inorganic Solid Electrolytes · Activation Energy · Interstitial Network · Deep Learning · Graph Convolutional Neural Network

1 Introduction

All-solid-state batteries (ASSBs), as a hot topic of research on electrochemical energy storage, have drawn widespread attention for various studies of performance characterization and prediction. One of the primary challenges in the

C. Cruz et al. (Eds.): IC 2023, CCIS 2036, pp. 153–164, 2024.
https://doi.org/10.1007/978-981-97-0065-3_11

advancement of inorganic solid electrolytes (ISEs) pertains to ion transport. Lower activation energy facilitates ion transport and also widens the range of operating temperatures [1]. Therefore, exploring materials with low activation energy (E_a) provides an effective approach to the development of high-performance ISEs.

Nowadays, there are two methods commonly used to derive activation energy. One is to conduct experiment, where electrical impedance spectroscopy is performed to obtain the conductivity data combined with the Arrhenius equation to calculate activation energy. The other is to carry out theoretical simulation, such as climbing-image nudged elastic band (CI-NEB) [2], molecular dynamic simulation (MD) [3], and others. Nevertheless, these methods have demanding requirements for experimental equipment and conditions. Also, they tend to incur high resource costs, which makes them unfit for large-scale search for new materials. Therefore, data-driven machine learning methods have attracted increasing attention from researchers due to their lower computational costs, faster processing speeds, and high predictive accuracy.

Machine learning (ML) has already been widely used for large-scale materials property prediction, particularly the prediction of activation energy [4,5]. Some research has been conducted to explore the approaches based on crystal structure descriptors. For example, Katcho et al. [6] used ion coordination numbers, ion site energies, polyhedral volumes, and compound density as descriptors and utilized a random forest method to model the structure-property relationship between these descriptors and activation energy. Liu et al. [7] performed feature engineering, with 30 descriptors relevant to activation energy prediction selected for ISEs, enabling accurate prediction of activation energy. However, traditional machine learning models are poor at predicting the properties of compounds with composition and structure varying in a wide range.

In recent years, graph neural network (GNN), such as deep learning models, has emerged as a powerful approach to the direct extraction of useful information from raw, unstructured graph data, which removes the need for laborious yet essential design of descriptors. Moreover, graph representation is a powerful method of non-Euclidean data representation that is commonly used to explore various complex properties of the material. Xie et al. [8] represented crystal structures by encoding atomic information and bonding interactions between atoms in the form of crystal graphs, and then proposed a crystal graph convolutional neural network (CGCNN) to accurately predict such properties as formation energy, bandgap, Fermi energy level, and elastic modulus. Based on the crystal structure graph representation, various GNNs have been proposed to improve prediction performance, such as ALIGNN [9], GATGNN [10], and DeeperGATGNN [11]. In addition, Gariepy et al. [12] developed an automatic graph representation algorithm (AGRA) tool that can be used for multi-phase catalysis to extract the local chemical environment of metallic surface adsorption sites. However, as far as we know, there is still no research conducted on the use of GNN to predict activation energy for ion transport in ISEs. Moreover, the mechanism that can affect activation energy is complicated. It is worth noting that the use of the representation methods in the studies cited above focuses on atomic and space information, which is insufficient to fully capture the underlying mechanisms of ion conduction. As for describing mechanisms of ion conduction, interstitial network (IN) is proposed to adequately characterize the essential geometric and topological attributes of ion transport properties, which are obtained from the network of transport paths formed by skeleton ions and bottle-neck connections in the crystal structure [14].

In this study, we propose a novel graph representation method based on the IN, to achieve an accurate prediction of ISEs activation energy. Then, to mine latent information comprehensively, a GNN model is designed and optimized according to the characteristics of IN. The results demonstrate that the IN representation method outperforms the crystal structure representation method by 10% in predicting activation energy, and it fits well with experimental and simulation results. This work can be adopted to substantially reduce the screening space required for high-throughput computations, contributing a novel solution to the development of machine·learning methods guided by domain knowledge.

2   Method

2.1   IN Calculation Method

IN calculations are derived from the screening platform for solid electrolytes (SPSE) [15]. Within ionic crystals, larger-radius anions form a relatively stable framework, while smaller-radius cations are distributed over the created voids. The IN is constructed by connecting the voids and this network is essential for investigating ion transport characteristics. The Voronoi decomposition algorithm represents a spatial partitioning technique widely applied in materials science to analyze crystal interstitial space [16] and is implemented in the Crystal structure Analysis by Voronoi Decomposition (CAVD) geometric analysis program, which can be used to efficiently calculated IN data for crystal structures [13].

The workflow of IN data calculation is shown in Fig. 1. For illustration, we used tetragonal $Li_7La_3Zr_2O_{12}$ (LLZO, icsd-246817) as an example. CAVD encodes the crystal structure using the material information from crystallographic information files (CIF) [17] , including chemical formulae, atomic valence states, and atomic positions. Through Voronoi decomposition, the voids within the crystal structure are transformed into an IN. Then, geometric and topological analyses are performed for this network, through which descriptors are obtained for the geometric structure of solid electrolytes, such as ion migration pathways topology and channel dimensions. Simultaneously, the activation energy values corresponding to the data for each compound are obtained through bond valence site energy (BVSE) [14]. The BVSE program uses CIF files to compute the energy potential field for migrating ions within the three-dimensional crystal structure space. Both CAVD and BVSE have been integrated into our independently developed SPSE. Users can retrieve or upload compounds of interest through the platform and generate the corresponding interstitial network files through CAVD calculations. Through the above calculations, the IN data and activation energy data are collected.

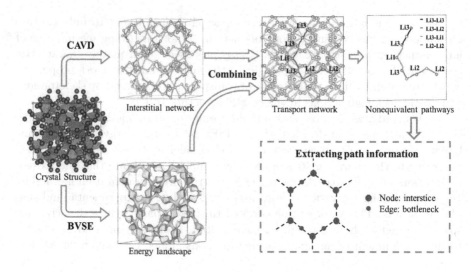

Fig. 1. The workflow of calculate the IN through SPSE. To obtain the IN data, CIF is processed through CAVD, and this information is combined with energy landscape calculated by BVSE to construct the transport network. Subsequently, a nonequivalent path calculation method is employed to filter out redundant data, allowing for the extraction of interstice and bottleneck information along the paths. This process resulted in the generation of IN data samples.

2.2   IN Representation Method

The IN representation method is based on the analysis of interstices and bottlenecks characteristics. In the Voronoi decomposition of the crystal structure, the interstices correspond to Voronoi polyhedra vertices or centers of Voronoi polyhedra faces. The bottlenecks correspond to the position of the minimum radius along each Voronoi polyhedra edge. After obtaining IN data, we utilized interstices as nodes and bottlenecks as edges. For each interstice or bottleneck, we choose the 12 nearest atoms to represent a local chemical environment. High-quality descriptors can greatly improve the accuracy of the model. Hence, we extracted the interstice size and bottleneck size from the IN file and added atomic information descriptors such as atomic numbers and electronegativity. In this process, the nonequivalent paths are calculated, and the bottlenecks and interstices located at the equivalent sites are filtered out to reduce the memory required for the calculation. The construction of the IN representation of the ISEs was then completed. The flowchart of the IN representation method is shown in Fig. 2.

The advantage of IN lies in the fact that the ion transport channels geometric characteristics, conduction thresholds, and ions transport descriptors (e.g., transport channel dimensionality) contained therein are critical to the ion transport performance. The geometric and topological features of the material are extracted and used to calculate the ion transport path information contained in the IN to predict the ion transport properties.

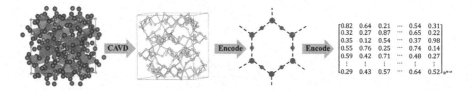

Fig. 2. IN representation method transformation flow. M means the number of nodes; N means the number of features.

2.3 Construction of GCN-Based Activation Energy Prediction Model

Graph Convolutional Neural Network. In this study, graph convolutional neural network (GCN) [18] is employed to extract features of nodes from interstitial network. Sketch of GCN architecture is shown in Fig. 3. For the nth layer in GCN, it takes adjacent matrix A of IN and hidden representation matrix $H_{(n+1)}$ as input, then the output of the next layer will be generated as follows:

$$H_{n+1} = \sigma(\tilde{D}^{-\frac{1}{2}}\tilde{A}\tilde{D}^{-\frac{1}{2}}H_nW^n) \tag{1}$$

where $\sigma(\cdot)$ is the sigmoid function and $H_0 = X$, $\tilde{A} = A + I$ is the adjacent matrix with self-connections. $\tilde{D}$ is the diagonal degree matrix of $\tilde{A}$, and $W^n \in \mathbb{R}^{d_n \times d_{n+1}}$ is a trainable weight matrix. For the ease of parameter tuning, we set output dimension $d_{n+1} = d_n = d$ for all layers.

Graph Pooling Operation. In order to select the key nodes that have great influence on ion transport performance, we adopted the graph pooling operation to simplify the structure of the IN. Through the graph pooling operation, we can eliminate some of the less significant nodes and retain the more valuable nodes. This strategy not only effectively reduces the redundancy of data, but also improves the computational efficiency and provides more refined graph structure data for subsequent analysis.

The Output Layer. After repeating the graph convolution and pooling operations for several times, the final graph level representation Z can be obtained. Finally, we fed the graph level representation into a Multilayer Perceptron (MLP) to perform the activation energy prediction task. The loss function is defined as the Mean Square Error of predictions over the targets:

$$\tilde{Y} = MLP(Z), \tag{2}$$

$$\mathcal{L} = \frac{\sum_i^N (\tilde{Y}_i - Y_i)^2}{N}, \tag{3}$$

where $\tilde{Y}_i$ represents the predicted activation energy values and Y_i is the Activation energy values calculated from SPSE. N denotes the training sets of graphs that have targets.

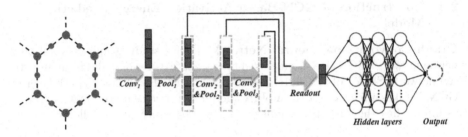

Fig. 3. Structure of GCN. On the left figure, green indicates interstitial; red indicates bottleneck. (Color figure online)

3   Experiments

3.1   Dataset

The IN dataset is collected from the SPSE, which comprises a total of 2150 lithium-containing compound entries. The CIF data is collected from the SPSE and Inorganic Crystal Structure Database (ICSD) [19]. Initial preparations involved organizing the CIF of target compounds and uploading to SPSE for calculation to get IN files. In addition, we collected data from some of the published literature for the comparison of results to validate our method for predicting activation energy in ISEs.

3.2   Experimental Setup

In all experiments, we employed mean squared error (MSE) loss function for training. To assess the accuracy of the model in predicting activation energy, we used determination coefficient (R^2) to evaluate the performance of model. The preprocessed interstitial network files and CIF files are divided into the training/testing/validation sets with a ratio of 8:1:1. We repeated this random splitting process 10 times, and the average performance is reported. Section 2.3 provides a comprehensive description of our model architecture. Table 1 details the parameter configuration.

Table 1. The parameter configuration of the model.

Parameter	Value
$Hidden1\_units$	128
$Hidden2\_units$	128
$Hidden3\_units$	128
$Pooling1\_ratio$	0.9
$Pooling2\_ratio$	0.8
$Pooling3\_ratio$	0.9
Epoch	150
Optimizer	Adam

Hidden unit means the output feature dimension after performing feature transformation and extraction on nodes in each hidden layer; Pooling ratio means the proportion of retained nodes compared to the total number of nodes after a pooling operation.

3.3 Experimental Results

The comparison between the proposed model and state-of-the-art models is shown in Fig. 4. The model based on IN significantly outperforms that based on crystal structures. The results of the models show that when the crystal structure is represented by atomic numbers as nodes, it is less effective information for predicting the activation energy. In contrast, the interstitial space is better representation of the crystal structure relevant for ionic transport, since IN contains additional geometric features and topological logic features, allowing the model to better identify their influence on the activation energy.

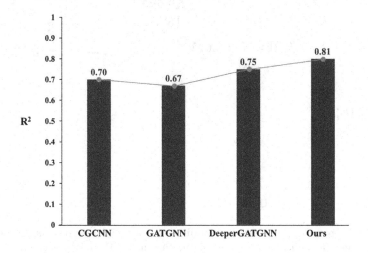

Fig. 4. The performance of different graph representation on the test set.

Partial prediction results are compared, as shown in Fig. 5. Under the guidance of materials domain knowledge, the data set is divided based on the activation energy value. From the accuracy of the activation energy predictions for various material subsets, it is evident that the model excels at accurately predicting materials with low activation energies. While the precision diminishes for materials with higher activation energies, the consistent predictive trends remain unaffected, ensuring the validity of the results. This validates the feasibility of the method for identifying materials with low activation energies. To demonstrate the accuracy of the model predictions clearer, we selected three crystals: Li_3N, $Li_7La_3Zr_2O_{12}$, and $LiFePO_4$, all of which are popular materials in study. As demonstrated in Table 2, a comparison is made among partial prediction results, BVSE results, and published data.

The time required for obtaining activation energy data using three different methods is as follows: Conventional experimental methods demand 48 h from

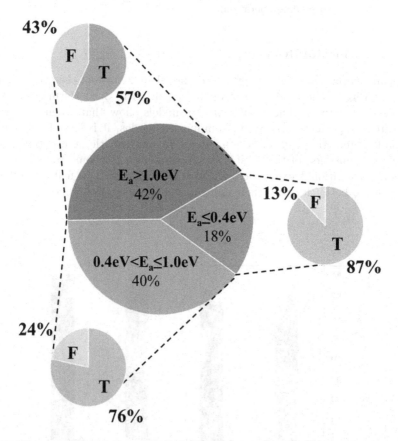

Fig. 5. The percentage of materials with different activation energies and the accuracy of predictions. "T" indicates errors below the threshold; "F" indicates errors above the threshold.

experimental setup preparation to testing for single activation energy value; calculations using the SPSE require 5 min, and lastly, our well-trained model completes the task within a matter of seconds. As a result of screening by this model, the material search space is considerably narrowed down, and experiment time is reduced by orders of magnitude.

Table 2. Comparison of predicted, BVSE-based, and published experimental E_a.

ID	Formula	Prediction(eV)	BVSE(eV)	Published(eV)	Ref
icsd_156898	Li_3N	0.21	0.21	0.13	[20]
icsd_183685	$Li_7La_3Zr_2O_{12}$	0.39	0.46	0.32	[21]
icsd_193797	$LiFePO_4$	0.38	0.39	0.48	[22]

3.4   Parameter Sensitivity Analysis

Performance and generalization ability of a deep learning model are significantly influenced by selection of hyperparameters. Proper hyperparameter choices can enhance model performance, while incorrect selections can result in issues like overfitting or underfitting. Two crucial hyperparameters are the learning rate and batch size. The learning rate dictates the magnitude of weight updates during model training, and an illsuited learning rate can hinder model convergence. On the other hand, batch size determines the number of samples processed in each training iteration, impacting training speed and convergence; an inappropriate batch size can result in slow convergence or training instability.

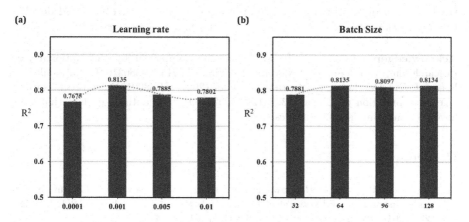

Fig. 6. The performance of the model on different hyperparameters. (a) Learning rate; (b) Batch size.

We conducted a sensitivity analysis on the hyperparameters of learning rate and batch size. Figure 6(a) illustrates how the performance of the model changes with learning rate. When the learning rate was set at 0.0001, the model exhibited its poorest performance, primarily due to the constrained magnitude of weight updates, making it unable to reach an optimal state within the same number of training iterations. Conversely, increasing the learning rate to 0.001 yielded the best performance. However, further increase to 0.005 and 0.01 led to a gradual performance deterioration. This was attributed to the larger learning rate, which induced substantial weight updates, causing the model to oscillate near the optimal point and impeding its convergence. Figure 6(b) illustrates how the performance of the model varied with changes in batch size. The model performed poorly when the batch size was set at 32. This can be ascribed to the smaller batch size, which increased sensitivity to effects of individual samples and reduced training speed. As the batch size increased, model performance gradually improved. Nonetheless, an excessively large batch size consumed more computational resources which may lead to potential scalability issues.

4   Conclusion

In summary, we proposed a graph representation method based on interstitial network (IN) and constructed a graph convolutional neural network (GCN), aiming to predict the activation energy of ISEs. By utilizing the geometric and topological features of the IN, we realized the high-accuracy prediction of activation energy through the graph representation method. Compared with the state-of-the-art crystal structure graph representation method, our method has better predictive accuracy and the efficiency of material screening is orders of magnitude higher compared with conventional calculations. In the future, we will extend the method to other migrating ions, realize the prediction of activation energy of multi-component inorganic solid-state electrolytes and develop the application of IN for predicting additional material properties.

Acknowledgement. This study was financially supported by the National Key Research and Development Program of China (2021YFB3802104), National Key Research and Development Program of China (2021YFB3802101), National Natural Science Foundation of China (92270124, 52073169). The authors declare that they have no competing financial interests.

References

1. Famprikis, T., Canepa, P., Dawson, J.A., Islam, M.S., Masquelier, C.: Fundamentals of inorganic solid-state electrolytes for batteries. Nat. Mater. **18**(12), 1278–1291 (2019). https://doi.org/10.1038/s41563-019-0431-3
2. Henkelman, G., Uberuaga, B.P., Jónsson, H.: A climbing image nudged elastic band method for finding saddle points and minimum energy paths. J. Chem. Phys. **113**(22), 9901–9904 (2000). https://doi.org/10.1063/1.1329672

3. Gao, Y., et al.: Classical and emerging characterization techniques for investigation of ion transport mechanisms in crystalline fast ionic conductors. Chem. Rev. **120**(13), 5954–6008 (2020). https://doi.org/10.1021/acs.chemrev.9b00747

4. Liu, Y., Guo, B., Zou, X., Li, Y., Shi, S.: Machine learning assisted materials design and discovery for rechargeable batteries. Energy Storage Mater. **31**, 434–450 (2020). https://doi.org/10.1016/j.ensm.2020.06.033

5. Liu, Y., Zhao, T., Ju, W., Shi, S.: Materials discovery and design using machine learning. J. Materiomics **3**(3), 159–177 (2017). https://doi.org/10.1016/j.jmat.2017.08.002

6. Katcho, N.A., et al.: An investigation of the structural properties of Li and Na fast ion conductors using high-throughput bond-valence calculations and machine learning. J. Appl. Crystallogr. **52**(1), 148–157 (2019). https://doi.org/10.1107/S1600576718018484

7. Liu, Y., Zou, X., Ma, S., Avdeev, M., Shi, S.: Feature selection method reducing correlations among features by embedding domain knowledge. Acta Mater. **238**, 118195 (2022). https://doi.org/10.1016/j.actamat.2022.118195

8. Xie, T., Grossman, J.C.: Hierarchical visualization of materials space with graph convolutional neural networks. J. Chem. Phys. **149**(17) (2018). https://doi.org/10.1063/1.5047803

9. Choudhary, K., DeCost, B.: Atomistic line graph neural network for improved materials property predictions. NPJ Comput. Mater. **7**(1), 185 (2021). https://doi.org/10.1038/s41524-021-00650-1

10. Louis, S.-Y., et al.: Graph convolutional neural networks with global attention for improved materials property prediction. Phys. Chem. Chem. Phys. **22**(32), 18141–18148 (2020). https://doi.org/10.1039/D0CP01474E

11. Omee, S.S., et al.: Scalable deeper graph neural networks for high-performance materials property prediction. Patterns **3**(5), 100491 (2022). https://doi.org/10.1016/j.patter.2022.100491

12. Gariepy Z, Chen Z, Tamblyn I, Singh C V, Tetsassi Feugmo C G. Automatic graph representation algorithm for heterogeneous catalysis. APL Machine Learning 1(3), (2023). https://doi.org/10.1063/5.0140487

13. He, B., Ye, A., Chi, S., et al.: CAVD, towards better characterization of void space for ionic transport analysis. Sci. Data **7**, 153 (2020). https://doi.org/10.1038/s41597-020-0491-x

14. He, B., et al.: A highly efficient and informative method to identify ion transport networks in fast ion conductors. Acta Materialia **203** (2021). https://doi.org/10.1016/j.actamat.2020.116490

15. Zhang, L.: A database of ionic transport characteristics for over 29 000 inorganic compounds. Adv. Func. Mater. **30**(35), 2003087 (2020). https://doi.org/10.1002/adfm.202003087

16. Willems, T.F., Rycroft, C.H., Kazi, M., Meza, J.C., Haranczyk, M.: Algorithms and tools for high-throughput geometry-based analysis of crystalline porous materials. Microporous Mesoporous Mater. **149**(1), 134–141 (2012). https://doi.org/10.1016/j.micromeso.2011.08.020

17. Hall, S.R., Allen, F.H., Brown, I.D.: The crystallographic information file (CIF): a new standard archive file for crystallography. Acta Crystallogr. Sect. A **47**(6), 655–685 (1991). https://doi.org/10.1107/S010876739101067X

18. Kipf T N, Welling M. Semi-supervised classification with graph convolutional networks. arxiv preprint arxiv:1609.02907 (2016)

19. Belsky, A., Hellenbrandt, M., Karen, V.L., Luksch, P.: New developments in the inorganic crystal structure database (ICSD): accessibility in support of materials research and design. Acta Crystallogr. Sect. B **58**(3 Part 1), 364–369 (2002). https://doi.org/10.1107/s0108768102006948

20. Kim, M.S., et al.: Revealing the multifunctions of Li3N in the suspension electrolyte for lithium metal batteries. ACS Nano **17**(3), 3168–3180 (2023). https://doi.org/10.1021/acsnano.2c12470

21. Liang, F., Sun, Y., Yuan, Y., Huang, J., Hou, M., Lu, J.: Designing inorganic electrolytes for solid-state Li-ion batteries: a perspective of LGPS and garnet. Mater. Today **50**, 418–441 (2021). https://doi.org/10.1016/j.mattod.2021.03.013

22. Sellami, M., Barre, M., Dammak, M., Toumi, M.: Local structure, thermal, optical and electrical properties of $LiFePO_4$ polycrystalline synthesized by co-precipitation method. Braz. J. Phys. **51**(6), 1521–1528 (2021). https://doi.org/10.1007/s13538-021-00977-6

AI for Medicine

KGCN-DDA: A Knowledge Graph Based GCN Method for Drug-Disease Association Prediction

Hongyu Kang[1,2], Li Hou[1], Jiao Li[1], and Qin Li[2(✉)]

[1] Institute of Medical Information, Chinese Academy of Medical Sciences and Peking Union Medical College, Beijing, China
[2] Department of Biomedical Engineering, School of Medical Technology, Beijing Institute of Technology, Beijing, China
liqin@bit.edu.cn

Abstract. Exploring the potential efficacy of a drug is a valid approach for drug discovery with shorter development times and lower costs. Recently, several computational drug repositioning methods have been introduced to learn multi-features for potential association prediction. A drug repositioning knowledge graph of drugs, diseases, targets, genes and side effects was introduced in our study to impose an explicit structure to integrate heterogeneous biomedical data. We revealed drug and disease embeddings from the constructed knowledge graph via a two-layer graph convolutional network with an attention mechanism. Finally, KGCN-DDA achieved superior performance in drug-disease association prediction with an AUC value of 0.8818 and an AUPR value of 0.5916, a relative improvement of 31.67% and 16.09%, respectively, over the second-best results of the four existing state-of-the-art prediction methods. Meanwhile, case studies have verified that KGCN-DDA can discover new associations to accelerate drug discovery.

Keywords: knowledge graph · drug repositioning · drug-disease · association prediction

1 Introduction

In recent decades, drug discovery techniques and biological systems have been intensively studied by multidisciplinary researchers. However, drug development is still a time-consuming, costly and labor-intensive process. Drug repositioning is a strategy for identifying new uses for approved or investigational drugs that are outside the scope of the original medical indications [1]. It could ease the drug development process, shorten the required time to 6.5 years, reduce costs to $300 million and reduce the risk of failure.

In recent years, computational drug repositioning methods [2] have attracted continuous attentions with explosive growth of large-scale genomic and phenotypic data. The previous computational methods can be roughly divided into three categories: complex network method [3], machine learning method [4], and deep learning method [5].

C. Cruz et al. (Eds.): IC 2023, CCIS 2036, pp. 167–173, 2024.
https://doi.org/10.1007/978-981-97-0065-3_12

Besides, the knowledge organization method [6], for example ontologies and knowledge graph, has also been gradually applied to the research of drug disease relationship prediction recently.

With the explosion of the total amount of drug discovery knowledge, the relationships between entities, such as drugs, diseases, targets, symptoms, etc., become progressively more complex. There is a wealth of associations hidden in literature, clinical guidelines, encyclopedias, and structured databases. Semi-structured and unstructured knowledge needs further exploration and exploitation. More hidden drug-disease associations can be found by fully utilizing public databases and literature knowledge related to drug development and disease treatment. This can reduce the risk of failure, shorten the time needed for research and development, and save money, manpower, and material resources. In this study, we first construct a drug repositioning knowledge graph and then propose a novel drug-disease association prediction method called KGCN-DDA based on multiple features in the knowledge graph and graph convolutional neural network. KGCN-DDA has achieved good performance in the prediction of unknown drug disease association. This method can find new indications of drugs, and also provide methodological reference and theoretical basis for drug relocation.

2   Methods and Materials

2.1   Dataset

Data for drug repositioning knowledge graph construction were primarily collected from various data sources including Comparative Toxicology Database (CTD), Drugbank, SIDER, MeSH and PubMed scientific literature from PubMed. Taking as a starting point, 269 drugs, 598 diseases and 18416 drug-disease associations originated from Comparative Toxicology Database (CTD). We extracted drug-target associations from Drugbank and drug-side effect associations from SIDER for drug repositioning knowledge graph construction. Biological semantic relationships between drugs, diseases, targets, genes, and side effects were also discovered from 12056 PubMed scientific literature which titles or abstracts containing drugs or diseases from the CTD dataset. Besides, drug chemical structures (represented by SMILES) from Drugbank, and diseases' tree numbers from MeSH served as entities attributes to in our study.

2.2   Drug–Disease Association Prediction Based on Knowledge Graph and GCN

In this study, we presented a comprehensive knowledge graph of drug repositioning with relevant drugs, diseases, targets, genes and side effects. Meanwhile, graph convolutional neural network worked as an efficient way to extract multi-features from the constructed knowledge graph. The workflow of KGCN-DDA was briefly shown in Fig. 1.

Drug Repositioning Knowledge Graph Construction. Our drug-centric knowledge graph data model comprised five types of entities includes drugs, diseases, and other entities that interact with the two entities, such as targets, side effects and genes. It curates and normalizes data from the four publicly available databases mentioned above, as well as information from PubMed publications based on a pre-training and fine-tuning BERT

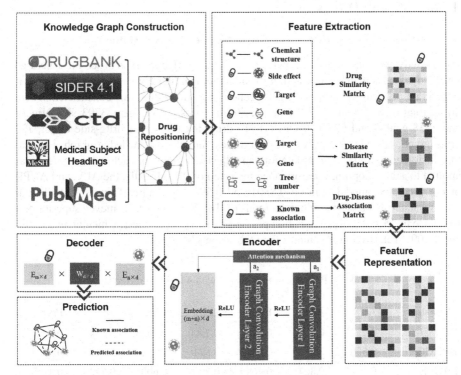

Fig. 1. The workflow of KGCN-DDA

model. The eight relationship types in drug repositioning knowledge graph include treat (between drugs and diseases), interact (between two drugs), cause (between drugs and side effects), target (between drugs and targets), associate (between drugs and genes), associate (between two genes), biomarker (between diseases and genes), and target (between diseases and targets).

Drug–Disease Feature Representation and Association Prediction. We calculated drug-drug similarities and disease-disease similarities based on multi features based on the drug repositioning knowledge graph, including: (1) drug-side effect associations, drug-target associations, drug-gene associations, drug molecular fingerprints, (2) disease-target associations, disease-gene target associations, disease MeSH tree-numbers. We then proposed this multi-feature fusion similarities and drug-disease associations in the knowledge graph to compute an association feature matrix. Finally, two GCN layers were applied to learn drug and disease embeddings of with an attention mechanism. An inner product decoder was used to discover unknown drug-disease associations.

3 Results and Discussion

3.1 Performances and Comparison with State-of-the-Art Methods

In this study, we constructed a drug repositioning knowledge graph based on structured knowledge and semantic information from biomedical literature. Specifically, a knowledge graph of drugs, diseases, targets, genes and side effects was constructed. There are in total of 8374 entities (269 drugs, 598 diseases, 266 targets, 3793 side effects, and 2938 genes) and 67350 triples (18416 drug-disease, 43508 drug-side effect, 722 drug-target, 4081 disease-gene, and 623 disease-target) in knowledge graph. For feature fusion and similarity computation, an adjusted weight for each measurement was applied to achieve optimal performance by a step of 0.01. Finally, the AUC and AUPR of our predictive model reached 0.8801 and 0.5961 optimality. Compared with four existing state-of-the-art prediction methods [7–10], KGCN-DDA achieved superior performance in drug-disease association prediction, shown in Table 1, which were 33.89% and 16.09% relative improvements than the second-best result.

Table 1. Performance compared with 4 baseline methods

Methods	AURP	AUC	F1	Acc	Rec	Spe	Pre
DDA-SKF	0.2521	0.7006	0.3281	0.7900	0.4478	0.8342	0.2591
DRHGCN	0.5063	0.8529	0.5013	0.8746	0.5503	0.9166	0.4604
LAGCN	0.5135	0.8045	0.4699	0.7966	0.6005	0.8220	0.4198
DRWBNCF	0.4552	0.8375	0.4739	0.8646	0.5321	0.9076	0.4280
KGCN-DDA	**0.5961**	**0.8818**	**0.5655**	**0.8885**	**0.6287**	**0.9224**	**0.5154**

Footnotes: The best results are in bold faces and the second-best results are underlined.

3.2 Case Study

To demonstrate KGCN-DDA's ability to discover new indications and new therapies, we conducted three case studies with validation from clinical indications already in use, Clinical Trials, CTD and public literature from PubMed: (1) Top 10 drug–disease associations, (2) Top 10 associated diseases for given drugs (Doxorubicin).

We listed the top 10 drug-disease associations predicted by KGCN-DDA in Table 2, and seven out of them can be demonstrated by the verification methods mentioned above. For example, we found olanzapine and fluoxetine together are more effective than duloxetine alone for treating severe depression in terms of improving physical and sleep quality [11]. Researchers examined how rosiglitazone inhibits hepatocellular carcinoma and showed that the medication can cause liver cancer cells to undergo apoptosis [12]. According to study from Johns Hopkins University in the United States, taking a certain amount of caffeine might enhance the body's memory function temporarily [13]. Cimetidine is a medication that can be used clinically to treat arrhythmia and chronic hepatitis B hepatitis. This therapeutic approach aligns with the expected management of inflammation and cardiac disease. Besides, several predictions have been confirmed effective by ClinicalTrials and CTD records.

Table 2. Predicted drug-disease association

No	Drug Name	Disease Name	Evidence
1	Olanzapine	Sleep wake disorders	PMID: 25062968
2	Rosiglitazone	Carcinoma, Hepatocellular	ClinicalTrials/PMID: 26622783
3	Docetaxel	Eosinophilia	ClinicalTrials/CTD
4	Venlafaxine Hydrochloride	Catalepsy	—
5	Caffeine	Amnesia	CTD/PMID: 24413697
6	Enalapril	Angina pectoris	ClinicalTrials/CTD
7	Propranolol	Urticaria	—
8	Cimetidine	Heart diseases	Clinical indications
9	Cimetidine	Inflammation	Clinical indications
10	Nifedipine	Anxiety disorders	—

The top 10 combinations in drug-disease prediction were examined from the viewpoint of a single medication, using doxorubicin as an example (Table 3). Doxorubicin is an anti-tumor medication that mostly inhibits DNA synthesis, but it can also limit RNA synthesis as well. It has a broad anti-tumor range and is mostly used in clinical practice to treat individuals with acute leukemia, including acute lymphocytic leukemia and acute myeloid leukemia. Combinations 1, 2, 3, 6, 8 [14–18] have been clinically treated and validated by literature, including doxorubicin, which has a certain ameliorative impact on non-small cell lung cancer, acute myeloid leukemia, trigeminal neuralgia, glioma, and osteosarcoma. Meanwhile, the remaining three combinations have not received much attention but have been predicted by the KGCN-DDA model. To some extent, this might give researchers fresh ideas for drug repositioning. As a result, it is feasible to predict drug-disease association by KGCN-DDA.

Table 3. Drug-disease association prediction for doxorubicin

Drug Name	No	Disease Name	Evidence
Doxorubicin	1	Carcinoma, Non-small-cell lung	ClinicalTrials/PMID: 33075540
	2	Leukemia	PMID: 32949646/Clinical indications
	3	Trigeminal neuralgia	CTD/PMID: 30235706
	4	Hemolytic-uremic syndrome	ClinicalTrials/CTD
	5	Cerebral hemorrhage	—
	6	Glioma	ClinicalTrials/CTD/PMID: 33475372
	7	Myocardial ischemia	—
	8	Osteosarcoma	ClinicalTrials/CTD/PMID: 31802872
	9	Atherosclerosis	—
	10	Vascular diseases	Clinical indications

4   Conclusions

In this study, we proposed a method called KGCN-DDA for drug-disease association prediction. Due to the huge amount of information contained in biomedical public databases and scientific literature, we constructed a drug repositioning knowledge graph and compute drug-drug and disease-disease similarities by knowledge graph multi-feature fusion. Two GCN layers were utilized to capture structural embeddings from association feature matrix. The proposed method achieved superior performance compared to four state-of-the-art methods, and we demonstrated its potential for identifying novel drug-disease associations in clinical practice.

However, there are still some limitations in our work that require an in-depth investigation. First, more association features should be further considered in our work. We can collect more prior biological knowledge from literature or datasets, such as drug-protein, drug-gene, disease-gene and drug-pathway from DisGeNET, Gene Ontology (GO) and so on, to improve similarity accuracy. Second, the two-layer GCN is a basic model for learning on graph-structured data, while some other graph neural network models are worth investigating in the future.

Above all, KGCN-DDA is able to learn scattered multidimensional information from heterogeneous networks and identify latent drug-disease associations. It gives researchers, pharmacologists, and pharmaceutical companies a tremendous opportunity to study and validate predictive associations that are more likely to exist. We expect KGCN-DDA to be an efficient approach that can improve drug repositioning in the future and shorten its cost and time.

Funding.   This work was supported by The National Social Science Fund of China (22CTQ024), Innovation Project of Chinese Academy of Medical Sciences (2021-I2M-1-001), The National Key Research and Development Program of China (2022YFB2702801).

References

1. Pushpakom, S., et al.: Drug repurposing: progress, challenges and recommendations. Nat. Rev. Drug Discov. **18**(1), 41–58 (2019)
2. Deng, J., Yang, Z., Ojima, I., Samaras, D., Wang, F.: Artificial intelligence in drug discovery: applications and techniques. Brief Bioinform. **23**(1), bbab430 (2022)
3. Wang, W., Yang, S., Zhang, X., Li, J.: Drug repositioning by integrating target information through a heterogeneous network model. Bioinformatics **30**(20), 2923–2930 (2014)
4. Napolitano, F., et al.: Drug repositioning: a machine-learning approach through data integration. J. Cheminform. **5**(1), 30 (2013)
5. Fatehifar, M., Karshenas, H.: Drug-Drug interaction extraction using a position and similarity fusion-based attention mechanism. J. Biomed. Inform. **115**(3), 103707 (2021)
6. Karim, M.R., Cochez, M., Jares, J., Uddin, M., Beyan, O., Decker, S.: Drug-drug interaction prediction based on knowledge graph embeddings and convolutional-LSTM network. ACM (2019). arXiv:1908.01288
7. Gao, C.Q., Zhou, Y.K., Xin, X.H., Min, H., Du, P.F.: DDA-SKF: predicting drug-disease associations using similarity Kernel fusion. Front. Pharmacol. **12**, 784171–784186 (2022)
8. Cai, L., et al.: Drug repositioning based on the heterogeneous information fusion graph convolutional network. Brief Bioinform. **22**(6), bbab319 (2021)
9. Yu, Z., Huang, F., Zhao, X., Xiao, W., Zhang, W.: Predicting drug-disease associations through layer attention graph convolutional network. Brief Bioinform. **22**(4), bbaa243 (2021)
10. Meng, Y., Lu, C., Jin, M., Xu, J., Zeng, X., Yang, J.: A weighted bilinear neural collaborative filtering approach for drug repositioning. Brief Bioinform. **23**(2), bbab581 (2022)
11. Qu, W., Gu, S., Luo, H., et al.: Effects of olanzapine-fluoxetine combination treatment of major depressive disorders on the quality of life during acute treatment period. Cell Biochem. Biophys. **70**(3), 1799–1802 (2014)
12. Bo, Q., Sun, X., Jin, L., et al.: Antitumor action of the peroxisome proliferator-activated receptor-γ agonist rosiglitazone in hepatocellular carcinoma. Oncol. Lett. **10**(4), 1979–1984 (2015)
13. Borota, D., Murray, E., Keceli, G., et al.: Post-study caffeine administration enhances memory consolidation in humans. Nature Neurosci. **17**(2), 201–212 (2014)
14. Ghosh, S., Lalani, R., Maiti, K., et al.: Synergistic co-loading of vincristine improved chemotherapeutic potential of pegylated liposomal doxorubicin against triple negative breast cancer and non-small cell lung cancer. Nanomedicine **31**(2), e102320 (2021)
15. Perry, J.M., Tao, F., Roy, A., et al.: Overcoming Wnt-β-catenin dependent anticancer therapy resistance in leukaemia stem cells. Nat. Cell Biol. **22**(6), 689–700 (2020)
16. Zheng, B., Song, L., Liu, H.: Gasserian ganglion injected with Adriamycin successfully relieves intractable trigeminal nerve postherpetic neuralgia for an elderly patient: a case report. Medicine (Baltimore) **97**(38), e12388 (2018)
17. Niu, W., Xiao, Q., Wang, X., et al.: A biomimetic drug delivery system by integrating grapefruit extracellular vesicles and doxorubicin-loaded heparin-based nanoparticles for glioma therapy. Nano Lett. **21**(3), 1484–1492 (2021)
18. Wei, H., Chen, J., Wang, S., et al.: A nanodrug consisting of doxorubicin and exosome derived from mesenchymal stem cells for osteosarcoma treatment in vitro. Int. J. Nanomed. **14**(1), 8603–8610 (2019)

Machine Learning for Time-to-Event Prediction and Survival Clustering: A Review from Statistics to Deep Neural Networks

Jinyuan Luo[1,3], Linhai Xie[2(✉)], Hong Yang[1(✉)], Xiaoxia Yin[1], and Yanchun Zhang[1,3(✉)]

[1] The Cyberspace Institute of Advanced Technology, Guangzhou University, Guangzhou, China
2112106041@e.gzhu.edu.cn, {hyang,xiaoxia.yin}@gzhu.edu.cn
[2] State Key Laboratory of Proteomics, National Center for Protein Sciences (Beijing), Beijing 102206, China
xielinhai@ncpsb.org.cn
[3] School of Computer Science and Technology, Zhejiang Normal University, Hangzhou, China
Yanchun.Zhang@vu.edu.au

Abstract. Survival analysis is a statistical method used in computational biology to investigate the time until the occurrence of an event of interest, such as death or disease recurrence. It plays a crucial role in analyzing and understanding time-to-event data in various medical and biological studies. Deep learning, as a subset of artificial intelligence, has shown remarkable success in diverse domains, such as image recognition and natural language processing. Its ability to automatically extract complex patterns and features from high-dimensional data makes it highly promising for enhancing survival analysis. While existing reviews have primarily focused on traditional statistical methods and conventional machine learning approaches in survival analysis, a critical aspect that has been largely overlooked is the integration of deep learning techniques. We not only considered traditional statistical methods and conventional machine learning approaches but also further incorporated deep learning methods. With this review article, we hope to provide researchers, doctors, and biologists with a comprehensive framework for understanding survival analysis and to foster the development of survival prediction and personalized medicine.

Keywords: Survival analysis · Time-to-event · Cluster

1 Introduction

Survival analysis is a statistical method used to analyze the probability and influencing factors of survival or experiencing a particular event for individuals

C. Cruz et al. (Eds.): IC 2023, CCIS 2036, pp. 174–192, 2024.
https://doi.org/10.1007/978-981-97-0065-3_13

or groups within a given time period [1]. In fields such as medicine, biology, and social sciences, survival analysis methods are widely applied in assessing treatment effectiveness, predicting risks, and studying disease progression, among others [2]. Through survival analysis, doctors and researchers can predict the survival time of patients, develop personalized treatment plans, and evaluate the effectiveness of different treatment approaches based on patient characteristics and clinical data [3]. Additionally, survival analysis can help researchers explore the patterns and pace of disease development, identify genetic variations or biomarkers associated with survival time, and provide scientific evidence for clinical decision-making [4].

However, despite the widespread application of survival analysis methods in various fields, there is relatively limited literature that simultaneously review two different branches of survival analysis, i.e. individualized time-to-event prediction and survival clustering. Except some non-parametric models such as a Kaplan-Meier model, most methods fall into the former branch. They formulate the survival prediction as a regression problem with censored data points and map from the covariates to the estimated risk of an individual, e.g. Cox proportional hazards models. Others, in contrast, are proposed as a unsupervised or semi-supervised clustering task which stratifies patients into subgroups with distinct survival distributions. Although these methods belong to totally different machine learning categories, an integrative overview that comprehensively review both branches of approach can provide deeper insights into survival analysis.

Furthermore, deep learning techniques have shown great promise in enhancing the accuracy and performance of predictive models, especially when dealing with complex and high-dimensional datasets. Its ability to automatically extract relevant features and patterns from raw data has opened up new opportunities for understanding the underlying mechanisms of survival outcomes and identifying critical prognostic factors. Unfortunately, previous reviews and analyses of the literature have provided limited coverage of state-of-the-art deep models in survival analysis.

Therefore, the objective of this review is to systematically review recent advance in survival analysis boosted by deep learning techniques from two different modeling perspectives, including "time-to-event" prediction and survival clustering. We will discuss the applications of statistical methods, traditional machine learning, and deep learning methods in each branch, and explore the advantages, limitations and suitable scenarios for various methods.

The organization of the subsequent sections of this article is as follows: Sect. 2 provides the necessary background knowledge. Section 3 provides an overview of predicting Time to Event. Section 4 introduces risk-based clustering methods. Section 5 concludes the entire article.

2 Preliminaries

In the study of survival analysis problems, it is possible that the event of interest is not observed in certain cases. This can occur due to limited observation time windows or other unobserved events that result in missing information. This concept is referred to as censoring [5]. Censoring can be classified into three groups based on the type of truncation of the event [6]: (I) right censoring, where the observed survival time is less than or equal to the true survival time; (II) left censoring, where the observed survival time is greater than or equal to the true survival time; and (III) interval censoring, where we only know that the event occurred within a given time interval. It is important to note that the true event times are unknown in all three cases. Among these, right censoring is the most common scenario in many practical problems [7]. Therefore, this article will primarily focus on the analysis of survival data with right-censored information.

For survival problems, the time to the event of interest (T) is precisely known only for those instances in which the event occurs during the study period. For the remaining cases, we can only observe a censoring time (C) as we may lose track of them within the observation time or their event occurrence time exceeds the observation time. In the context of survival analysis, they are considered censored instances. In other words, for any given instance i, we can only observe either the survival time (T_i) or the censoring time (C_i), but not both simultaneously. The dataset is referred to as right-censored only when $y_i = min(T_i; C_i)$ can be observed during the study. In right-censored survival problems, since the samples are entered into the study randomly, and the randomness of the censoring time, the censoring time of the sample is also a random variable. Therefore, in this article, we assume that censoring occurs randomly in survival problems. For brevity, this article refers to randomly occurring right-censored instances as censored instances.

Problem Statement: For a given instance i, it is represented by a triplet $(x_i; y_i; \delta_i)$, where $x_i \in \mathbf{R}^{1*P}$ is the feature vector, δ_i is a binary event indicator such that $\delta_i=1$ for uncensored instances and $\delta_i=0$ for censored instances, and y_i represents the observation time, which equals the survival time T_i for uncensored instances and the censoring time C_i for censored instances. Write

$$y_i = \begin{cases} T_i & \text{if } \delta_i = 1 \\ C_i & \text{if } \delta_i = 0 \end{cases} \tag{1}$$

In survival analysis, the objective is to estimate the time from a new instance j, characterized by feature predictors represented by X_j, to an event of interest, T_j. X_j is the variable input to the model. It can be data of various types such as transcriptomics data, medical images, and so on. It is important to note that in survival analysis, the value of T_j is both non-negative and continuous.

3   Time to Even Prediction

Time-to-event prediction, as an application of survival analysis, aims to forecast the probability of a specific event (e.g., death, failure, cure) occurring within a future period. This time frame can be measured from the start of a study or from a particular event's occurrence. Time-to-event prediction holds significant practical significance as it assists in understanding the probability and timing of an event, enabling individuals to make more informed decisions.

3.1   Statistics Based Methods

Statistical methods in survival analysis can be classified into three major categories: (I) parametric methods, (II) non-parametric methods, and (III) semi-parametric methods. Parametric methods are highly effective and accurate in predicting event times when assuming that the data set follows a specific distribution. For instance, if the time in the examined data set follows a well-known theoretical distribution such as the exponential distribution, it is straightforward to use it for estimating event durations. However, in real-life data sets, it is often challenging to obtain data that precisely conforms to known theoretical distributions. In such cases, non-parametric methods can be employed, as they do not rely on assumptions about the underlying distribution of event times. The Kaplan-Meier (Kaplan and Meier, 1958) method is one of the most popular approaches in this category [8]. The Nelson-Aalen (NA) estimator is another non-parametric estimator based on modern counting process techniques [9]. The log-rank test is an application of the Kaplan-Meier method for interval-grouped survival data [10]. The third category comprises a combination of parametric and non-parametric methods. Similar to non-parametric methods, semi-parametric models do not require knowledge of the underlying distribution of event times. The Cox proportional hazards model (Cox, 1972) is the most widely used semi-parametric method in survival analysis [11]. It assumes that the attributes have a multiplicative effect in the hazard function and remain constant over time. In the Cox proportional hazards model, it is assumed that the hazard ratio between two covariates is independent of time and is defined as follows:

$$\lambda\left(t|x\right) = \lambda_0\left(t\right)\exp\left(h\left(x\right)\right) \tag{2}$$

$$h(x) = \theta^T x \tag{3}$$

where $\lambda_0(t)$ is the baseline risk. $h(x)$ is the risk function, which represents the impact of covariates on an individual's risk of death. $\theta = \left(\theta_1, \theta_2, ..., \theta_n\right)$ can be estimated using the maximum partial likelihood function. Partial likelihood is the product of the probabilities of individual i being at risk at each event time T_i and experiencing an event. The Cox partial likelihood function is parameterized by θ and defined as follows:

$$L(\theta) = \prod_{i:E=1} \frac{\exp\left(\hat{h}\left(x_i\right)\right)}{\sum_{j \in R(T_i)} \exp\left(\hat{h}\left(x_j\right)\right)} \tag{4}$$

where T_i, E_i, and x_i represent the event time, event indicator, and covariates, respectively, for the i-th observation. The risk set $R(t)$ represents the set of patients who are still at risk of death at time t. The survival function proposed by Cox is highly influential and widely incorporated in subsequent studies within the field of computational biology.

Table 1. Dataset

Dataset	Acronym
Worcester Heart Attack Study	WHAS
Study to Understand Prognoses Preferences Outcomes and Risks of Treatment	SUPPORT
The Molecular Taxonomy of Breast Cancer International Consortium	METABRIC
The Cancer Genome Atlas	TCGA
National Lung Screening Trial	NLST

3.2   Traditional Machine Learning Based Methods

Although statistical techniques aim to characterize the distribution of event times and the statistical properties of each (statistical) model's parameters, machine learning methods seek to predict event occurrences at given time points. The decision tree algorithm (Bou-Hamad et al., 2011) is based on recursive partitioning with specific splitting criteria applicable to survival analysis [12]. Due to the key feature of this algorithm being the splitting criterion, there has been some research focused on finding effective splitting criteria for survival analysis [13]. Random survival forests (RSF) employ the log-rank test as the splitting criterion to construct random forests. It calculates the cumulative hazards of leaf nodes and averages them over all elements [14]. The LASSO-COX model, utilizing the least absolute shrinkage and selection operator (LASSO), applies feature selection to choose relevant subsets for cancer prediction [15]. SVRc improves the loss function to handle truncated data [16]. It leverages the advantages of standard Support Vector Regression (SVR) and makes it applicable to censored cases through an updated asymmetric loss function that considers both uncensored and censored instances in the model.

Bayesian analysis is one of the fundamental principles in statistics, linking posterior probabilities with prior probabilities. Some studies employ this model to predict the probability of the events of interest [17], benefiting from the desirable properties of Bayesian modeling, such as interpretability [18]. Features in Bayesian networks can be interrelated at different levels and can be represented graphically to depict the theoretical distribution of a set of variables. Bayesian networks provide an intuitive representation of all relationships among variables, making them interpretable for end-users. Knowledge information can be acquired by estimating the network structure and parameters from a given dataset. (Fard et al., 2016) introduced a novel framework that combines the representational power of Bayesian networks with the accelerated failure time (AFT) model by extrapolating the prior probabilities to future time points [19]. The computational complexity of these Bayesian methods primarily depends on the type of Bayesian techniques used in the model.

Support Vector Machine (SVM) is also a crucial class of machine learning algorithm [20]. It can be used for both classification and regression tasks and has been successfully applied to survival analysis problems [21]. (Van et al., 2007) studied a learning machine designed for predictive modeling of survival data with independent correct censoring by introducing a health index as a proxy between instance covariates and outcomes [22]. (Van et al., 2011) presented an SVR-based method that combines ranking and regression approaches within the context of survival analysis [23]. On average, these methods have a time complexity of $O(N^3)$, which is comparable to the time complexity of standard Support Vector Machines.

3.3   Neural Network Based Methods

Currently, deep learning has emerged as a highly successful technique in machine learning. It has demonstrated the ability to train complex models and extract advanced features from real-world datasets. In deep learning, generative networks can capture intricate relationships between features through deep neural network structures, thereby enhancing the accuracy of predictions.

With the increasing prevalence and integration of various data types, such as genomics, transcriptomics, and tissue pathology data, cancer treatment is shifting towards precision medicine [24]. Utilizing and interpreting multiple high-dimensional data types in translational research or clinical tasks requires a significant amount of time and expertise. This necessitates modeling algorithms capable of learning from a multitude of complex features. Excitingly, deep learning models have the potential to leverage this complexity to provide meaningful insights and identify relevant granular features from diverse data types [25,26]. Whether it is tabular data or image data, the application of deep learning-based survival analysis models can be achieved by constructing appropriate model architectures and training processes. It is crucial to select suitable model structures based on specific data types and application scenarios, while also performing feature engineering and model optimization to obtain accurate survival predictions. We discuss deep learning-based survival analysis models from two perspectives: utilizing omics-data and image data. In Table 1, we have compiled the datasets used in this study. The WHAS dataset comprises trends in cardiac event occurrences, patient characteristics, and the quality of healthcare in the region. The SUPPORT study aims to understand the communication and decision-making processes among patients, their families, and healthcare providers. It examines medical decisions related to various types of treatment interventions and assesses the impact of these decisions on patient survival and quality of life. METABRIC aims to investigate the classification and molecular characteristics of breast cancer through molecular biology and genetics approaches. TCGA aims to improve the prevention, diagnosis, and treatment of various cancer types by conducting in-depth research on the genomes and molecular characteristics of cancer. TCGA data provides extensive information related to tumor genomics, mutations, gene expression, methylation, and more. NLST aims to assess the effectiveness of lung CT scans in lung cancer screening.

Table 2. Summary of omics-dataset.

Publication	Type of Disease	Dataset	Variable of input
[27]	Heart attack, Breast(BRCA)	WHAS, SUPPORT, METABRIC	Age, sex, body-mass-index(BMI), left heart failure complications (CHF), and order of MI (MIORD)
[28]	Pan-glioma (LGG/GBM), BRCA, Pan-kidney (KIPAN)	TCGA	Gene expression features, protein expression, copy number, age and stage
[29]	BRCA	METABRIC	Tumor size, number of positive lymph nodes
[30]	Comprehensive	CLINIC	Physiologic variables
[31]	BRCA	TCGA	Gene expression data, miRNA data, Copy number burden (CNB), Tumor mutation burden and clinical information
[32]	BRCA	TCGA	methylation and mRNA data

Omics-Data. Transcriptomics data is currently the most commonly used type of Omics-data. Table 2 provides an overview of the omics data used in related works. Transcriptomic analysis can be employed to assign cancer to clinically meaningful molecular subtypes with diagnostic, prognostic, or therapeutic relevance. Standard computational methods for cancer subtyping, such as SVM or k-Nearest Neighbors (kNN), may be prone to batch effects and can be error-prone, relying only on a small number of feature genes while disregarding important biological information. Deep learning algorithms can overcome these limitations by learning patterns across the entire transcriptome. DeepSurv, based on the semi-parametric Cox proportional hazards model, uses a deep neural network instead of a linear network for prediction [27]. DeepSurv has demonstrated the significant role of deep neural networks in survival analysis tasks. However, this prediction method has limitations when learning from high-dimensional profiles generated from these platforms and relies on expert manual selection of a few features to train the predictive model. SurvivalNet demonstrates how deep learning and Bayesian optimization methods can be applied to predict cancer outcomes, which have been successful in general high-dimensional prediction tasks [28]. SurvivalNet is a Bayesian optimization-based deep survival model that successfully transfers information across diseases to improve prognostic accuracy. However, these models rely on strong parametric assumptions, which are often violated in practice. DeepHit, on the other hand, does not make any assumptions about underlying stochastic processes and allows for the possibility of time-varying relationships between covariates and risks [29]. Moreover, DeepHit handles competing risks. It introduces a ranking loss in the loss function to address these challenges. DeepHit effectively improves the prediction of correct order pairs by optimizing this loss function:

$$A_{k,i,j} \equiv \mathbf{1}(k^{(i)} = k, s^{(i)} < s^{(j)})$$ (5)

$$L = -\sum_{k=1}^{K} \alpha_k \sum_{i \neq j} A_{k,i,j} \cdot \eta \left(F_k \left(s^{(i)} | x^{(i)} \right), F_k \left(s^{(i)} | x^{(j)} \right) \right)$$ (6)

In this context, where k represents the occurrence of an event, s denotes the time of event occurrence, x represents patient covariates, and $F()$ is the prediction function, $F(s|x)$ represents the estimated risk of an event at time s for a given patient. During the model training process, this function aims to make the model predict patients with longer survival times and lower risk of death as indicated by the labels.

Few studies have considered sequence patterns within the feature space. To address this, (Ren et al., 2019) proposed a deep recursive survival analysis model that utilizes deep learning for fine-grained conditional probability prediction of the data, while incorporating survival analysis to address censoring issues [30]. This approach models the conditional probability of events for each sample, capturing temporal dependencies to predict the likelihood of real event occurrences. It also estimates the survival rates over time for censored data, thereby improving prediction accuracy.

Numerous studies have made further contributions to the richness of input data, allowing the model training to encompass not only single-source data such as gene expression data or biomarkers but also incorporate multiple omics data. SAMMON has been proposed to aggregate and simplify gene expression data and cancer biomarkers for prognostic prediction [31]. Experimental results demonstrate that performance is improved when more omics data is used in model construction. However, incorporating more data into the model also presents challenges. Gene data is characterized by high dimensionality, non-linearity, and sparsity. Directly training the model with these high-dimensional data does not yield optimal results. (Bichindaritz et al., 2021) employed the local maximum quasi-cliques merging (lmQCM) algorithm to reduce the dimensions of mRNA and methylation features and extract clustered feature genes [32]. They introduced an auxiliary ordinal loss on top of the original Cox model to enhance the optimization learning process during training and regularization. The auxiliary loss helps mitigate the problem of vanishing gradients in early layers and aids in reducing the loss of the primary task. Finally, they constructed an ordered Cox risk model for survival analysis and employed the Long Short-Term Memory (LSTM) approach to predict patients' survival risks.

Image Data. Convolutional neural network (CNNs) have been extensively applied in the intersection of computer science and medicine. These end-to-end

Table 3. Summary of image dataset.

Publication	Type of Disease	Dataset	Variable of input
[34]	Lung and brain cancers	NLST, TCGA	Whole Slide Histopathological Images(WSIs)
[35]	Lung cancer	NLST	The screening radiography arm of the NLST
[36]	Lung and colorectal cancers	NLST	WSIs
[37]	BRCA	TCGA	WSIs
[38]	Heart failure	Clinical records	The clinical information, cine MR images, and clinic outcome
[39]	Glioblastoma multiforme (GBM), Lung squamous cell carcinoma (LUSC)	TCGA	WSIs

deep neural networks possess stronger feature extraction capabilities compared to traditional methods [33]. Unlike manual feature selection, CNNs can automatically extract the most discriminative features from images. The Table 3 provides an overview of the imaging data used in the relevant studies. Moreover, they can leverage pre-trained models trained on ImageNet during training, enabling them to be quickly trained and optimized on medical image datasets with limited data, yielding satisfactory results. However, in a high-resolution medical image, only a small portion typically contains the relevant features. The prevalence of negative samples significantly outnumbering positive samples can lead to suboptimal model performance in such scenarios. (Zhu et al., 2017) proposed an effective whole slide image-based survival analysis framework (WSISA) to overcome the aforementioned challenges [34]. To leverage WSIs for survival discrimination, they first extracted hundreds of patches from each WSI through adaptive sampling and then grouped these images into different clusters. Subsequently, they trained an ensemble model to make patient-level predictions based on the cluster-level deep convolutional survival (DeepConvSurv) predictions. This framework efficiently mines and utilizes all discriminative patterns within WSIs to predict patient survival status.(Lu et al., 2019) presented the prediction of patients' mortality risk using X-ray images [35]. Traditional image-based survival prediction models rely on annotated discriminative markers, which limits their scalability to large datasets. When there are no available annotations for the classification task, the multiple instance learning (MIL) framework proves useful for histopathological images. Unlike existing image-based survival models that are restricted to extracting key patches or clusters obtained from whole slide images (WSIs), DeepAttnMISL [36]effectively learns imaging features from WSIs by introducing attention-based MIL pooling and aggregates WSI-level information into patient-level predictions. In current survival models, attention-based aggregation offers more flexibility and adaptability compared to traditional aggregation techniques. (Liu et al., 2022) proposed and experimentally evaluated a multi-resolution deep learning approach for breast cancer survival analysis [37]. This method integrates multiple resolution image data with tumor, lymphocyte, and nucleus segmentation results from deep learning models. The results demonstrate that this approach significantly improves the performance of deep learning models compared to using only raw image data. However, conventional images capture two-dimensional spatial information as they only contain information within the plane. They cannot capture the temporal trends of patients over time. (Guo et al., 2023) proposed a predictive model based on cardiac images that incorporates temporal sequence information [38]. By fusing the boundary information of cardiac images with the motion field of the heart, they obtained cardiac motion information, which improved the survival probability prediction for heart failure patients. (Li et al., 2023) introduced a patch sampling strategy based on image information entropy and constructed a multi-scale feature fusion network (MSFN) using a self-supervised feature extractor [39]. Specifically, this work employed image information entropy as a criterion to select representative sampling patches, thereby avoiding noise interference caused by random sampling in blank regions. Additionally, a pre-training process was performed on the feature extractor using self-supervised learning mechanisms to enhance feature extraction efficiency. Furthermore, a global-local feature fusion prediction network based on attention mechanisms was constructed to improve the survival prediction of WSIs with comprehensive multi-scale information representation.

Table 4. Summary of multimodal dataset.

Publication	Type of Disease	Dataset	Variable of input
[40]	Lower-Grade Glioma (LGG) and Glioblastoma (GBM)	TCGA	Whole-slide images, clinical and genomic data
[41]	20 cancer types	TCGA	MicroRNAs, gene expression data, and clinical data
[42]	Lung and colorectal cancers	NLST	CT scans, age, gender, and smoking history

Multimodal Data. Multimodal fusion refers to the process of integrating information from two or more modalities to make predictions. In prediction tasks, individual modalities often do not contain all the necessary information required for accurate prediction results. The process of multimodal fusion combines information from multiple modalities to complement each other, expand the coverage of information contained in the input data, enhance the accuracy of prediction results, and improve the robustness of the prediction model (Fig. 1).

Simultaneously combining Omics-data and image data is a key factor in further improving model performance. The Table 4 provides an overview of the multimodal data used in the related studies.(Mobadersany et al., 2018) demonstrated a computational approach that utilizes deep learning to learn patient outcomes from digital pathology images by combining adaptive machine learning algorithms with traditional survival models [40]. This work showcased how survival convolutional neural networks (SCNNs) integrate information from histopathological images and genomic biomarkers into a unified framework to predict time-to-event outcomes. It showed superior prediction accuracy for overall survival

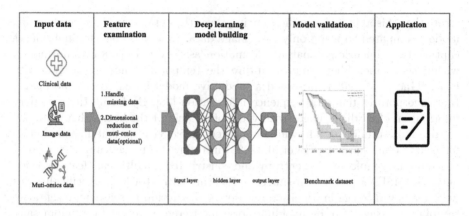

Fig. 1. Workflow for establishing a deep learning model for prognosis prediction.

of glioma patients beyond current clinical models. (Cheerla et al., 2019) further developed a model based on a multimodal neural network that utilizes clinical data, mRNA expression data, microRNA expression data, and WSIs to predict survival of patients across 20 different cancer types [41]. They devised an unsupervised encoder that compresses the four data modalities into a single feature vector per patient, handling missing data using an elastic, multimodal dropout approach. The encoding method is tailored for each data type - using deep highway networks to extract features from clinical and genomic data and employing convolutional neural networks to extract features from WSIs. (Lu et al., 2023) proposed a hybrid CNN-RNN approach to investigate the long-term survival rates of subjects in lung cancer screening studies [42]. This method utilizes a CNN model to capture imaging features from CT scans and employs an RNN model to examine the time series aspect, thereby capturing global information. By combining time series information with multimodal data, the performance of the model is effectively improved.

4   Cluster Based Risk Profile

Clustering is an unsupervised learning method that can uncover hidden patterns and structures within data. Clustering is a valuable tool for data-driven disease discovery and classification. The objective of survival clustering is to map subjects (e.g., users in a social network, patients in medical research) into K clusters ranging from low-risk to high-risk.

4.1   Statistics Based Methods

(Li et al., 2004) proposed an approach to examine survival data by extending the partial least squares (PLS) regression to the framework of the Cox model [43]. They presented a parallel algorithm for constructing latent components.

This algorithm utilized residual iterative least squares fitting and Cox regression fitting to construct predictive components. These components could then be used to build useful survival prediction models and also employed for clustering survival data since the principal components were simultaneously constructed.

(Bair et al., 2004) were the first to explore clustering methods for survival data, introducing a technique known as semi-supervised clustering (SSC) [44]. In their study, they proposed using risk scores from univariate Cox regression as a preselection step to choose variables and then applied k-means clustering to the selected subset of features to discover patient subgroups. In the second part of the method, they employed only clinical data to test the clustering assignments. Using the clinical data, they treated the clustering assignments as the outcome variable and applied a classification algorithm. The classification algorithm performed well, indicating correct identification of the clustering assignments. However, both the regression and survival models utilized principal components. Since principal components may not capture a large portion of the variance present in the data, these methods cannot guarantee that these components are relevant to the outcomes. Therefore, (Bair et al. 2006) proposed a semi-supervised approach called supervised principal components (SPC) [45]. In this method, they computed univariate Cox scores for each feature and selected the most important features by choosing only those with the best Cox scores.

Sparse clustering methods and semi-supervised clustering approaches rely on the number of features that have been characterized as "salient," and therefore, these methods have notable limitations. (Gaynor et al., 2017) proposed an enhanced method called pre-weighted sparse clustering to overcome the limitations of sparse clustering [46]. Its objective is to overcome the limitations of traditional sparse clustering by identifying features that have different means within clusters. This approach can identify features that exhibit variations in their average values across clusters.

4.2   Traditional Machine Learning Based Methods

(Zhang et al. 2016) employed a hybrid approach consisting of statistical and machine learning methods, focusing on cluster discovery in clinical and gene expression data [47]. They utilized penalized logistic regression and penalized proportional hazards models along with an expectation-maximization algorithm to select only the most relevant clinical features associated with the event of interest. This approach allowed them to identify the most important clinical features specifically related to the event of interest. (Mouli et al., 2017) proposed a decision tree-based method aimed at achieving survival clustering [48]. The ultimate goal was to cluster examination data and identify two or more populations with different risk levels. The objective was to determine clusters where the survival distributions differed. (Ahlqvist et al., 2018) utilized Cox regression to explore differences between subgroups of diabetes patients discovered through k-means and hierarchical clustering [49]. This method divided patients into five subgroups based on distinct disease progression and risk of diabetes complications. This novel subtyping could potentially contribute to personalized and targeted early treatments for patients who would benefit the most, representing an initial step towards precision medicine in diabetes.

Table 5. Summary of cluster dataset.

Publication	Type of Disease	Dataset	Variable of input
[50]	Nondisease	Friendster	age, gender, marital status, occupation, and interest
[51]	Acute coronary syndrome	Private	disease types, demographics, personal disease history, comorbidities, habits, laboratory test results, and procedures
[52]	Type2 diabetes, heart dise	FLCHAIN, SUPPORT, SEER, EHR, SLEEP, FRAMINGHAM	serum data, breast cancer subcohort and clinical data
[53]	BRCA	SUPPORT, METABRIC, SEER	Age, gender, race, education, income, physiological measurements, co-morbidity information, gene expressions, and clinical variables
[54]	Alzheimer's disease(AD)	EHR	demographic information, laboratory tests, and diagnoses and symptoms
[55]	Cardiovascular disease	ALLHAT-A, ALLHAT-B, ACCORD	Clinical information on the use of different concentrations of drugs
[56]	22 high-grade glioma patients, non-small cell	SUPPORT, HGG, Hemodialysis, FLChain, NSCL	Observational cohort and CT image

4.3   Neural Network Based Methods

Deep neural networks play a significant role in the advancement of clustering tasks. They are commonly used to handle large-scale and high-dimensional data, such as images, speech, and textual data, in the field of computational biology. The Table 5 provides an overview of the data used in deep learning clustering research. Traditional survival methods assume the existence of explicit end-of-life signals or introduce them artificially using predefined timeouts (Mouli et al., 2018). They proposed a deep clustering approach that distinguishes long-term and short-term survivors based on a modified Kuiper statistic, even in the absence of end-of-life signals [50]. In their study, they introduced a loss function that utilizes an enhanced Kuiper statistic to differentiate the empirical survival distributions of clusters. By optimizing this loss, a deep neural network is learned to softly cluster users into survival groups. (Xia et al., 2019) employed a multi-task learning approach for outcome-driven clustering of patients with acute coronary syndrome [51]. The proposed method utilized an attention-based multi-task neural network as the modeling framework, which includes patient state learning, cluster analysis, and feature importance analysis.

However, traditional survival analysis methods estimate risk scores or personalized event time distributions that depend on covariates. In practice, due to (unknown) subpopulations having different risk profiles or survival distributions, there often exists substantial population-level phenotypic heterogeneity. Therefore, in survival analysis, there is an unmet need to identify subgroups with distinct risk profiles while simultaneously considering accurate personalized event time predictions. Methods addressing this need may improve the characterization of individual outcomes by leveraging the regularities within subpopulations, thereby accounting for population-level heterogeneity. (Chapfuwa et al., 2020) proposed a Bayesian nonparametric method that represents observations (subjects) in a clustering latent space and encourages accurate time-to-event predictions and clustering (subpopulations) with distinct risk profiles [52]. (Nagpal et al., 2021) have explored similar techniques, introducing a finite mixture of Weibull distributions known as the Deep Survival Machine (DSM) [53]. DSM fits a survival regression model mixture on the representations learned by an encoder neural network. From a modeling perspective, the aforementioned methods focus on outcome-driven clustering, where they fully recover clusters with distinct survival distribution characteristics. In this work, the Deep Cox Mixture (DCM) is introduced, which jointly fits a VAE and Cox regression mixture without specifying a generative model. The loss of DCM is derived by combining the VAE loss and likelihood estimation of survival time.

Previous research has primarily utilized imaging or cognitive data, with limitations in data breadth and sample size. Data-driven models have not been able to perform well in these cases. Certain diseases exhibit a high degree of heterogeneity, such as Alzheimer's disease (AD), where different trajectories and outcomes are observed in clinical populations. (Alexander et al., 2021) identified AD patients using a previously validated rule-based phenotype algorithm from the Clinical Practice Research Datalink (CPRD), which contains primary care electronic health records [54]. They extracted and incorporated a range of comorbidities, symptoms, and demographic features as patient characteristics, thus expanding the breadth of data. However, this approach did not consider the evaluation of treatment effects concerning clinical interventions involving continuous time-to-event outcomes, such as time to death, readmission, or composite events that may be subject to review. In such cases, counterfactual inference is required to disentangle the effects of confounding physiological features affecting baseline survival rates from the effects of the interventions being evaluated. (Nagpal et al., 2022) proposed a latent variable approach to simulate heterogeneous treatment effects, suggesting that an individual can belong to one of several latent clusters with different response characteristics [55]. Experimental results demonstrate that this latent structure can modulate baseline survival rates and help determine the effects of interventions. However, clustering of survival data remains an underexplored problem. In this scenario, only a few methods have been proposed, either with limited scalability in high-dimensional unstructured data or focused on discovering purely outcome-driven clusters, i.e., clusters solely based on survival time as the defining feature. The latter may fail in applications where individual survival distribution information alone is insufficient for stratifying the population. For example, patient groups with similar survival outcomes may exhibit vastly different responses to the same treatment. To address these challenges, (Manduchiy et al., 2022) introduced a novel survival data clustering approach called Variational Deep Survival Clustering (VaDeSC), which discovers patient groups with distinct characteristics in terms of the underlying mechanisms generating survival outcomes [56]. It extends previous variational methods used for unsupervised deep clustering by incorporating survival models specific to each cluster within the generative process. VaDeSC focuses not only on survival but also captures the heterogeneity in the relationship between covariates and survival outcomes.

5   Conclusions and Future Directions

Despite the numerous scientific reports on the application of machine learning techniques for time-to-event prediction, there has been relatively less research on survival clustering techniques. Survival clustering techniques are particularly useful when there is a need to identify unknown subpopulations within an entire dataset. They can discover clusters with significantly different survival capabilities, which cannot be achieved by traditional clustering techniques. These techniques focus on finding clusters with distinct survival distributions, providing a unique perspective to understand the characteristics of the dataset.

Many research works in the field of survival analysis are currently exploring the application of deep learning methods, which have powerful modeling capabilities and predictive performance. However, a major limitation of deep learning models is their lack of interpretability. This means that although deep learning models can generate accurate prediction results, it is often challenging to explain how the model arrives at those predictions and what features and patterns it relies on for decision-making. Further more, the research efforts in exploring key biomarkers are also relatively scarce in the field.

Enhancing Model Interpretability: One significant challenge in the application of machine learning models to survival analysis lies in their inherent complexity and the difficulty in understanding how they arrive at specific predictions. As the use of machine learning in this domain continues to grow, there is a pressing need to develop interpretable models that can provide meaningful insights into the underlying biological mechanisms governing survival outcomes. Future research efforts should focus on incorporating techniques such as feature importance analysis, attention mechanisms, and visualizations to shed light on the decision-making processes of these models. By achieving better interpretability, researchers can gain a deeper understanding of the relationships between genomic, clinical, and imaging features, ultimately leading to more reliable and clinically actionable predictions.

Identifying Novel Biomarkers: While machine learning techniques have shown great promise in survival analysis, there remains untapped potential in the discovery of novel and robust biomarkers that can accurately predict patient outcomes. Leveraging multi-omics data integration and advanced feature selection methods can facilitate the identification of previously unrecognized biomarkers with strong prognostic significance. Additionally, collaborative efforts between computational biologists, bioinformaticians, and domain experts can drive the development of innovative approaches to uncover hidden patterns and relationships within complex biological datasets. The integration of these novel biomarkers into clinical practice has the potential to revolutionize patient risk stratification, enabling tailored treatment strategies and personalized medicine.

By addressing these challenges, it not only contributes to enhancing the feasibility and acceptability of survival analysis methods in medicine and other fields but also promotes the advancement of survival analysis techniques to meet broader application needs. Furthermore, it provides more reliable decision support for clinical practice and disease management.

References

1. Guo, S.: Survival Analysis. Oxford University Press (2010)
2. Emmert-Streib, F., Dehmer, M.: Introduction to survival analysis in practice. Mach. Learn. Knowl. Extr. **1**(3), 1013–1038 (2019)
3. Nezhad, M.Z., Sadati, N., Yang, K., et al.: A deep active survival analysis approach for precision treatment recommendations: application of prostate cancer. Expert Syst. Appl. **115**, 16–26 (2019)
4. Singer, J.D., Willett, J.B.: Modeling the days of our lives: using survival analysis when designing and analyzing longitudinal studies of duration and the timing of events. Psychol. Bull. **110**(2), 268 (1991)
5. Klein, J.P., Moeschberger, M.L.: Survival Analysis. SBH, Springer, New York (2003). https://doi.org/10.1007/b97377
6. Lee, E.T., Wang, J.: Statistical Methods for Survival Data Analysis. John Wiley & Sons (2003)
7. Marubini, E., Valsecchi, M.G.: Analysing Survival Data from Clinical Trials and Observational Studies. John Wiley & Sons (2004)
8. Kaplan, E.L., Meier, P.: Nonparametric estimation from incomplete observations. J. Am. Stat. Assoc. **53**(282), 457–481 (1958)
9. Andersen, P.K., Borgan, O., Gill, R.D., et al.: Statistical Models Based on Counting Processes. Springer Science & Business Media (2012)
10. Cutler, S.J., Ederer, F.: Maximum utilization of the life table method in analyzing survival. J. Chronic Dis. **8**(6), 699–712 (1958)
11. Lin, D.Y., Wei, L.J., Ying, Z.: Checking the Cox model with cumulative sums of martingale-based residuals. Biometrika **80**(3), 557–572 (1993)
12. Bou-Hamad, I., Larocque, D., Ben-Ameur, H.: A review of survival trees (2011)
13. Ciampi, A., Chang, CH., Hogg, S., McKinney, S.: Recursive partition: a versatile method for exploratory-data analysis in biostatistics. In: MacNeill, I.B., Umphrey, G.J., Donner, A., Jandhyala, V.K. (eds.) Biostatistics. The University of Western Ontario Series in Philosophy of Science, vol. 38. Springer, Dordrecht (1987). https://doi.org/10.1007/978-94-009-4794-8_2

14. Ishwaran, H., Kogalur, U.B., Blackstone, E.H., et al.: Random survival forests (2008)
15. Shahraki, H.R., Salehi, A., Zare, N.: Survival prognostic factors of male breast cancer in Southern Iran: a LASSO-Cox regression approach. Asian Pac. J. Cancer Prev. **16**(15), 6773–6777 (2015)
16. Khan, F.M., Zubek, V.B.: Support vector regression for censored data (SVRc): a novel tool for survival analysis. In: 2008 Eighth IEEE International Conference on Data Mining, pp. 863–868. IEEE (2008)
17. Kononenko, I.: Inductive and Bayesian learning in medical diagnosis. Appl. Artifi. Intell. Int. J. **7**(4), 317–337 (1993)
18. Ibrahim, J.G., Chen, M.H., Sinha, D., et al.: Bayesian survival analysis. Springer, New York (2001). https://doi.org/10.1007/978-1-4757-3447-8
19. Fard, M.J., Wang, P., Chawla, S., et al.: A Bayesian perspective on early stage event prediction in longitudinal data. IEEE Trans. Knowl. Data Eng. **28**(12), 3126–3139 (2016)
20. Van Belle, V., Pelckmans, K., Suykens, J.A.K., et al.: Support vector machines for survival analysis. In: Proceedings of the Third International Conference on Computational Intelligence in Medicine and Healthcare (CIMED2007), pp. 1–8 (2007)
21. Shivaswamy, P.K., Chu, W., Jansche, M.: A support vector approach to censored targets. In: Seventh IEEE International Conference on Data Mining (ICDM 2007), pp. 655–660. IEEE (2007)
22. Van Belle, V., Pelckmans, K., Suykens, J.A.K., et al.: Support vector machines for survival analysis. In: Proceedings of the Third International Conference on Computational Intelligence in Medicine and Healthcare (CIMED 2007), pp. 1–8 (2007)
23. Van Belle, V., Pelckmans, K., Van Huffel, S., et al.: Support vector methods for survival analysis: a comparison between ranking and regression approaches. Artif. Intell. Med. **53**(2), 107–118 (2011)
24. Tran, K.A., Kondrashova, O., Bradley, A., et al.: Deep learning in cancer diagnosis, prognosis and treatment selection. Genome Med. **13**(1), 1–17 (2021)
25. Wainberg, M., Merico, D., Delong, A., et al.: Deep learning in biomedicine. Nat. Biotechnol. **36**(9), 829–838 (2018)
26. Zou, J., Huss, M., Abid, A., et al.: A primer on deep learning in genomics. Nat. Genet. **51**(1), 12–18 (2019)
27. Katzman, J.L., Shaham, U., Cloninger, A., et al.: DeepSurv: personalized treatment recommender system using a Cox proportional hazards deep neural network. BMC Med. Res. Methodol. **18**(1), 1–12 (2018)
28. Yousefi, S., Amrollahi, F., Amgad, M., et al.: Predicting clinical outcomes from large scale cancer genomic profiles with deep survival models. Sci. Rep. **7**(1), 11707 (2017)
29. Lee, C., Zame, W., Yoon, J., et al.: DeepHit: a deep learning approach to survival analysis with competing risks. In: Proceedings of the AAAI Conference on Artificial Intelligence, vol. 32, no. 1 (2018)
30. Ren, K., Qin, J., Zheng, L., et al.: Deep recurrent survival analysis. In: Proceedings of the AAAI Conference on Artificial Intelligence. vol. 33, no. 01, pp. 4798–4805 (2019)
31. Huang, Z., Zhan, X., Xiang, S., et al.: SALMON: survival analysis learning with multi-omics neural networks on breast cancer. Front. Genet. **10**, 166 (2019)

32. Bichindaritz, I., Liu, G., Bartlett, C.: Integrative survival analysis of breast cancer with gene expression and DNA methylation data. Bioinformatics **37**(17), 2601–2608 (2021)

33. Raza, R., Zulfiqar, F., Tariq, S., et al.: Melanoma classification from Dermoscopy images using ensemble of convolutional neural networks. Mathematics **10**(1), 26 (2022)

34. Zhu, X., Yao, J., Zhu, F., et al.: WSISA: making survival prediction from whole slide histopathological images. In: Proceedings of the IEEE Conference on Computer Vision and Pattern Recognition, pp. 7234–7242 (2017)

35. Lu, M.T., Ivanov, A., Mayrhofer, T., et al.: Deep learning to assess long-term mortality from chest radiographs. JAMA Netw. Open **2**(7), e197416–e197416 (2019)

36. Yao, J., Zhu, X., Jonnagaddala, J., et al.: Whole slide images based cancer survival prediction using attention guided deep multiple instance learning networks. Med. Image Anal. **65**, 101789 (2020)

37. Liu, H., Kurc, T.: Deep learning for survival analysis in breast cancer with whole slide image data. Bioinformatics **38**(14), 3629–3637 (2022)

38. Guo, S., Zhang, H., Gao, Y., et al.: Survival prediction of heart failure patients using motion-based analysis method. Comput. Methods Programs Biomed. **236**, 107547 (2023)

39. Li, L., Liang, Y., Shao, M., et al.: Self-supervised learning-based Multi-Scale feature Fusion Network for survival analysis from whole slide images. Comput. Biol. Med. **153**, 106482 (2023)

40. Mobadersany, P., Yousefi, S., Amgad, M., et al.: Predicting cancer outcomes from histology and genomics using convolutional networks. Proc. Natl. Acad. Sci. **115**(13), E2970–E2979 (2018)

41. Cheerla, A., Gevaert, O.: Deep learning with multimodal representation for pancancer prognosis prediction. Bioinformatics **35**(14), i446–i454 (2019)

42. Lu, Y., Aslani, S., Zhao, A., et al.: A hybrid CNN-RNN approach for survival analysis in a Lung Cancer Screening study (2023). arXiv preprint arXiv:2303.10789

43. Li, H., Gui, J.: Partial Cox regression analysis for high-dimensional microarray gene expression data. Bioinformatics **20**(suppl_1), i208–i215 (2004)

44. Bair, E., Tibshirani, R.: Semi-supervised methods to predict patient survival from gene expression data. PLoS Biol. **2**(4), e108 (2004)

45. Bair, E., Hastie, T., Paul, D., et al.: Prediction by supervised principal components. J. Am. Stat. Assoc. **101**(473), 119–137 (2006)

46. Gaynor, S., Bair, E.: Identification of relevant subtypes via preweighted sparse clustering. Comput. Stat. Data Anal. **116**, 139–154 (2017)

47. Zhang, W., Tang, J., Wang, N.: Using the machine learning approach to predict patient survival from high-dimensional survival data. In: 2016 IEEE International Conference on Bioinformatics and Biomedicine (BIBM), pp. 1234–1238. IEEE (2016)

48. Mouli, S.C., Naik, A., Ribeiro, B., et al.: Identifying user survival types via clustering of censored social network data (2017). arXiv preprint arXiv:1703.03401

49. Ahlqvist, E., Storm, P., Käräjämäki, A., et al.: Novel subgroups of adult-onset diabetes and their association with outcomes: a data-driven cluster analysis of six variables. Lancet Diab. Endocrinol. **6**(5), 361–369 (2018)

50. Mouli, S.C., Ribeiro, B., Neville, J.: A Deep Learning Approach for Survival Clustering without End-of-life Signals (2018)

51. Xia, E., Du, X., Mei, J., et al.: Outcome-driven clustering of acute coronary syndrome patients using multi-task neural network with attention. In: MedInfo, pp. 457–461 (2019)

52. Chapfuwa, P., Li, C., Mehta, N., et al.: Survival cluster analysis. In: Proceedings of the ACM Conference on Health, Inference, and Learning, pp. 60–68 (2020)
53. Nagpal, C., Li, X., Dubrawski, A.: Deep survival machines: fully parametric survival regression and representation learning for censored data with competing risks. IEEE J. Biomed. Health Inform. **25**(8), 3163–3175 (2021)
54. Alexander, N., Alexander, D.C., Barkhof, F., et al.: Identifying and evaluating clinical subtypes of Alzheimer's disease in care electronic health records using unsupervised machine learning. BMC Med. Inform. Decis. Mak. **21**(1), 1–13 (2021)
55. Nagpal, C., Goswami, M., Dufendach, K., et al.: Counterfactual phenotyping with censored time-to-events. In: Proceedings of the 28th ACM SIGKDD Conference on Knowledge Discovery and Data Mining, pp. 3634–3644 (2022)
56. Manduchi, L., Marcinkevičs, R., Massi, M.C., et al.: A deep variational approach to clustering survival data (2021). arXiv preprint arXiv:2106.05763

Label-Independent Information Compression for Skin Diseases Recognition

Geng Gao[1], Yunfei He[1], Li Meng[1], Jinlong Shen[1], Lishan Huang[1], Fengli Xiao[2], and Fei Yang[1(✉)]

[1] School of Biomedical Engineering, Anhui Medical University, Hefei, China
yangfei@ahmu.edu.cn
[2] Department of Dermatology of First Affiliated Hospital, and Institute of Dermatology, Anhui Medical University, Hefei, China

Abstract. Skin diseases have a widespread impact on people's lives, making their identification crucial in medical data analysis. Convolutional neural networks (CNNs) are widely used for skin disease recognition, yielding success. However, the current research overlooks the crucial issue of compressing label-independent information. Key information in skin disease images exists in small symptomatic patches, while the rest is extraneous. Regrettably, prevailing CNNs-based methods embed redundancy in skin disease features across convolutional layers, reducing accuracy. This paper introduces an Information Bottleneck theory-based algorithm for selective information propagation. The Hilbert-Schmidt independence criterion (HSIC) is used to calculate the dependency between variables in the algorithm. This algorithm enhances the independence of CNN's convolutional layers and the input features of the skin disease image during training while improving the dependence of convolutional layers and the label value of the image. This confines superfluous data spread and boosts vital information propagation. Experiments conducted on both a self-collected hypopigmentation dataset and the publicly available ISIC2018 dataset, utilizing ResNet-50, DenseNet-169, Inception-v4, and ConvNeXt-B, confirm the effectiveness of the algorithm. The experimental results demonstrate significant enhancements in accuracy following the application of the algorithm. These four CNNs achieve accuracy improvements of 2.68%, 7.63%, 4.20%, and 3.44% on hypopigmentation dataset, and 0.86%, 0.33%, 2.84%, and 1.19% on ISIC2018 dataset. Thus, this algorithm effectively compresses label-independent information, enhancing skin disease recognition.

Keywords: Convolutional neural networks · Skin diseases · Information bottleneck

1 Introduction

Skin diseases are a highly prevalent class of illnesses, significantly impacting both physical and mental well-being. For instance, vitiligo can lead to the appearance

C. Cruz et al. (Eds.): IC 2023, CCIS 2036, pp. 193–204, 2024.
https://doi.org/10.1007/978-981-97-0065-3_14

of white patches on a person's skin, greatly affecting people's appearance [25]. Skin cancer poses a threat to human life [15]. Central to this issue is skin disease recognition. Skin disease recognition not only enhances cure rates but also prevents deterioration [3,7]. Deep learning provides new opportunities for skin disease recognition, which relies on skin disease images combined with relevant algorithms. Among them, the most popular and representative method is convolutional neural networks (CNNs) and their variants [5,21,29].

CNNs excel in extracting crucial image information using convolutional layers. Especially in the medical domain, CNNs have achieved remarkable success in effectively classifying skin conditions such as melanoma and psoriasis. [28,31]. Some studies have explored enhancing CNNs classification performance by employing ensemble methods and altering CNNs architectures. [1,18,32]. Some other studies have proposed the integration of clinical and dermoscopic images for comprehensive multimodal classification [13,24]. As an instance, Yang et al. [30] introduced a multi-scale fully shared fusion network for classifying dermatoscopic and clinical images using multiple modes. Their work emphasized the significance of capturing intra-modal relationships between clinical and dermatoscopic images. Additionally, a study has investigated the integration of dermatologic images and patient information for multimodal diagnostic [6]. Significantly, researchers have devised CNN-based models for skin disease recognition and pitted them against dermatologist evaluations. Interestingly, these models frequently achieve comparable or even superior accuracy when compared to dermatologist assessments [2,26]. While these studies have achieved significant advancements in skin disease recognition, they overlook the imperative task of eliminating redundant (i.e., label-independent) information from skin disease images. This excessive redundant information considerably impedes the ability of CNNs to identify pivotal features. The extent to which the convolutional layers relies on the input layer (i.e., input image) and target layer (i.e., label value) during the training process significantly impacts the CNN's performance. If the convolutional layers displays a strong dependent with the input layer but a weak dependent with the target layer, it indicates an excess of redundant information and a lack of discernible feature extraction, ultimately diminishing the effectiveness of skin disease recognition.

Information Bottleneck(IB) [8] is a good method to remove redundant information. The IB theory is a concept in information theory, initially proposed by Tishby et al. [22]. It is used to understand and explain phenomena that occur in data compression and feature selection. The core idea of this theory is that when dealing with data, a system needs to retain essential information while eliminating unnecessary redundancy to achieve efficient data representation and compression. The main task of IB is to find a mapping that maximizes the compression of input variables as much as possible to retain the information related to the label value, i.e., minimizing the mutual information(MI) of the input layers and the convolutional layers, and maximizing the MI of the label value and the convolutional layers. The MI is a measure of the mutual dependency between two variables. Since the calculation of MI is very complicated and inconvenient to use in IB, researchers have proposed alternative methods [19,27].

Hilbert-Schmidt independence criterion(HSIC) was first proposed by Gretton et al. [10]. It is similar to MI, which is also a nonparametric independence criterion that measures the correlation between two variables, and the smaller its value, the stronger the correlation between the two variables. The computational effort of HSIC is much smaller compared to MI, and its accuracy is also very high. Researchers have tried to introduce HSIC into deep learning for optimization [19]. For example, Greenfeld and Shalit [9] utilized HSIC as a loss function to learn robust regression and classification models by satisfying noise independent of the distribution of random variables. Considering the complexity of MI computation and the advantages of HSIC, researchers have tried to use HSIC instead of MI in IB [19,27]. For example, Wu et al. [27] used HSIC [11] instead of MI to reconstruct and predict self-coding. Ma et al. [19] proposed a method to optimize the HSIC bottleneck for DNNs. Inspired by these studies, we introduce the IB based on HSIC as a loss function to eliminate redundant information from skin disease images, thereby enhancing CNN's ability to identify skin diseases.

This study presents an algorithm named CIH rooted in the IB theory to tackle the challenges at hand. The IB theory is a cornerstone of information theory, delving into how to extract crucial information from data. This theory was proposed by Tishby and his colleagues [23], aiming to address the balance between compression and retention of information during information transmission. In the CIH algorithm, the IB theory is harnessed to orchestrate the behavior of convolutional layers, effectively streamlining the network by discarding label-independent information while retaining meaningful information. The original IB theory involves using MI to compute dependency between variables. However, the computation of MI can be intricate and unwieldy. Consequently, HSIC [19] is ingeniously employed as a proxy to calculate the dependency among variables within the IB. Through the application of the IB and HSIC, the CIH algorithm masterfully navigates the trade-off between information preservation and redundancy elimination, leading to an enhanced capacity to recognize valuable information in the skin disease images.

The main contributions of this paper are as follows:

- This paper applies the information bottleneck theory to skin disease diagnosis. HSIC is used in this theory to calculate the correlation between variables.
- This paper constructs an algorithm named CIH for removing label-independent information based on CNN and information bottleneck theory. The algorithm improves the performance of CNN in recognizing skin diseases.
- Multiple experiments on two datasets and four classical CNNs validate the validity of CIH.

2   Method

2.1   Information Bottleneck

The essence of the IB concept lies in channeling image feature information to the label value through a defined "bottleneck". This bottleneck typically encompasses concealed attributes within a neural network, which can equivalently be represented by the convolutional layers in CNNs. The theoretical definition of IB is as follows:

$$\min : \mathcal{L}_{IB} = M(\boldsymbol{X},\boldsymbol{C}) - \beta \cdot M(\boldsymbol{C},\boldsymbol{Y}) \tag{1}$$

where $M(\boldsymbol{X},\boldsymbol{C})$ means the MI between the input image $\boldsymbol{X}$ and the convolutional layers $\boldsymbol{C}$, $M(\boldsymbol{C},\boldsymbol{Y})$ means the MI between the convolutional layers $\boldsymbol{C}$ and the label value $\boldsymbol{Y}$, σ means the tradeoff ratio between $M(\boldsymbol{X},\boldsymbol{C})$ and $M(\boldsymbol{C},\boldsymbol{Y})$. MI is a statistical tool employed to gauge the dependency between two variables. Higher MI values signify a more dependency between the variables. The IB facilitates the assessment of the dependency between a specific convolutional layer and both the input layer and the label value. The key strategy to curtail label-independent information within the convolutional layers entails ensuring a diminished dependency with the input layer while fostering a potent dependency with the label value.

2.2   Hilbert-Schmidt Independence Criterion

Given the complexity and potential inaccuracies inherent in practical MI calculations, the HSIC emerges as a pragmatic alternative for quantifying dependency. The theoretical definition of HSIC is as follows: Given any two sets of variables, $\boldsymbol{X} = [\boldsymbol{X}_1, \cdots, \boldsymbol{X}_N]$ and $\boldsymbol{Y} = [\boldsymbol{Y}_1, \cdots, \boldsymbol{Y}_N]$. $\boldsymbol{X}$ and $\boldsymbol{Y}$ have a corresponding relationship. The value of HSIC between $\boldsymbol{X}$ and $\boldsymbol{Y}$ is defined as follows:

$$HSIC(\boldsymbol{X},\boldsymbol{Y}) = \frac{1}{(N-1)^2} tr(\boldsymbol{K_X H K_Y H}) \tag{2}$$

where $\boldsymbol{H} = \boldsymbol{I} - \frac{1}{N}\boldsymbol{1}\boldsymbol{1}^T$, $\boldsymbol{I}$ is the identity matrix, $\boldsymbol{1}$ is a column vector with all elements being 1. And $\boldsymbol{K_X}$ is the kernel matrix with elements $\boldsymbol{K_X}(\boldsymbol{X}_i, \boldsymbol{X}_j) = exp(\frac{\|\boldsymbol{X}_i - \boldsymbol{X}_j\|_2^2}{2\sigma^2})$, where σ is the bandwidth. Moreover, the definition of $\boldsymbol{K_Y}$ is similar to $\boldsymbol{K_X}$.

2.3   The Proposed CIH Algorithm

Throughout the training phase, skin disease images encompass a surplus of label-independent elements, including background noise, clutter and so on, which can profoundly impede CNN's extraction of pivotal features. To counteract this redundancy and retain the essence of importance, the CIH algorithm emerges as a strategic solution for selectively advancing convolutional layers during training. According to the CIH algorithm, the primary focus lies in diminishing the dependency between the convolution layers and the input image, while concurrently optimizing the dependency between the convolution layers and the label value. The HSIC stands as the pivotal tool for calculating this dependency. The architecture of the CIH algorithm is illustrated in Fig. 1. During CNN training, it minimizes the dependency between input image feature and each convolutional layer, while maximizing the dependency between label value and each convolutional layer. This process ensures that the final predictions gradually approach the label value. The implementation principles of this algorithm will be elaborated below.

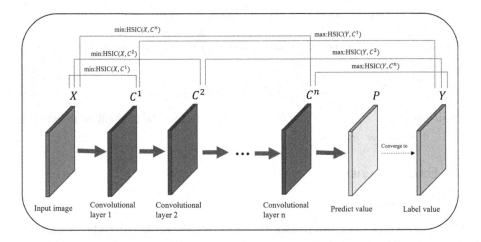

Fig. 1. Architecture of the CIH Algorithm.

Firstly, The fundamental notion underpinning CNN's propagation process is akin to a progressive feature compression, gradually filtering out inessential aspects from the original image. Consequently, minimizing the dependency between the convolution layers and the input image becomes imperative. Given the skin disease image X and the convolutional layers $\{C_1, ..., C_n\}$, the objective of reducing their dependency can be formulated as follows:

$$\min : \mathcal{L}_X = \sum_{i=1}^{n} \text{HSIC}\left(X, C^i\right) \tag{3}$$

where the calculation method of HSIC$(\cdot, \cdot)$ is in Eq. (2).

Secondly, CNN's propagation process can be likened to the preserving features. This endeavor aims to bring the convolution layers in closer alignment with the characteristics of the label value, thereby guaranteeing the extraction of pertinent information. In essence, maximizing the dependency between the convolution layers and the label value Y remains pivotal. This objective is concisely encapsulated in the following expression of the corresponding objective function:

$$\min : \mathcal{L}_Y = \sum_{i=1}^{n} \text{HSIC}\left(C^i, Y\right) \tag{4}$$

where the calculation of HSIC$(\cdot, \cdot)$ is also in Eq. (2).

Finally, combining Eq. (3) and Eq. (4) according to the IB calculation Eq. (1) forms an application method for using HSIC-based IB in CNN, called CIH algorithm, as shown below:

$$\min : \mathcal{L}_{CIH} = \mathcal{L}_X - \lambda \cdot \mathcal{L}_Y \tag{5}$$

where the λ means the tradeoff ratio between $\mathcal{L}_X$ and $\mathcal{L}_Y$.

During CNN training, $\mathcal{L}_{CIH}$ acts as a loss function. In this study, it is used in conjunction with the cross-entropy loss function. Combining $\mathcal{L}_{CIH}$ with $\mathcal{L}_{Cross-entropy}$ can be expressed as:

$$\min : \mathcal{L}_{Loss} = \mathcal{L}_{CIH} + \gamma \cdot \mathcal{L}_{Cross-entropy} \tag{6}$$

where the $\mathcal{L}_{Loss}$ is the final loss function, the γ means the tradeoff ratio between $\mathcal{L}_{CIH}$ and $\mathcal{L}_{Cross-entropy}$.

3   Experiments

3.1   Datasets

We used two datasets to validate the effectiveness of the CIH algorithm. The first is a self-collected hypopigmentation dataset(HD dataset), and the second is a publicly ISIC2018 dataset [4]. The HD dataset was captured in clinical settings using smartphones and digital cameras by dermatologists. It comprises 2,613 images encompassing six skin types, including vitiligo (VI), pityriasis alba (PA), pityriasis versicolor (PV), achromic nevus (AN), idiopathic guttate hypomelanosis (IGH), and normal skin (NS). Remarkably, VI, PA, PV, AN, and IGH all belong to the category of hypopigmented skin diseases. The ISIC2018 dataset contains seven skin diseases, Melanoma(MEL), Melanocytic nevus(MV), Basal cell carcinoma(BCC), Actinic keratosis/intraepithelial carcinoma(AKIEC), Benign keratosis(BKL), Dermatofibroma(DF) and Vascular lesion(VASC), for a total of 11,702 images. The detailed division of training, validation and test data on the two datasets is shown in Table 1.

3.2   Implementation Details

To verify the effectiveness of the proposed CIH algorithm, experiments were performed on HD dataset and ISIC2018 dataset using four classical CNNs: ResNet-50 [12], DensNet-169 [14], Inception-v4 [20], and ConvNeXt-B [17]. The four CNNs were trained with CIH and not with CIH, and then the resulting models were tested using the same test dataset. In addition, this paper reproduces an intelligent diagnostic model for skin diseases on multiple views. The model is trained using ResNet-50 network for RGB, HSV and YCbCr images of skin diseases separately and finally the probability values of the prediction results are averaged for all three images [16]. The model is named ResNet-50+multi-view. In this paper, the ResNet-50+multi-view model is compared with ResNet-50+CIH in the same test environment.

Table 1. The statistical information of the datasets.

Datasets	Class	Training	Validation	Test	Total
HD Dataset	VI	784	98	98	980
	PA	339	42	42	423
	PV	136	17	17	170
	AN	238	30	30	298
	IGH	237	30	30	297
	NS	355	45	45	445
ISIC2018 Dataset	MEL	1,101	30	171	1,302
	NV	6,788	30	909	7,727
	BCC	497	30	92	619
	AKIEC	305	30	43	378
	BKL	1,089	30	217	1,336
	DF	86	30	44	160
	VASC	115	30	35	180

The deep learning framework used in the experiment is Pytorch (version 1.10.2) and the graphics card model is NVIDIA GeForce RTX 3090Ti. The hyperparameters are set as follows: Optimizer Type, Stochastic Gradient Descent; Learning rate, 0.01; Batch size, 32; Epochs, 100. Additionally, we have configured the three crucial parameters σ, λ, and γ in Sect. 2.3 to be 2, 3, and 1, respectively. The input image sizes for the ResNet-50, DenseNet-169, Inception-v4, and ConvNeXt-B networks were 224×224, 224×224, 299×299, and 288×288, respectively. To enrich the datasets and improve the robustness of the models, random clipping and mirror flipping data augmentation methods are used in the data preprocessing stage. Using accuracy, precision, recall, and f1-score as metrics to test the performance of CNNs. Visualize the model's test results using the confusion matrix.

3.3   Results

Table 2 presents the performance test results of four classical CNNs on the HD dataset before and after implementing the CIH algorithm. The table reveal that for ResNet-50, there were increases of 2.68%, 3.93%, 3.39%, and 3.57% in accuracy, precision, recall, and f1-score respectively. For DenseNet-169, these metrics experienced growth by 7.63%, 5.75%, 7.92%, and 6.96% respectively. For Inception-v4, the metrics exhibited growth by 4.2%, 5.13%, 4.08%, and 4.98%. As for ConvNeXt-B, the metrics showed growth by 4.2%, 3.7%, 4.66%, and 5.09% respectively.

Table 3 displays the performance test results of four classical CNNs on the ISIC2018 dataset before and after implementing the CIH algorithm. The results reveal that for ResNet-50, there were increases of 0.86%, 0.45%, 1.18%, and

1.33% in accuracy, precision, recall, and f1-score respectively. As for DenseNet-169, these metrics experienced growth by 0.33%, 0.52%, 1.79%, and 1.14% respectively. In the case of Inception-v4, the metrics exhibited growth by 2.84%, 4.35%, 5.80%, and 5.78%. For ConvNeXt-B, the metrics showed growth by 1.19%, 0.99%, 3.12%, and 1.63% respectively.

Table 2. The experimental results(%) of four classical CNNs on the HD dataset before and after using the CIH algorithm.

Models	Accuracy	Precision	Recall	F1-score
ResNet-50	87.40	84.79	83.71	83.96
ResNet-50+CIH	**90.08**	**88.72**	**87.10**	**87.53**
ResNet-50+multi-view	88.93	87.32	85.24	85.64
DenseNet-169	84.35	82.80	82.94	82.36
DenseNet-169+CIH	**91.98**	**88.55**	**90.86**	**89.32**
Inception-v4	90.46	87.42	90.15	88.35
Inception-v4+CIH	**94.66**	**92.55**	**94.23**	**93.24**
ConvNeXt-B	91.22	88.82	88.40	87.57
ConvNeXt-B+CIH	**94.66**	**92.52**	**93.06**	**92.66**

Table 3. The experimental results(%) of four classical CNNs on the ISIC2018 dataset before and after using the CIH algorithm.

Models	Accuracy	Precision	Recall	F1-score
ResNet-50	79.55	73.58	62.23	66.12
ResNet-50+CIH	**80.41**	**74.03**	**63.41**	**67.45**
ResNet-50+multi-view	80.68	77.33	62.83	68.25
DenseNet-169	76.44	68.47	56.58	61.12
DenseNet-169+CIH	**76.77**	**68.99**	**58.37**	**62.26**
Inception-v4	82.20	76.01	66.40	69.61
Inception-v4+CIH	**85.04**	**80.36**	**72.20**	**75.39**
ConvNeXt-B	81.67	76.06	66.97	70.49
ConvNeXt-B+CIH	**82.86**	**77.05**	**70.09**	**72.12**

From these two tables, it can be seen that the accuracy, precision, recall and f1-score of CNNs for recognizing skin diseases can be significantly improved by using the CIH algorithm in CNNs. This implies the importance of removing label-independent information and the effectiveness of the CIH algorithm. In addition, it can be seen in the test results of both datasets that ResNet-50+CIH performs better compared to the ResNet-50+multi-view model. This demonstrates that the introduction of the CIH algorithm into CNNs can achieve results that are superior to current good skin disease diagnosis models.

Then, compared with other models, DensNet-169 has deeper convolution layers but poor performance. This indicates that label-independent information in skin disease images is easily propagated and confused with deeper networks, potentially threatening the diagnosis performance. On the other hand, this phenomenon proves the necessity and effectiveness of removing label-independent information.

Figure 2 displays the confusion matrices of the four classical CNNs before and after implementing the CIH algorithm on the HD dataset. The top four figures illustrate confusion matrices without the CIH algorithm, while the bottom four figures show matrices with the algorithm. Notably, (a) and (e) reveal CIH's enhanced performance for ResNet-50 across 3 categories. Similarly, (b) and (f) demonstrate improved predictions with CIH for Densenet-169 across 5 categories. Likewise, (c) and (g) highlight CIH's predictive boost for Inception-v4 across 5 categories. Furthermore, (d) and (h) underscore CIH's contribution to improved ConvNeXt-B predictions across 4 categories.

In summary, there was a distinct improvement in the predictive performance of most diseases for each CNN model through the application of the CIH algorithm. This emphasizes the efficacy of the proposed CIH algorithm, effectively enhancing the CNN's capability to identify skin disease images by removing label-independent information.

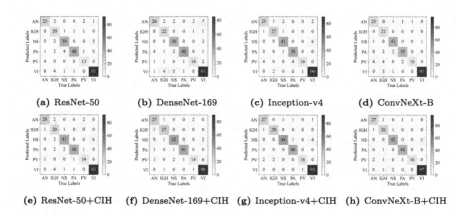

(a) ResNet-50   (b) DenseNet-169   (c) Inception-v4   (d) ConvNeXt-B

(e) ResNet-50+CIH   (f) DenseNet-169+CIH   (g) Inception-v4+CIH   (h) ConvNeXt-B+CIH

Fig. 2. Confusion matrices of the four CNNs on HD dataset before and after applying the CIH algorithm.

4   Conclution

In this paper, we have proposed an algorithm for removing label-independent information based on the Information Bottleneck theory, applied within the training process of CNNs for skin diseases image recognition. The algorithm employs HSIC to calculate the dependency between variants, aiming to minimize the dependency between convolutional layers and the input image while maximizing the dependency between convolutional layers and label value. This approach is utilized to constrain the propagation of label-independent information. We validated the algorithm using four well-established CNNs on two datasets. The experimental results strongly indicate that the proposed algorithm leads to significant improvements in the testing metrics of four CNNs.

As we look forward, we are considering ways to reduce the complexity of the algorithm and enhance training efficiency. Additionally, we plan to explore the application of this algorithm beyond the realm of skin diseases, extending into other domains of medical image recognition, to evaluate its effectiveness in diverse scenarios.

Acknowledgement. This work is supported by the National Natural Science Foundation of China (62306011,81972926), the Natural Science Foundation of Anhui Province of China (2108085MH303), and the Key Scientific Research Foundation of Education Department of Anhui Province (2023AH050631). The numerical calculations in this paper have been done on the Medical Big Data Supercomputing Center System of Anhui Medical University.

References

1. Al-masni, M.A., Kim, D.H., Kim, T.S.: Multiple skin lesions diagnostics via integrated deep convolutional networks for segmentation and classification. Comput. Methods Programs Biomed. **190**, 105351 (2020). https://doi.org/10.1016/j.cmpb. 2020.105351
2. Ba, W., et al.: Convolutional neural network assistance significantly improves dermatologists' diagnosis of cutaneous tumours using clinical images. Eur. J. Cancer **169**, 156–165 (2022). https://doi.org/10.1016/j.ejca.2022.04.015
3. Balch, C.M., et al.: Final version of 2009 ajcc melanoma staging and classification. J. Clin. Oncol. **27**, 6199 (2009)
4. Codella, N., et al.: Skin lesion analysis toward melanoma detection 2018: a challenge hosted by the international skin imaging collaboration (isic) (2019)
5. El Saleh, R., Bakhshi, S., Amine, N.A.: Deep convolutional neural network for face skin diseases identification. In: 2019 Fifth International Conference on Advances in Biomedical Engineering (ICABME), pp. 1–4. IEEE (2019)
6. da F. Mendes, C.F.S., Krohling, R.A.: Deep and handcrafted features from clinical images combined with patient information for skin cancer diagnosis. Chaos Solitons Fractals **162**, 112445 (2022). https://doi.org/10.1016/j.chaos.2022.112445
7. Felsten, L.M., Alikhan, A., Petronic-Rosic, V.: Vitiligo: a comprehensive overview: part ii: treatment options and approach to treatment. J. Am. Acad. Dermatol. **65**, 493–514 (2011)

8. Gedeon, T., Parker, A.E., Dimitrov, A.G.: The mathematical structure of information bottleneck methods. Entropy **14**, 456–479 (2012). https://doi.org/10.3390/e14030456

9. Greenfeld, D., Shalit, U.: Robust learning with the hilbert-schmidt independence criterion. In: International Conference on Machine Learning, pp. 3759–3768. PMLR (2020)

10. Gretton, A., Bousquet, O., Smola, A., Schölkopf, B.: Measuring statistical dependence with hilbert-schmidt norms. In: Jain, S., Simon, H.U., Tomita, E. (eds.) ALT 2005. LNCS (LNAI), vol. 3734, pp. 63–77. Springer, Heidelberg (2005). https://doi.org/10.1007/11564089_7

11. Gretton, A., Bousquet, O., Smola, A.J., Schölkopf, B.: Measuring statistical dependence with hilbert-schmidt norms. In: Proceedings of the 16th International Conference on Algorithmic Learning Theory, ALT, Singapore, vol. 3734, pp. 63–77 (2005)

12. He, K., Zhang, X., Ren, S., Sun, J.: Deep residual learning for image recognition. In: 2016 IEEE Conference on Computer Vision and Pattern Recognition (CVPR), pp. 770–778 (2016). https://doi.org/10.1109/CVPR.2016.90

13. He, X., Wang, Y., Zhao, S., Chen, X.: Multimodal skin lesion classification in dermoscopy and clinical images using a hierarchical attention fusion network. J. Invest. Dermatol. **141**, S52 (2021)

14. Huang, G., Liu, Z., Van Der Maaten, L., Weinberger, K.Q.: Densely connected convolutional networks. In: 2017 IEEE Conference on Computer Vision and Pattern Recognition (CVPR), pp. 2261–2269 (2017). https://doi.org/10.1109/CVPR.2017.243

15. Huang, J., et al.: Global incidence, mortality, risk factors and trends of melanoma: a systematic analysis of registries. Am. J. Clin. Dermatol. (2023). https://doi.org/10.1007/s40257-023-00795-3

16. Liu, F., Yan, J., Wang, W., Liu, J., Li, J., Yang, A.: Scalable skin lesion multi-classification recognition system. Mater. Continua Comput. (2020)

17. Liu, Z., Mao, H., Wu, C.Y., Feichtenhofer, C., Darrell, T., Xie, S.: A convnet for the 2020s. In: 2022 IEEE/CVF Conference on Computer Vision and Pattern Recognition (CVPR), pp. 11966–11976 (2022). https://doi.org/10.1109/CVPR52688.2022.01167

18. Liu, Z., Xiong, R., Jiang, T.: Ci-net: clinical-inspired network for automated skin lesion recognition. IEEE Trans. Med. Imaging **42**, 619–632 (2023). https://doi.org/10.1109/TMI.2022.3215547

19. Ma, W.D.K., Lewis, J., Kleijn, W.B.: The hsic bottleneck: deep learning without back-propagation. In: Proceedings of the AAAI Conference on Artificial Intelligence, pp. 5085–5092 (2020)

20. Szegedy, C., Ioffe, S., Vanhoucke, V., Alemi, A.: Inception-v4, inception-resnet and the impact of residual connections on learning. In: Proceedings of the AAAI Conference on Artificial Intelligence, vol. 31 (2017)

21. Thieme, A.H., et al.: A deep-learning algorithm to classify skin lesions from mpox virus infection. Nat. Med. **29**, 738–747 (2023)

22. Tishby, N., Pereira, F.C., Bialek, W.: The information bottleneck method (2000)

23. Tishby, N., Zaslavsky, N.: Deep learning and the information bottleneck principle. In: 2015 IEEE Information Theory Workshop (ITW), pp. 1–5. IEEE (2015)

24. Wang, Y., et al.: Adversarial multimodal fusion with attention mechanism for skin lesion classification using clinical and dermoscopic images. Med. Image Anal. **81**, 102535 (2022). https://doi.org/10.1016/j.media.2022.102535

25. Whitton, M.E., et al.: Interventions for vitiligo. Cochrane Database Systemat. Rev. **36** (2015)

26. Winkler, J.K., et al.: Assessment of diagnostic performance of dermatologists cooperating with a convolutional neural network in a prospective clinical study: human with machine. JAMA Dermatol. **159**, 621–627 (2023). https://doi.org/10.1001/jamadermatol.2023.0905

27. Wu, D., Zhao, Y., Tsai, Y.H., Yamada, M., Salakhutdinov, R.: Dependency bottleneck in auto-encoding architectures: an empirical study. CoRR arxiv:1802.05408 (2018)

28. Xin, C., et al.: An improved transformer network for skin cancer classification. Comput. Biol. Med. **149** (2022). https://doi.org/10.1016/j.compbiomed.2022.105939

29. Yanagisawa, Y., Shido, K., Kojima, K., Yamasaki, K.: Convolutional neural network-based skin image segmentation model to improve classification of skin diseases in conventional and non-standardized picture images. J. Dermatol. Sci. **109**, 30–36 (2023)

30. Yang, Y., et al.: Skin lesion classification based on two-modal images using a multi-scale fully-shared fusion network. Comput. Methods Programs Biomed. **229**, 107315 (2023). https://doi.org/10.1016/j.cmpb.2022.107315

31. Zhu, C.Y., et al.: A deep learning based framework for diagnosing multiple skin diseases in a clinical environment. Front. Med. **8** (2021). https://doi.org/10.3389/fmed.2021.626369

32. Zhuang, D., Chen, K., Chang, J.M.: CS-AF: a cost-sensitive multi-classifier active fusion framework for skin lesion classification. Neurocomputing **491**, 206–216 (2022). https://doi.org/10.1016/j.neucom.2022.03.042

AI for Civil Aviation

3D Approach Trajectory Optimization Based on Combined Intelligence Algorithms

Li Lu(✉), Juncheng Zhou, Chen Li, Yuqian Huang, Jiayi Nie, and Junjie Yao

School of Air Traffic Management, Civil Aviation Flight University of China, Sichuan 618307, Guanghan, China
421361599@qq.com

Abstract. Nowadays, with the global warmth spread all around the world, thunderstorms, tsunami, turbulence and other natural disaster weather frequently occur around us, these kind of weather can cause hazardous disaster to aviation, especially thunderstorms, for example, when aircraft flight during approach segment, it will be very dangerous if the thunderstorm appears. This research adopt two intelligence algorithms, the Rapidly-Exploring Random Tree (RRT) algorithm combined with Artificial Potential Field (APF) method to design the trajectory of approaching aircraft. At first, this algorithm can guide the aircraft fly from start point to the destination, by the meanwhile, the trajectory can avoid the thunderstorms, finally, the result shows that the trajectory of aircraft simulated can evade moving thunderstorm smoothly, and the profile trajectory meet the approach procedure standard, it is proved that the algorithm is of great value in aviation industry.

Keywords: Thunderstorm weather · Approach flight segment · Trajectory optimization · Combined intelligence algorithms

1 Introduction

In recent years, due to the global warmth, there have been more and more unsafe incidents of aircraft flight caused by unstable airflow. Thunderstorms, as such unstable airflow, posing a huge safety hazard to aircraft flight, especially during the approach segment when the aircraft needs to execute corresponding flight procedures in each segment. At this time, the aircraft's configuration may change, such as flaps and landing gear should be in the landing position, If the aircraft encounters the thunderstorm at this time, it will create a huge safety hazard to the landing. In 2022, Lul et al. conducted a detailed analysis of the hazards of thunderstorms during the approach process and designed a deviation trajectory using dynamic window algorithms. W Fan et al. and Y. Su et al. analyzed the safety threats of severe weather such as thunderstorms to aviation in 2018 and 2019, respectively [1–3].

C. Cruz et al. (Eds.): IC 2023, CCIS 2036, pp. 207–217, 2024.
https://doi.org/10.1007/978-981-97-0065-3_15

How solve the problem depicted above? Since 2016, many domestic or abroad schol-ars have begun to explore artificial intelligence algorithms and computer applications to research in path optimization area, for example, using algorithms to optimizing the trajectory of unmanned vehicles, including RRT algorithm, APF algorithm, and Astar algorithm, which have made important groundwork for subsequent research on path optimization [4–11]. In 2021, CAO Kai et al. proposed a solution that combines RRT algorithm and AFP algorithm to solve the path optimization problem of robots. This article provides a theoretical research approach for this study [12].

In this research, the author use the regress and Monte Carlo algorithm to predict the dynamic weather movement, and than, combine the RRT (Rapidly-Exploring Random Tree) with APF (artificial potential field) to solve the trajectory design.

2   Thunderstorm Movement Prediction

This part focuses on the processing of thunderstorm clouds, as shown in the Fig. 1. The main idea is to extract the dangerous weather color blocks in the radar echo images, extracting the coordinate point data matrix in the color blocks, and then predict the future time of thunderstorm clouds.

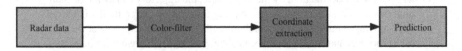

Fig. 1. The processing of thunderstorm clouds

Using graphic algorithms to extract dangerous red and yellow color blocks from radar echoes, which are very dangerous for aircraft flight during approach segment. Therefore, collecting radar echoes and extract thunderstorm color blocks from the images as shown Fig. 2.

Fig. 2. The extracted color blocks of thunderstorm

The next step is extracting the envelope of thunderstorm color blocks, and the enclosed area formed by the envelope serves as a moving obstacle in the terminal area of the airport. This research focuses on three-dimensional path planning, treating the height of thunderstorm clouds as infinite, as shown in the Fig. 3, which is the closed envelope of thunderstorm clouds.

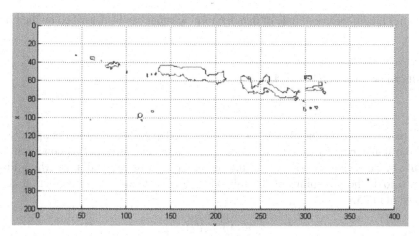

Fig. 3. The envelope of thunderstorm

Next, collecting the data extracted by different times echo maps when thunderstorm occurs. All the points in the thunderstorm envelope include the position information, therefore, the thunderstorm should be expressed as Eq. 1:

$$X = \begin{pmatrix} x_{11} & \cdots & x_{1n} \\ \vdots & \ddots & \vdots \\ x_{m1} & \cdots & x_{mn} \end{pmatrix} \tag{1}$$

As to the thunderstorm points matrix, m and n represent the maximum length and width values of the thunderstorm range, taking the shape of the thunderstorm cloud is not fixed and can change at any time into consideration, m and n represent the entire airport terminal area airspace. If there is no thunderstorm at a certain point in the airspace, the x_{ij} is 0 at that position, conversely, it will be 1. Thus defining a dynamic thunderstorm obstacle. Based on the collected thunderstorm echo maps at different times, the dynamic trajectory of the next thunderstorm can be predicted. This study uses multiple linear regression algorithm to predict the movement trend of thunderstorm.

The following algorithm is to predict the thunderstorm. Firstly, anticipate the thunderstorm moving direction:

Using the regress algorithm to deal with this problem, overlap the thunderstorm samples, find the intersection part, these parts can sure predict the movement of thunderstorm, and constitute by many points, one points move to the other points can be used by a vector to predict the next step which thunderstorm will move, every vector can be described by the following formula.

Assuming the position of each thunderstorm cloud point in the thunderstorm cloud matrix is $p_{storm}(x_i, y_i)$. Therefore, it can be thought that the short-term movement trend of thunderstorms is uniform linear movement. Shown in Eq. 2:

$$y_i = k_i x_i + b_i \tag{2}$$

k_i and b_i stands for the coefficient of the regress formula.

After time t, the coordinate of thunderstorm weather can be described by the Eq. 3.

$$y_{it} = k_i(x_i + t) + b_i \tag{3}$$

It can be transferred to the Eq. 4.

$$y_{it} = y_i + k_i t \tag{4}$$

Apart from the original thunderstorm, when overlap the thunderstorm samples, some random thunderstorms may be occurred in the complement area, this condition also should be taken into consideration, the possibility can be referred as p, which range from 0 to 1, the possibility of the following random thunderstorm prediction is p_a. Shown in Eq. 5.

$$p_a = Ap + A'(1 - p) \tag{5}$$

$A = rand(map_x, map_y)$ stands for the occurrence of the random thunderstorm currently, $A' = rand(map_x, map_y)$ stands for the occurrence of the random thunderstorm next time.

Define a random value $temp$, which obey uniform distribution U(0,1), will decide whether some random thunderstorm will occur, can be depicted as Eq. 6.

$$\begin{cases} p_a < tmp \, (no \, random \, thunderstorm) \\ p_a > tmp \, (random \, thunderstorm \, occurs) \end{cases} \tag{6}$$

Finally, predict the trajectory of aircraft when approach under thunderstorm weather, this study adopts RRT and APF algorithm to deal with this problem.

The cloud map of the next moment of thunderstorm predicted by the above algorithm is shown in the Fig. 4:

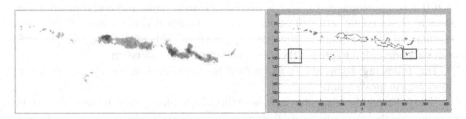

Fig. 4. The prediction of thunderstorm

From Fig. 4, it can be seen that the location of future thunderstorm clouds predicted based on multiple linear regression combined with Monte Carlo algorithm is highly consistent with the true thunderstorm cloud color block at that time. However, the thunderstorm points selected in the red box in the simulation figure are newly added thunderstorm cloud points calculated based on Monte Carlo algorithm, which also confirms that this algorithm can be used as a method for predicting the trajectory of thunderstorm clouds and of high application value.

3   Trajectory Prediction Algorithm of Approaching Aircraft

The main idea of this research is to solve the trajectory optimization problem of thunderstorms encountered during aircraft approach. The main research idea is shown in the Fig. 5. Starting from the starting point of the state space, the next path point is searched. If there are no obstacles, the shortest geometric distance path is selected as the optimal trajectory. If there are obstacles, the next path point is searched again, The trajectory algorithm used in this study is a combination of RRT and APF algorithm.

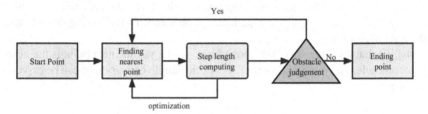

Fig. 5. The trajectory optimization flow chart

3.1   Artificial Potential Field Algorithm

The APF algorithm is divided into repulsive field and gravitational field, where attractive and repulsive force serve as target points and obstacles to attract and repel moving objects, respectively, This algorithm assure that the flight trajectory will not contact the thunderstorms, which can be seen as obstacles and will produce repulsion to aircraft, however, the destination position will produce attraction to aircraft as Eq. 7.

$$\begin{cases} F_{att} = \mu \times dg \times n_{rg}\,(no\ random\ thunderstorm) \\ F_{att} = 0\,(other\ condition) \end{cases} \tag{7}$$

The thunderstorms will produce repulsion to aircraft as Eq. 8.

$$\begin{cases} F_{rep} = 10 \times \left[\varepsilon \times R \times dg \times \left(\frac{1}{d} - \frac{1}{\rho}\right)^2 \times n_{or} - \varepsilon \times R \times \left(\frac{1}{d} - \frac{1}{\rho}\right) \times \frac{dg^2}{d^2} \times n_{rg} \right] (d \in [\tau, 2\tau)) \\ F_{rep} = 100 \times \left[\varepsilon \times R \times dg \times \left(\frac{1}{d} - \frac{1}{\rho}\right)^2 \times n_{or} - \varepsilon \times R \times \left(\frac{1}{d} - \frac{1}{\rho}\right) \times \frac{dg^2}{d^2} \times n_{rg} \right] (d \in (0, \tau)) \\ F_{rep} = 0\,(other condition) \end{cases} \tag{8}$$

d stands for the distance between the position now and the next new position, ρ stands for the efficient distance of repulsion, $dg = \left(x_p - x_g\right)^2 + \left(y_p - y_g\right)^2$, This research use 0.4 for R.

The APF algorithm can be used as a trajectory planning guidance in multi-objective environments, with strong advantages in local short distance path guidance. However, for overall path planning, this algorithm is difficult to find the optimal path, and in high-density moving obstacle environments, it may create obstacles that cannot be avoided.

3.2 Rapidly-Exploring Random Tree Algorithm

The RRT algorithm is used to find the optimal path in complex environments, initialize the starting point, and randomly select adjacent nodes from the established state space. If there are obstacles in the path planning process, the next path node will be changed. After multiple iterations, an optimal path can be obtained. The following is the function of the RRT algorithm.

Imaging that the aircraft is in the position $p_n = (x_n, y_n)$, the next point is the nearest without obstacle, assuming $p_{n+1}=(x_{n+1}, y_{n+1})$, the length of every step can be expressed by the Eq. 9:

$$L_{n+1} = \sqrt{(x_{n+1} - x_n)^2 + (y_{n+1} - y_n)^2} \tag{9}$$

When n approaches infinity, it can be assumed that $\overline{v_n}$ follows a uniform distribution, shown in Eq. 10:

$$\overline{v_n} = \frac{d(p_{n+1} - p_n)}{dt} \tag{10}$$

For RRT optimization theory, the path taken is the shortest sum of paths after iterative operation, Shown in Equation 11:

$$\begin{cases} L_{total} = min \sum_{n=0}^{iterations} L_{n+1}, p_n + \overline{v}t \neq p_{storm} + \overline{v_o}t \\ p_{n+1} is\ not\ available,\ p_n + \overline{v}t = p_{storm} + \overline{v_o}t \end{cases} \tag{11}$$

In this research, the following chart is the pseudo-code of the RRT algorithm:

Algorithm: RRT

	Input: x_{start}, x_{end} *Result: A path from* x_{start} *to* x_{end}
	for $i=0$ *to n do*
	$x_{rand} \leftarrow$ *choose target* $(x_{rand},\ x_{end},\ P\ )$
	$x_{near} \leftarrow$ *Near* (x_{rand})
	$x_{new} \leftarrow$ *Steer* $(x_{rand},\ x_{near},\ Stepsize)$
	End *Edge* $(x_{new},\ x_{near})$
	If there are no obstacles *Add node* *If* $x_{new} = x_{end}$ *successful*

The RRT algorithm has advantages in overall path planning, but it also has drawbacks. Each step of calculation exploration seeks the optimal path for the next step, resulting in an overall path that is not smooth and does not conform to the design principles of the approach flight program.

3.3 The Combined Algorithm

The Table 1 shows the three algorithms' performance from the aspects of path length, nodes explored and elapsed time:

Table 1. The performance of the three algorithms

Algorithm	Path Length	Nodes Explored	Elapsed Time (s)
RRT	519.3913	1000	0.0057233
Artificial Potential Field	520.2005	1000	0.0031489
Combined	376.2294	1000	0.002784

The Fig. 6 shows the search complexity of three algorithms in seeking paths in state space. It is clear that the combined algorithm can achieve optimal path search in space and time. Therefore, the algorithm selected in this research is the algorithm combined RRT with APF.

Path Visualization

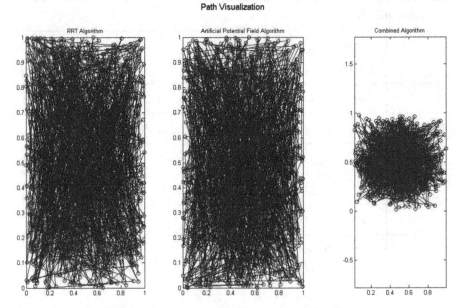

Fig. 6. The search complexity of the three algorithms

This research adopts a combined algorithm of RRT and APF. As can be seen from Fig. 6, this kind of algorithm has significant advantages in terms of time and space utilization. The combination of the two algorithms has the following advantages:

1. Effectively avoiding obstacles in dynamic and complex environments;
2. The RRT algorithm can compensate for the oscillations caused by attractive and repulsive force of APF algorithm in the local space. Therefore, the combined algorithm is beneficial for overall trajectory optimization;
3. The trajectory generated by the RRT algorithm is optimized by each iteration, and the heading generated by each iteration is inconsistent, which does not conform to the smooth design of the trajectory. Therefore, adding the APF algorithm makes the path smoother;
4. From the figure, it can be seen that the combined trajectory algorithm can save more space and time, and has better path calculation performance.

4   Case Study

The case study used in this study is based on the terminal area of Chongqing Jiangbei Airport, as shown in the Fig. 7, which shows the radar echo map of June 5, 2023. Starting from 15:00 pm, meteorological data is collected every 6 min for one hour to optimize the approach trajectory during thunderstorm weather.

Fig. 7. The thunderstorm echo map of Chongqing

The simulation parameters are set as Table 2:

Table 2. Data settings

Start point	(50 km, 150 km, 4500 m)
Ending point	(350 km, 20 km, 0 m)
Initial approach descent gradient	4%
Intermediate approach descent gradient	0
Final approach descent gradient	5.2%
Speed	300km/h

The simulation results using MATLAB are shown in the Fig. 8 and Fig. 9, and the program generates two optimized trajectories during thunderstorm weather:

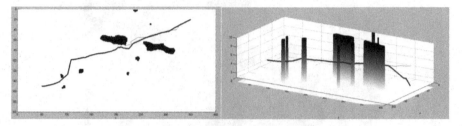

Fig. 8. The first trajectory under thunderstorm weather

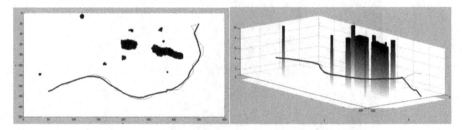

Fig. 9. The second trajectory under thunderstorm weather

According to the results, it can be clearly seen that both aircraft approach trajectories meet the design requirements of the flight procedure. It can be seen that in the initial approach segment, the turning angle of the route is less than 120 degrees, and the descent gradient in the middle and final approach segments is less than 6%. Finally, it can ensure the stable landing of the aircraft at the target airport.

From the results, it can be seen that this study can provide optional temporary diversion trajectories for aircraft approach safety in thunderstorm weather, which has practical application value. However, some of the content in this research can be further optimized. In future research, precise prediction of meteorological data will be conducted in depth, making track prediction more reasonable.

Funding Statement. This research is supported by the China Scholarship Council [Grant Nos. 202108510115].

References

1. Lu, L., Liu, C.: Research on trajectory planning in thunderstorm weather based on dynamic window algorithm during approach segment. Sci. Program. **2022**, 1–10 (2022). https://doi. org/10.1155/2022/7031928
2. Fan, W.: Research on the Influence of Thunderstorm on Aviation Flight. Chengdu University of Information Technology, Chengdu (2018)

3. Su, Y., Pei, Z., Chang, X.: Impact of climate change on civil aviation operation. Civil Aviat. Manag. **11**, 60–62 (2019)
4. González, D., Pérez, J., Milanés, V., et al.: A review of motion planning techniques for automated vehicles. IEEE Trans. Intell. Transp. Syst. **17**(4), 1135–1145 (2016)
5. Chen, Y., Luo, G., Mei, Y., et al.: UAV path planning using artificial potential field method updated by optimal control theory. Int. J. Syst. Sci. **47**(6), 1407–1420 (2016)
6. Ren, Y., Zhao, H.B.: Improved robot path planning based on artificial potential field method. Comput. Simulat. **37**(2), 360–364 (2020)
7. Pan, H., Guo, C., Wang, Z.D.: Research for path planning based on improved a start algorithm. In: Proceedings of the 4th International Conference on Information, Cybernetics and Computational Social Systems, Dalian, 24–26 July 2017, pp. 225–230. IEEE, Piscataway (2017)
8. Zhang, K., Liu, P., Kong, W., Lei, Y., Zou, J., Liu, M.: An improved heuristic algorithm for UCAV path planning. In: Gong, M., Pan, L., Song, T., Zhang, G. (eds.) BIC-TA 2016. CCIS, vol. 682, pp. 54–59. Springer, Singapore (2016). https://doi.org/10.1007/978-981-10-3614-9_7
9. Mac, T.T., Copot, C., Tran, D.T., et al.: Heuristic approaches in robot path planning: a survey. Robot. Auton. Syst. **86**, 13–28 (2016)
10. Wang, X., Shi, Y., Ding, D., et al.: Double global optimum genetic algorithm particle swarm optimization-based welding robot path planning. Eng. Optim. **48**(2), 299–316 (2016)
11. Choudhury, S., Gammell, J.D., Barfoot, T.D., et al.: Regionally accelerated batch informed trees (RABIT*): a framework to integrate local information into optimal path planning. In: Proceedings of the 2016 International Conference on Robotics and Automation, Stockholm, 16–21 May 2016, pp. 4207–4214. IEEE, Piscataway (2016)
12. Cao, K., Chen, Y., Gao, S., Gao, J.: Vortex artificial-potential-field guided RRT* for path planning of mobile robot. J. Front. Comput. Sci. Technol. **15**(4), 723–732 (2021). https://doi.org/10.3778/j.issn.1673-9418.2004037

A-SMGCS: Innovation, Applications, and Future Prospects of Modern Aviation Ground Movement Management System

Jiahui Shen⬡, Lin Lin(✉)⬡, and Runzong Shangguan⬡

Civil Aviat ion Flight University of China, Guanghan 618307, China
1836792382@qq.com

Abstract. This paper provides an overview of the Advanced Surface Movement Guidance and Control System (A-SMGCS) as a pivotal feature of modern aviation ground movement management. The C/S structure, distributed, and open nature of A-SMGCS suit the demands of complex airport environments. The system comprises multiple modules, including data communication, surveillance data processing, flight data processing, route planning, human-machine interface, digital clearance, and data recording. Data processing and fusion facilitate accurate information aggregation, while surveillance and positioning modules locate aircraft and vehicles through multi-source data analysis. Route planning and conflict detection optimize aircraft taxi paths, enhancing ground traffic efficiency. Aircraft state machines and multi-zone segmentation reduce the complexity of intrusion judgment, enabling early runway intrusion alerts. Furthermore, this paper presents the application of Petri net technology in taxiway lighting control. The synergistic effect of these key technologies empowers A-SMGCS to maintain high levels of safety, efficiency, and controllability under diverse conditions, providing robust support for modern airport operations.

Keywords: A-SMGCS · Petri net · aircraft state machine

1 Introduction

As the aviation industry continues to thrive, airport ground movement management faces increasingly serious challenges, especially in high-traffic and complex airport layouts. Traditional manual coordination and monitoring methods are gradually struggling to meet the demands of increasing airport traffic and stringent safety standards. The Advanced Surface Movement Guidance and Control System (A-SMGCS) is an innovative technology designed to address these issues. In 2004, the International Civil Aviation Organization (ICAO) released the "A-SMGCS Manual."

A-SMGCS, through the integration of data from multiple sources and intelligent algorithms, enables real-time monitoring and conflict detection, significantly

Supported by organization Civil Aviation Flight University of China.

enhancing the safety of airport ground operations. Furthermore, A-SMGCS optimizes the flow of ground movements at airports, reducing waiting times, increasing airport operational capacity, minimizing delays, and generating higher benefits for all stakeholders. It efficiently allocates ground resources, enhances overall airport efficiency, and plays a crucial role in handling the growing volume of air traffic.

In addition, A-SMGCS reduces fuel consumption and environmental emissions by optimizing aircraft taxiing paths and parking processes, contributing positively to sustainability. A-SMGCS holds an indispensable position in modern airport ground movement management, bringing significant improvements in safety, efficiency, and sustainability. Internationally, A-SMGCS has been researched and implemented at many airports. Various countries and regions have adopted A-SMGCS systems to enhance the efficiency, safety, and controllability of airport ground traffic.

Airports such as Amsterdam Schiphol Airport (Netherlands), Paris Charles de Gaulle Airport (France), Frankfurt Airport (Germany), Stockholm Arlanda Airport (Sweden), major airports in China, and airports like Abu Dhabi International Airport and Dubai International Airport in the United Arab Emirates were among the early adopters of A-SMGCS.

2   A-SMGCS Overview

2.1   Concept of A-SMGCS

A-SMGCS, which stands for Advanced Surface Movement Guidance and Control System, is a system that integrates various technologies, including radar, sensors, automation systems, and intelligent algorithms, with the aim of comprehensive management of ground activities at airports. A-SMGCS offers a wide range of functionalities, covering real-time monitoring, conflict detection, path planning, guidance, and control, among others. Through these functionalities, A-SMGCS aims to enhance the safety, efficiency, and capacity of airport ground operations. This is of significant importance in the modern aviation industry, as A-SMGCS aids in optimizing ground movement processes, reducing potential conflicts and delays, and adapting to complex and dynamic airport environments. By providing comprehensive ground traffic management, A-SMGCS makes a crucial contribution to the sustainable development of the aviation industry. According to Mao Huijia (2014) [3], as per the provisions in the "A-SMGCS Manual" and "ICAO A-SMGCS Operational Requirements," the fundamental functions of the A-SMGCS system include monitoring, path planning, guidance, and control. The monitoring aspect requires accurate position information, identity recognition, and labeling, covering all moving and stationary aircraft or vehicles while ensuring functionality under adverse conditions. Path planning encompasses assigning routes, handling destination changes, and addressing complex requirements. Guidance functions ensure the maintenance of situational awareness and path changes while indicating restricted areas. Control functions cover conflict detection, runway incursion and taxiway incursion alerts, as well as the resolution of conflicts in critical and emergency areas.

2.2   Technological Evolution of A-SMGCS

In 1986, ICAO published Doc9476, the "SMGCS Manual." In 2004, ICAO released Doc9830, the "A-SMGCS Manual." ICAO classifies A-SMGCS into five levels: Level I for monitoring functions, Level II for alert functions, Level III for route planning, Level IV for surface guidance, and Level V for on-board and on-vehicle driver assistance guidance. Munich Airport in Germany became the first airport to install A-SMGCS in 1997, utilizing a Level 2 system that provided basic functionalities like real-time monitoring, conflict detection, and collision avoidance. Frankfurt Airport installed a Level 3 A-SMGCS system in 2003, which introduced advanced features and complexity, including automatic route planning, conflict resolution, and flight coordination. In 2019, China's Beijing Daxing International Airport installed an A-SMGCS system and received a Level 4 usage permit from the Civil Aviation Administration of China in 2021. However, the deployment of A-SMGCS Level 4 systems is not widespread globally. A-SMGCS Level 4 represents the highest level of the system, offering the highest degree of automation and conflict resolution capabilities to support the safety and efficiency of airport ground operations. Nonetheless, due to its high complexity and cost, Level 4 system deployment is relatively limited and primarily used for research and pilot projects. Similarly, A-SMGCS Level 5 has few practical applications due to system complexity, high construction costs, and low reliability.

3   Components of A-SMGCS

A-SMGCS (Advanced Surface Movement Guidance and Control System) is a modern aviation ground movement management system designed to enhance the safety, efficiency, and controllability of airport ground traffic. Drawing from comprehensive research by Wang Wenzai (2023) [4], Liu Jiachen (2021) [5], Xing Jin (2020) [6], and Yu Qian (2021) [7], it can be concluded that the system adopts a C/S (Client/Server) architecture, incorporating distributed and open characteristics to meet the demands of complex airport environments. A-SMGCS is composed of multiple key components and functional modules aimed at achieving real-time monitoring, guidance, and management of ground aircraft and vehicles.

Data Communication Processing Module (DCP). This module is responsible for interfacing with external systems, receiving and pre-processing various data from external sources, such as clock data, radar data, ADS-B data, flight plan data, weather data, etc. DCP ensures the accuracy and consistency of data, providing essential information for subsequent processing.

Surveillance Data Processing Module (SDP). SDP serves as the core processor of the A-SMGCS system and is responsible for handling surveillance data. It integrates data from multiple sources, including surface surveillance radar, air traffic control radar, multilateration radar, etc. Through data fusion and analysis, it achieves the positioning and identification of aircraft and vehicles on the surface. The SDP module supports functions like front-end processing of surveillance data, data processing servers, and path planning processing.

Flight Data Processing Module (FDP): The FDP module is responsible for processing flight plan data and weather data. It connects with external flight planning systems and weather systems, cross-referencing flight plans with actual operations to ensure the reasonableness and safety of flight processes.

Route Planning Processing Module (RTP). The RTP module is responsible for planning the ground taxi routes for aircraft. It considers the airport's layout and operational conditions to plan the optimal taxi routes for aircraft, avoiding conflicts, and improving ground traffic efficiency.

Human-Machine Interface Module (HMI). This module provides the interface for operators to interact with the A-SMGCS system. This interface may include ground control positions, tower control positions, and enables operators to monitor the scene and perform tasks such as flight clearances and taxi guidance.

Digital Clearance System (DCL). The digital pre-departure clearance system enables bidirectional communication between pilots and tower controllers through ground-air data links, facilitating digital clearances and enhancing departure efficiency and process management.

Data Recording and Replay Module (DRF): The DRF module is used to record and replay the operations and scenarios of the A-SMGCS system, aiding in post-analysis, training, and troubleshooting.

Incorporating the functions of the above modules, the A-SMGCS system achieves critical functionalities such as real-time surveillance, conflict detection, flight guidance, and taxi route planning for ground movements. The application of data fusion from multiple sources and intelligent algorithms elevates the safety, efficiency, and controllability of airport ground traffic, providing robust support and tools for modern airport operations and management (Fig. 1).

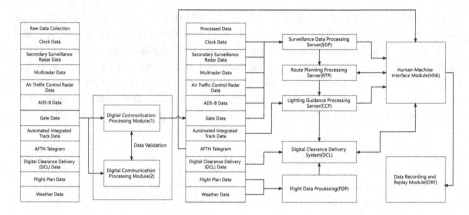

Fig. 1. A-SMGCS System Module Diagram

4   Key Technologies of A-SMGCS

4.1   Data Fusion and Processing

One of the cornerstones of A-SMGCS is data integration and processing. A-SMGCS gathers data from various sources, including clock data, radar data, ADS-B data, flight plan data, and weather data. Researchers like Zhao Lin [10] have employed algorithms such as the Interacting Multiple-Model Probabilistic Data Association (IMM-PDA) and Interacting Multiple-Joint Probabilistic Data Association (IMM-JPDA) to handle multi-tracked target trajectory data in A-SMGCS scenes. This endows the A-SMGCS system with the capability to process complex scene data, including dealing with weather fronts and clutter reflections while effectively avoiding target decay issues. This multi-sensor target data fusion system represents the latest advancements in target tracking technology, providing an effective approach for real-time A-SMGCS target detection, tracking, and data fusion, addressing challenges such as track loss, target classification, handling close encounters, high clutter environments, multi-sensor calibration, and more. The system combines target fusion filters and the Joint Probabilistic Data Association (JPDA) algorithm, enabling it to simultaneously track different types of targets, thereby reducing the risk of false targets entering the environmental assessment model. A-SMGCS's environmental assessment module is used to monitor environmental changes and adapt to evolving conditions by predicting environmental parameters. The track initiation and maintenance module is responsible for establishing new target tracks and maintaining existing ones. Additionally, track management encompasses track classification and gap bridging. The IMM filter serves as a crucial tool for extrapolating and updating the A-SMGCS system's track states. It has the ability to simultaneously track highly maneuverable targets and targets in high clutter environments. It utilizes branch-and-merge algorithms to enhance tracking accuracy and robustness [11]. Lastly, track termination is performed in cases of continuity or inadequate accuracy, parameterized based on known clutter areas or regions with low detection probabilities. Track management also includes track classification and gap bridging to maintain track identification and differentiate between different types of targets. In summary, the multi-sensor target data fusion system efficiently handles complex target tracking tasks through a variety of functionalities and algorithms. It enhances system performance and robustness, providing a solid foundation for efficient ground movement management and monitoring in the A-SMGCS system.

4.2   Surveillance Data Analysis and Positioning

A key technology involves the use of the Surveillance Data Processing Module (SDP) to locate and identify aircraft and vehicles on the ground. Multiple sources of surveillance data, such as surface movement radar, air traffic control radar, multilateration systems (MLAT), and Automatic Dependent Surveillance-Broadcast (ADS-B), are fused and analyzed to allow the system to precisely determine the positions of aircraft and vehicles on the ground and provide a real-time view of ground targets. A-SMGCS primarily relies on surface movement radar, multilateration systems, and ADS-B to monitor airport movements. However, there may be some discrepancies between these data sources, which can impact the accuracy and reliability of surveillance data.

To address this issue, Wang Zhenfei (2023) proposed a scene surveillance enhancement method based on video recognition data fusion, introducing innovative technology and equipment to upgrade and improve airport surveillance systems.

The core of this method involves the introduction of a panoramic gimbal camera, which uses deep learning video recognition technology based on the YOLO v5 network model to enhance the monitoring and identification capabilities of aircraft and vehicles on the airport ground. Here are the main points and steps of this method:

1. Multi-Source Surveillance Data Fusion: First, this method fuses and analyzes data from various surveillance sources through the Surveillance Data Processing Module (SDP). These sources include surface movement radar, air traffic control radar, multilateration radar, and ADS-B. By integrating and analyzing this data, the system can more accurately determine the positions of aircraft and vehicles on the ground, providing a real-time view of ground targets.

2. Video Recognition Technology: This method introduces a panoramic gimbal camera, which uses deep learning video recognition technology based on the YOLO v5 network model to identify and track objects in the airport scene. This includes identifying targets like aircraft, vehicles, and pedestrians. After identification, the system generates data including the target's identification, category, location, and confidence level.

3. Preliminary Comparison Using the Hungarian Algorithm: Received surveillance data and panoramic video recognition data are initially compared and

analyzed using the Hungarian algorithm. This step helps match and confirm targets, leading to a better understanding of the accuracy of surveillance data.

4. Gimbal Camera Target Capture: For exceptional targets, the system employs the gimbal camera to capture target details, providing more detailed target information. This process is conducted to further confirm and analyze targets that are questionable or anomalous.

5. Secondary Comparison Analysis: After the initial comparison analysis and gimbal camera target capture data, the system performs a secondary comparison analysis. This step helps further validate the accuracy of targets and enhances the confidence in target confirmation.

6. Comprehensive Track Data Correlation Fusion: Finally, the processed data is fed back to the A-SMGCS system for comprehensive track data correlation fusion processing. This means that video recognition data and traditional surveillance data are combined, offering a more complete understanding of the status and location of targets in the airport scene.

This method was validated and applied at Huai'an Airport, demonstrating its effectiveness [12]. Through the introduction of this technology, false targets were reduced by 90%, and monitoring of vehicles without onboard ADS-B equipment was increased. This innovative method improves the accuracy and reliability of airport scene surveillance, providing better monitoring and management support to airport tower controllers and enhancing flight safety. This approach not only utilizes AI-based video recognition technology but also addresses the limitations of traditional surveillance data through secondary comparison analysis, improving the performance of existing surveillance systems.

The scene surveillance enhancement method based on video recognition data fusion represents a significant advancement in the field of airport surveillance systems. It integrates traditional surveillance data with state-of-the-art video recognition technology, providing more robust monitoring and identification capabilities for airport ground operations. This not only enhances flight safety but also improves operational efficiency at airports, reducing errors and confusion. The successful implementation of this technology opens new avenues for future airport surveillance systems, making ground operations safer, more reliable, and efficient. This is forward-looking work that brings substantial potential and opportunities to our aviation industry and airport management.

4.3   Flight Data Processing and Matching

The Advanced Surface Movement Guidance and Control System (A-SMGCS) is a crucial tool for ensuring the safety and reasonableness of aviation operations. In addition to monitoring aircraft, the system connects with external flight planning and meteorological systems through the Flight Data Processing Module (FDP) for information exchange. By connecting to external flight planning systems, A-SMGCS can real-time cross-reference actual flight plans with operations to ensure the reasonableness and safety of plans. Furthermore, the system connects to meteorological systems to obtain real-time weather information, aiding flight planning adjustments based on meteorological conditions to ensure flight safety. FDP acts as the system's manager, responsible for ensuring flight safety and reasonableness. The comprehensive functionality of A-SMGCS supports efficient flight management, providing the necessary tools to ensure highly safe and reasonable flight operations. This system plays a critical role in ensuring the safety and smoothness of operations in the aviation sector, ensuring that every flight is conducted efficiently, reasonably, and safely.

4.4   Path Planning

The Path Planning Processing Module (RTP) plays a crucial role in the A-SMGCS system, responsible for planning the ground movement paths of aircraft while considering the airport's layout and operational conditions to avoid conflicts and improve ground traffic efficiency. This is achieved through intelligent algorithms, allowing the system to accurately calculate paths to ensure safe passage.

Petri nets, as described by An Hongfeng (2011), serve as one of the core components of the A-SMGCS system. This mathematical modeling tool, through state transitions and time-triggered events, helps the system capture various states and events during aircraft movement on the ground, providing precise state information to better identify potential conflicts and make corresponding decisions. By combining Petri nets with multiple protected zones, flight state machines, intrusion detection, logic mutual constraint controllers, and guidance light control command mapping, the A-SMGCS system effectively enhances the safety and efficiency of aircraft ground movement. In testing, the application of Petri nets has demonstrated excellent performance and contributed to the successful performance of the A-SMGCS system. However, further research and improvements are still necessary to continuously enhance the adaptability and reliability of Petri nets in different airports and under various conditions.

Additionally, the logic mutual constraint controller plays a critical role in the A-SMGCS system, ensuring that the system does not enter undefined states, effectively preventing repeated switching of the aircraft's flight state to avoid chaotic situations. This controller is built upon the concept of forbidden arcs, using mutex constraints to prevent potential conflicts between different states. In this way, the system can stably guide aircraft, reducing unexpected incidents and ensuring high safety during aircraft ground movement. The logic mutual constraint controller works in conjunction with other technologies, such as Petri nets, providing a solid foundation for the comprehensive performance of the A-SMGCS system. However, this also underscores the need for ongoing research and improvement to ensure the effectiveness of the logic mutual constraint controller in different environments and applications. The synergy of these two key components enhances the capabilities of the A-SMGCS system, ensuring efficient and safe aircraft ground movement [1].

Yunfan Zhou (2019) proposed the use of DQN (Deep Q-Network) learning to address conflicts in hot spot areas in A-SMGCS systems. Hot spot areas refer to locations on the ground where aircraft may cross paths and potentially conflict at intersections. DQN learning dynamically adjusts the ground speed of aircraft based on the current environment and conflict situations, reducing wait times and improving airport ground traffic efficiency. The application of the DQN model has shown promise in experiments, especially when resolving conflicts between aircraft with different speeds. This suggests that DQN has the potential to be effectively applied in A-SMGCS systems to improve ground movement safety and efficiency.

However, further research and evaluation are required to assess the applicability of DQN in scenarios involving multiple aircraft and variations in the workload of air traffic controllers. This highlights that while DQN performs well in specific conflict situations, further in-depth research and testing are needed in more complex airport environments and with a higher number of aircraft involved [13].

When it comes to solving aircraft ground routing problems in A-SMGCS systems, Tang Yong (2014) introduced an innovative approach using free time windows and a Multi-Agent System (MAS) to achieve more efficient and safe ground routing. This task is particularly critical in large airports where managing the ground movements of numerous aircraft amid complex traffic conditions is essential.

The paper first clearly defines aircraft ground routing plans, including sequences of resource nodes and time windows from the starting point to the destination. These plans must meet various conditions, including start times, physical connections between resource nodes, and time window lengths.

The paper introduces the concept of free time windows, representing time intervals on resource nodes that are not scheduled for use. These free time windows are fully utilized during the planning process to ensure smooth aircraft movements. Reachability conditions are crucial to ensure that the departure time of one free time window overlaps with the arrival time of the next one, preventing conflicts.

To better manage and search for free time windows, the paper introduces the concept of a free time window graph, describing the reachability relationships between different time windows, and utilizes classic algorithms such as A* and Dijkstra to search this graph.

To address aircraft ground routing planning problems, the paper introduces a Multi-Agent System (MAS), including routing management agents, resource node agents, aircraft agents, and environmental objects. This system coordinates the interactions and collaboration among various agents to accomplish the routing planning task.

The operation of the MAS system includes initialization, interactions between agents, the routing planning process, and the final output of the optimal routing plan. Through MAS, aircraft ground routing planning is carried out sequentially according to priority, ensuring efficient overall ground movement.

Experimental results demonstrate that the MAS system effectively improves the overall operational efficiency of the scene, meeting A-SMGCS requirements for real-time and optimized aircraft routing planning. In conclusion, the paper provides a method based on a Multi-Agent System that can be applied to aircraft ground routing planning at large airports, enhancing efficiency and safety. Future research could explore optimization methods to further improve the algorithm's efficiency [14].

4.5 Conflict Detection and Alerting

Jing Li (2016) introduced a runway intrusion detection system for airport runway safety, which plays a crucial role in ensuring the safety and compliance of aircraft during takeoff and landing processes. The core concept of the system is based on the concept of multiple protected areas, such as approach area, lineup area, runway area, and exit area, to identify potential conflicts. Here are the key highlights [2]:

1. Multiple Protected Areas: The system introduces multiple intrusion zones to reduce the preprocessing complexity of intrusion detection. These areas include the approach area, lineup area, runway area, and exit area, covering various airport situations and potential conflicts.
2. Target Classification and State Machine: Each flight target is divided into nine flight states, including approach state, landing state, taxiing state, lineup state, rolling state, departure state, exit state, and unknown state. This classification simplifies the detection of complex conflict situations.
3. Event-Driven Alerts: The system features event-driven alert functionality, where events correspond to the arrival of new surveillance data. Once new data is available, the system performs conflict detection through a multi-step algorithm, including data decoding, target storage, target scanning, and conflict detection.
4. Human-Machine Interface (HMI): Alert results are displayed on the HMI, managed by the Air Traffic Control (ATC) controller. Controllers can confirm conflicts through alert windows, ensuring timely action. Alerts are marked in yellow or red based on different types of conflicts.
5. Performance and Validation: The system was tested and validated at Chengdu Shuangliu International Airport, showing promising performance metrics, including detection probability, false alarm rate, and alert response time. Results indicate the system's potential in early detection and alerting for runway safety.

This runway intrusion detection system provides airports with an important tool for detecting and addressing potential aircraft conflicts, thus enhancing runway safety. While it has demonstrated good performance in testing, further development and testing are needed to adapt to various intrusion scenarios and weather conditions. Future research will focus on enhancing and expanding the system to meet evolving airport requirements.

4.6 Human-Machine Interface and Operational Support

A-SMGCS provides an intuitive human-machine interface for operators, including ground control positions and tower control positions. These interfaces allow operators to monitor real-time scenarios, carry out tasks such as flight clearances and taxiing guidance, ensuring the coordination and management of the entire process.

4.7   Digital Pre-departure Clearance System (DCL)

This technology enables two-way communication between pilots and air traffic controllers via the ground-to-air data link, facilitating digital pre-departure clearances. This helps enhance takeoff efficiency and streamline the departure process.

Through the integration and synergy of these key technologies, A-SMGCS achieves critical functions such as real-time ground movement monitoring, conflict detection, flight guidance, and taxi route planning. The application of these technologies ensures that airport ground traffic maintains a high level of safety, efficiency, and manageability in various situations, providing robust support and tools for modern airport operations.

5   Application and Benefits of A-SMGCS

In a case study presented by Gao Zanhua (2023), we can take Dubai Airport as an example. This airport faced challenges with only two closely spaced parallel runways. As flight volumes sharply increased, the airport was operating beyond its capacity, with peak hours seeing an impressive 43 movements, and only 17 during low visibility conditions [8]. However, in 2016, Dubai Airport introduced the A-SMGCS Level IV system, a decision that had a significant impact.

The introduction of A-SMGCS not only increased the airport's capacity but also improved on-time performance of flights. Airlines benefited from reduced taxi times, leading to reduced carbon emissions. Peak-hour movements during low visibility conditions increased from 17 to 49, a critical enhancement for operational flexibility. This successful case not only improved the airport's operational efficiency but also had a positive impact on the passenger throughput, reaching 88.4 million passengers in Dubai Airport in 2018.

Through the Dubai Airport case, we can clearly see how the A-SMGCS system helped the airport overcome challenges and improve safety, efficiency, and controllability in practical applications. This success story serves as a model for other airports, particularly those facing operational pressures and peak-hour challenges.

Another airport case is Beijing Daxing International Airport in China, which is the country's first airport planned and built according to A-SMGCS Level IV standards. According to the trial operation in September 2019, this system was developed by the 28th Research Institute of China Electronics Technology Group. Within the airport's movement area, there are 309 stop bar light groups and 2,053 centerline light segments, totaling 17,980 centerline lights, all controlled by the A-SMGCS system. The system automatically plans the optimal taxi route for aircraft and illuminates green centerline lights in front of the air-craft, allowing pilots to follow these lights to taxi to their respective parking stands or runway departure positions. The introduction of this technology significantly improves the safety of aircraft taxiing while also reducing taxi times.

The example of Beijing Daxing Airport illustrates that the application of A-SMGCS systems can effectively ensure the safety of aircraft taxiing and improve airport operations' efficiency. After two years of testing and operation, the A-SMGCS lighting guidance system operates around the clock and has reached international advanced levels. It plays a crucial role in ensuring the safety of runway operations and improving operational efficiency.

It is worth noting that based on data from the Civil Aviation Administration of China (CAAC) shown in Fig. 2 and Fig. 3 [9], we can clearly see the annual passenger volume changes at Beijing Capital Airport and Chengdu Shuangliu Airport. Beijing Capital Airport installed an A-SMGCS Level 3 system in 2010, a substantial improvement that significantly increased the airport's efficiency and airspace capacity, eventually allowing the airport to achieve an astonishing passenger capacity of 100 million in 2018. A similar situation occurred at Chengdu Shuangliu Airport, which installed an A-SMGCS system in 2012 to cope with the growing passenger and flight demands. As a result, by 2019, the annual passenger volume at the airport had reached nearly 56 million.

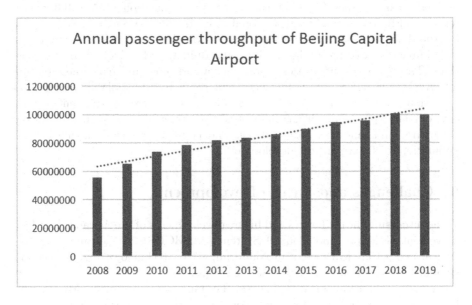

Fig. 2. Beijing Capital Airport throughput data

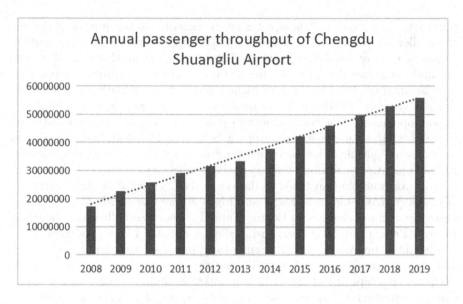

Fig. 3. Chengdu Shuangliu Airport throughput data

These case studies highlight the successful application of A-SMGCS systems in both domestic and international airports. These systems have not only improved airport operational efficiency but also enhanced safety and had a positive impact on taxiing guidance, flight scheduling, and reducing carbon emissions. Through data integration, intelligent algorithms, and real-time monitoring, A-SMGCS has elevated the safety, efficiency, and controllability of ground operations. This system can effectively address challenges in different airport environments, providing airport managers with powerful tools to ensure on-time departures and landings, reduce delays and conflicts, all while laying a solid foundation for future airport development.

6   Challenges and Future Development

When discussing the challenges and future developments of the Advanced Surface Movement Guidance and Control System (A-SMGCS), we encounter several critical issues that will have a profound impact on technology, regulations and standards, and the direction of future development.

In terms of technological challenges, the implementation of A-SMGCS may face issues related to data integration, complex system integration, and the development of intelligent algorithms. Integrating multisource data from different sensors into consistent and accurate information, ensuring data consistency and integrity, will be a challenge. Additionally, effectively integrating different subsystems to ensure their seamless cooperation will require overcoming complex systems engineering challenges. Furthermore, designing and implementing intelligent algorithms capable of effectively predicting and avoiding ground conflicts will be a technological challenge.

On the application front, A-SMGCS has been installed and used at many airports worldwide, with most airports operating at Level 3, and only a few airports implementing Level 4 A-SMGCS systems. Level 5 A-SMGCS systems are still in the exploration process.

Regulations and standards play a crucial role in the development of A-SMGCS. The International Civil Aviation Organization (ICAO) and national aviation authorities continuously refine regulations and standards related to A-SMGCS. These regulations and standards not only specify the performance and requirements of A-SMGCS but also impose strict requirements on safety and compliance during its implementation. Therefore, the practical application of A-SMGCS needs to closely adhere to regulations and standards to ensure the reliability and safety of the system.

Looking ahead to the future development of A-SMGCS, automation and artificial intelligence will be essential directions. As technology advances, A-SMGCS will leverage automation to achieve a higher degree of autonomous ground operations, thereby reducing operational risks. Artificial intelligence will be applied to decision support, data analysis, and forecasting, enabling A-SMGCS to intelligently handle complex scenarios. Furthermore, communication technologies with aircraft and vehicles, such as 5G and the Internet of Things, are likely to introduce more innovative features to A-SMGCS, further enhancing its performance and efficiency.

In summary, the challenges and future development of A-SMGCS call for interdisciplinary cooperation and innovation. Overcoming technological challenges and adhering to regulations and standards while actively exploring frontiers in automation and artificial intelligence will ensure the continued development of A-SMGCS, bringing a higher level of safety, efficiency, and sustainability to future air transportation.

7 Conclusion

In summary, we have drawn important conclusions regarding the Advanced Surface Movement Guidance and Control System (A-SMGCS) based on the comprehensive content of this paper. The paper extensively discusses the technology, applications, benefits, challenges, and future developments of A-SMGCS. A-SMGCS plays a crucial role in modern aviation operational management by integrating advanced data processing and intelligent algorithms to achieve real-time monitoring, guidance, and control of ground-based aircraft and vehicles. The significance of A-SMGCS cannot be overstated. It not only enhances the safety of airport ground traffic but also significantly improves efficiency and controllability. By reducing conflicts, optimizing flight guidance, and implementing digital clearances, A-SMGCS has had a profound impact on the aviation industry. It has the potential to reduce delays, improve resource utilization, and provide airport managers with more precise real-time data support, making air transportation more smooth and efficient. In terms of future research and development directions, we recommend continuing to focus on several key areas. Firstly, technological innovation is of paramount importance, including higher-precision data integration, intelligent decision-making algorithms, and seamless integration with other systems. Secondly, given the rapid development of the aviation industry, attention should be continuously paid to updates in regulations and standards to ensure the implementation and compliance of A-SMGCS. Additionally, we encourage in-depth exploration in areas such as automation and artificial intelligence to achieve more intelligent ground operations management. In conclusion, A-SMGCS has brought about significant changes in airport ground operations management and opened up vast prospects for the future. Through ongoing innovation and collaboration, we are confident that A-SMGCS can be developed into a safer, more efficient, and more intelligent aviation ground operations management system, making a positive contribution to the development of the aviation industry and the transportation systems of modern society.

References

1. An, H., Tang, X., Zhu, X., et al.: Research on taxiway lighting control method based on petri nets. J. Traffic Inf. Saf. **29**(4), 28–32 (2011)
2. Li, J., Wang, G.Q., Zhu, P., Lu, X., Su, J.: A runway incursion detection approach based on multiple protected areas and flight status machine for A-SMGCS. In: MATEC Web of Conferences, vol. 44, p. 01084 (2016). https://doi.org/10.1051/matecconf/20164401084
3. Mao, H.: Research on advanced surface movement guidance and control system (A-SMGCS) of airports. Technol. Forecast. (2014)
4. Wang, W.: Design and implementation of A-SMGCS system test platform for Beijing Daxing international airport. Comput. Knowl. Technol. **19**(7), 109–111+119 (2023). https://doi.org/10.14004/j.cnki.ckt.2023.0449
5. Liu, J.: Analysis of digital clearance function in A-SMGCS system of Beijing Daxing international airport. Mod. Inf. Technol. **5**(22), 130–132+136 (2021). https://doi.org/10.19850/j.cnki.2096-4706.2021.22.037
6. Xing, J.: Processing of surveillance data in A-SMGCS advanced ground surveillance and guidance system of Beijing Daxing international airport. Digital Commun. World (2020)
7. Yu, Q.: Design and analysis of A-SMGCS system. Inf. Technol. Informatizat. (2021)
8. Gao, Z.: Discussion on the construction scheme of airport A-SMGCS system. J. Civil Aviat. Flight Univ. China **34**(3), 68–72 (2023)
9. Civil Aviation Administration of China: National Civil Transport Airport Production Statistics Bulletin (2013). http://www.caac.gov.cn/XXGK/XXGK/TJSJ/index_1216.html. Accessed 16 Mar 2023
10. Zhao, L.: Data fusion of airport surface based on A-SMGCS with multiple sensors (2015)
11. Zhang, Q., Wang, Y.: Discussion on advanced surface movement guidance and control systems. China Civil Aviat. **2**(74), 46–49 (2007)
12. Wang, Z., Huang, Y., Wang, L., et al.: Enhanced scene monitoring method based on video recognition data fusion. Aviat. Comput. Technol. **53**(1), 113–117 (2023)

13. Zhou, Y., Liu, W., Li, Y., Jiang, B.: Taxiing speed intelligent management of aircraft based on DQN for A-SMGCS. J. Phys: Conf. Ser. **1345**(4), 042015 (2019). https://doi.org/10.1088/1742-6596/1345/4/042015

14. Tang, Y., Hu, M., Huang, R., et al.: Aircraft taxi route planning for A-SMGCS based on free time windows and multi-agent systems. Acta Aeronautica et Astronautica Sinica **36**(5), 1627–1638 (2015)

AI for High Energy Physics

An Intelligent Image Segmentation Annotation Method Based on Segment Anything Model

Jiameng Zhao[1,2], Zhengde Zhang[2(✉)], and Fazhi Qi[2(✉)]

[1] School of Computer and Artificial Intelligence, ZhengZhou University, Zhengzhou, China

[2] Institute of High Energy Physics, Chinese Academy of Sciences, Beijing, China
`{zhaojm,qfz}@ihep.ac.cn`

Abstract. Training of supervised neural network models requires a large amount of high-quality datasets with true values. In computer vision tasks such as object detection and image segmentation, the process of annotating a large number of original two-dimension data segments is extremely costly, which greatly affects the application rate of AI for HEP (High Energy Physics). The SAM(Segment Anything Model) based on transformer provides a promising solution to this problem. This paper proposes an intelligent image segmentation annotation method based on the SAM, by which the annotation efficiency can be increased by 50 times. Examples of annotations, the API (Application Programming Interfaces), and GUI (Graphical User Interfaces) are also provided. The use of this tool will greatly accelerate the process of transforming high-energy physics image-style data from raw data to AI-Ready data.

Keywords: SAM · Segmentation Annotation

1 Introduction

The task of image segmentation is fundamental in computer vision and has been applied in various fields, including medicine, autonomous driving, and high-energy physics etc. In recent years, the development of deep learning has brought image segmentation to a new level. However, since deep learning is a data-driven algorithm, it requires high-quality, large-scale annotated datasets for accurate model training. The data annotation is a costly process that significantly affects the speed of the application of deep learning algorithms.

Researchers have made substantial efforts to address the image segmentation annotation problem. They have developed annotation software, such as LabelMe [1], LabelImg [2], and CVAT [3], and designed semi-automatic annotation systems [4,5] that iteratively combine model annotation and manual correction. However, these methods or tools still have limitations, such as the need for manual annotation, difficult installation, and slow model iteration.

C. Cruz et al. (Eds.): IC 2023, CCIS 2036, pp. 237–243, 2024.
https://doi.org/10.1007/978-981-97-0065-3_17

In 2023, the SAM (Segment Anything Model) [6], a large-scale segmentation model, was proposed in the computer vision field. The SAM has learned the features of general objects and has strong transferability and zero-shot generalization capabilities, providing a promising solution to the high cost of segmentation annotation.

In this paper, we propose an intelligent image annotation method based on the SAM model (Fig. 1), and deploy it on the HepAI platform. Examples of X-ray image segmentation demonstrate that the method can effectively be used for image annotation and accelerate AI development. Our contributions are as follows:

(1) Based on the SAM model, we developed SAM intelligent segmentation annotation workers on the distributed deployment architecture of the HepAI platform, providing segmentation annotation API interfaces.
(2) We developed a GUI interface based on QT that can be used for intelligent segmentation annotation.
(3) We shared the code in gitlab [8].

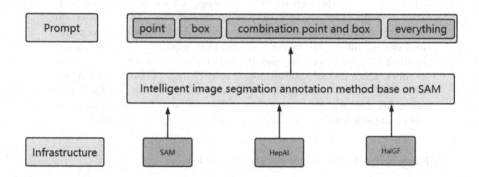

Fig. 1. Intelligent Image Segmentation Annotation Framework Based on the SAM.

2   Method

2.1   Segment Anything Model

The Segment Anything Model (SAM) originated from the 2023 Segment Anything (SA) project. SAM is a fundamental model in the field of image segmentation, unifying the entire image segmentation task. The research path is shown in Fig. 2 and is divided into tasks, models, and data engines.

The SAM model structure includes image encoders, cue encoders, and mask decoders, which can accept images and various types of cues (such as points, boxes, masks, text) as input and output predicted segmentation masks. SAM was trained on more than 11 million images to obtain over 1 billion masks, learning features of common objects and having powerful zero-shot transfer ability.

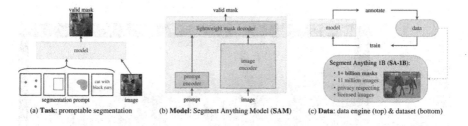

Fig. 2. Research path of segment anything model

2.2 SAM Worker

The HepAI platform adopts a distributed deployment architecture to deploy AI models and provide AI model services. The main structure of the architecture is designed with a controller and workers.

The controller serves as the gateway for publishing services and the medium for users to request AI services. Users apply for AI model services through HTTP requests. In addition to this, the controller is also responsible for registering and monitoring workers. Once a worker registers with the controller, it can provide services outside of the worker. However, if a worker becomes inactive, it is removed from the model list maintained by the controller, and the worker no longer provides the model service.

On the other hand, the worker is the instance for deploying the model and the party that actually provides the service. The worker deploys the AI model and provides service interfaces. It registers with a controller, and the controller is responsible for forwarding request parameters and service results.

We integrated the SAM model into the HepAI platform and deployed it in the form of a worker to provide model services to the public. SAM is deployed in the form of a worker and encapsulates the inference function, providing segmentation services under different types of prompts, such as point prompts, box prompts, and panoramic segmentation prompts. The controller injects the required inference parameters and returns the segmentation result in a unified format.

2.3 GUI

The GUI interface, written in QT [7], helps users to perform segmentation annotation more conveniently (Fig. 3). It includes a resource bar for loading image folders, a main sidebar for adding different types of prompts or annotation functions, and a central control for displaying images and drawing annotation prompts. With the help of the GUI interface, image segmentation annotation can be performed interactively.

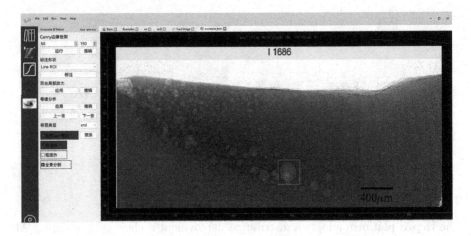

Fig. 3. Label-Tool GUI

3   Usage and Application Cases

3.1   API Usage

Based on the concept of model-as-a-service on the HepAI platform, SAM model services are provided through the HepAI platform, and SAM can be used to assist in image segmentation annotation through the API interface.

In Step1, to use SAM model services, the API key needs to be obtained, which is the permission control and can be obtained at https://ai.ihep.ac.cn/.

In Step 2, the Hepai library is installed in the local environment to list available models through the API key.

In Step 3, the SAM segmentation service is called by inputting the model name, image, and segmentation prompts as parameters. The segmentation results are processed based on actual annotation requirements to obtain the segmentation annotation data.

3.2   GUI Usage

In addition to the API interface, a GUI interface is also provided to use the intelligent segmentation annotation service. The process is shown in Fig. 4, which includes five steps:

Step 1 involves using the resource browser in the HaiGF core function bar to load the folder containing the dataset and viewing the file directory list of the dataset image in the main sidebar. The dataset images are displayed in the central control by using double-click, previous image, and next image function keys, preparing for annotation.

In Step 2, the label plugin is enabled, and the SAM button is clicked to enable the intelligent annotation model. The SAM button is used to control the annotation model, i.e., the AI intelligent annotation model and the manual annotation mode.

Step 3 involves using the label plugin to provide four types of prompts in the main sidebar: point prompts, box prompts, a combination of point and box prompts, and panoramic segmentation prompts. According to the characteristics of the dataset image and the segmentation target, the corresponding prompt type is selected for annotation, and the prediction button is clicked.

In Step 4, HepAI returns the annotation results, and the next step is processed. Finally, Step 5 involves further processing the annotation results according to different purposes, such as saving the bounding box of the mask as yolo or voc format, saving the mask as png or jpg files, etc.

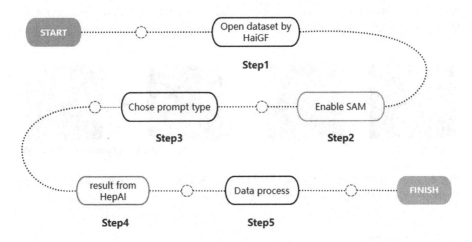

Fig. 4. Path to segmentation annotation via GUI

3.3   Use Cases

X-Ray Image Segmentation. X-ray additive manufacturing bubble defect detection analysis guides the optimization and design of metal additive manufacturing, supporting the research and rapid manufacturing of strategic high-end equipment in fields such as medicine, aviation, aerospace, and defense. The bubble defect dataset is shown in Fig. 5(a), and Fig. 5(b) shows the results of applying intelligent annotation methods to annotate the dataset. By processing the annotation results, high-quality annotation data can be produced, which can be used to train AI models, as shown in Fig. 5(c).

Fluorescence Image Segmentation. Fluorescence image segmentation requires the separation of biological samples from the image, and traditional denoising algorithms may lose target information. The use of AI models requires solving the problem of annotated data for training. Segmentation based on the

Fig. 5. Example of X-ray image segmentation. (a) Sample of X-ray bubble detection dataset. (b) Detection result from SAM. (c) Yolo format via detection result.

SAM model can effectively segment images that are enhanced by ROI, threshold segmentation, histogram equalization and obtain the cerebellar region (Fig. 6). By adopting this technology, training data can be accumulated to train specialized object detection models, which can achieve fluorescence image segmentation with polar resolution.

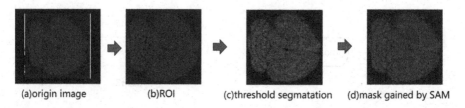

(a)origin image   (b)ROI   (c)threshold segmatation   (d)mask gained by SAM

Fig. 6. Fluorescence Image Segmentation. (a) origin image. (b) apply ROI. (c) apply threshold segmentation. (d) get mask via SAM

3.4 Comparison

A comparison is made between this method and other annotation methods on the additive manufacturing bubble dataset. Each image in this dataset contains 30 to 50 bubble instances with different sizes. At the same time, most bubbles are very small in size.

Manual Annotation. Traditional manual annotation is conducted with the help of annotation software like LabelMe and LabelImag. Due to the small size of bubbles in additive manufacturing, it requires high concentration to annotate them accurately. It would take at least 5 min to finish bounding box annotation for one image, and at least half an hour for pixel-level annotation of one image. With the intelligent annotation method proposed in this paper, both bounding box annotation and pixel-level annotation can be controlled within 20 s.

Semi-automatic Annotation. Existing semi-automatic annotation methods all follow the process of manually annotating data, training a model with the data, using the model to help annotate data, and using new data to improve model performance. In the early stage of this process, a large amount of manpower and effort is still needed to annotate data. Meanwhile, the workload of human verification is also large during the process of improving model performance. In contrast, the SAM model used in this paper can be directly used to help annotation without fine-tuning, greatly reducing the time cost and manpower cost.

4 Conclusion

This paper proposes a method for intelligent image segmentation annotation based on the SAM model. By using interactive prompts of different types, high-quality image segmentation data can be obtained quickly, thus speeding up the process of accumulating training data. The feasibility and development prospects of the proposed method for image segmentation annotation were confirmed through its application to the X-ray additive manufacturing bubble defect detection dataset. This method can be widely applied to 2D image segmentation in various fields. We have deployed this method on the HepAI platform, allowing users to use the model directly without having to consider algorithm, weight, computing power, and other issues. Proper use of this tool will significantly improve research efficiency and promote a paradigm shift in scientific research.

Acknowledgements. This work was supported by the Network Security and Informatization Project of the Chinese Academy of Sciences [Grant Number: CAS-WX2022SF-0104] and "From 0 to 1" Original Innovation Project of IHEP [E3545PU2]. Special thanks to Mr. Binbin Zhang, Ms. Chunxia Yao, and Mr. Yu Hu for providing support with the original experimental data.

References

1. Labelme. https://github.com/wkentaro/labelme
2. Labelimage. https://github.com/rachelcao277/LabelImage
3. Computer vision annotation tool (cvat). https://github.com/openvinotoolkit/cvat
4. Hao, Y., Liu Y, Chen Y, et al.: EISeg: An Efficient Interactive Segmentation Annotation Tool based on PaddlePaddle. arXiv preprint arXiv:2210.08788 (2022)
5. Hu, M., Xia, X., Yang, C., Cao, J., Chai, X.: Design and implementation of semi-supervised image annotation system based on deep learning. J. China Agricultural Univ. (05), 153–162 (2021)
6. Kirillov, A., et al.: Segment anything. arXiv preprint arXiv:2304.02643 (2023)
7. HaiGF. https://code.ihep.ac.cn/zdzhang/hai-gui-framework
8. HaiGF-LabelTool. https://code.ihep.ac.cn/1746163329/haigf-labeltrain

ParticleNet for Jet Tagging in Particle Physics on FPGA

Yutao Zhang[1]([✉]), Yaodong Cheng[2,3], and Yu Gao[2,3]

[1] School of Computer and Artificial Intelligence, ZhengZhou University, Zhengzhou, China
z_zhangyt@163.com
[2] Institute of High Energy Physics, Chinese Academy of Sciences, Beijing, China
[3] University of Chinese Academy of Sciences, Beijing, China

Abstract. Jet tagging is a crucial classification task in particle physics experiments. In recent years, the introduction of deep learning methods has significantly improved the accuracy of jet tagging classification tasks, with graph neural networks like ParticleNet demonstrating outstanding performance in this domain. Regarding model deployment, common hardware options include CPUs, GPUs, FPGAs, and ASICs. Presently, due to FPGA's advantages such as low power consumption, low latency, and hardware programmability, it has become a cutting-edge research focus for accelerating AI deployment. Compared to CPUs, FPGAs can achieve better parallel operations, while compared to GPU platforms, using FPGAs can enhance computation efficiency and reduce power consumption. Hence, porting and optimizing ParticleNet on FPGAs can enable rapid and low-power execution of classification tasks in particle physics, thereby reducing economic costs and expediting data processing in particle physics research. This study will adopt the CPU+FPGA heterogeneous computing model, offloading compute-intensive tasks to the FPGA for faster execution and reduced algorithm latency.

Keywords: Jet tagging · Particle physics · Deep learning · FPGAs · Acceleration

1 Introduction

The Large Hadron Collider (LHC) is a high-energy physics device used for accelerating and colliding protons. Collisions occur in the LHC every 25 nanoseconds, producing data at the TB level per second. Real-time processing of the data generated from high-energy proton collisions is a challenging task in the trigger system, which aims to retain events of interest while ensuring low latency and accuracy.

This research has not been published in any other journal and has not been presented at any other conference or undergone peer review in any other journal.

C. Cruz et al. (Eds.): IC 2023, CCIS 2036, pp. 244–253, 2024.
https://doi.org/10.1007/978-981-97-0065-3_18

In recent years, the introduction of deep learning has transformed the handling of jet tagging tasks in particle physics and significantly improved their accuracy. Among them, ParticleNet network model stands out as one of the state-of-the-art methods, achieving first place in the 2020 High Energy Particle Classification Challenge with a top tagging dataset accuracy of 94% [1], making it crucial for the trigger system.

In the trigger system, accuracy and power consumption are important considerations for jet tagging classification tasks. Common hardware platforms include CPUs, GPUs, and FPGAs. However, CPUs may struggle to meet real-time processing demands, while GPUs consume more power as deployment platforms. Hence, using FPGAs is a favorable choice, offering faster computation and lower power consumption compared to CPUs. In recent years, there has been considerable research on implementing jet tagging tasks on FPGAs. For instance, [2,3] achieved 75% accuracy by implementing an MLP network for jet tagging on an FPGA. Additionally, [4] achieved 80% accuracy by implementing an optimized JEDInet graph neural network on an FPGA.

The ParticleNet is one of the state-of-the-art methods, we aim to explore its implementation onto an FPGA. ParticleNet possesses a complex network structure, making it challenging to deploy the entire model on an FPGA. The ParticleNet architecture consists of three EdgeConv blocks, one aggregation layer, and two fullyconnected layers. We simply divide ParticleNet into convolution and other modules. Through time analysis, we found that the convolution module accounts for approximately 70.8% of the time, while the remaining modules account for 29.2%. Thus, we plan to implement the convolution module using an FPGA and the remaining modules using a CPU, forming a CPU + FPGA cooperative computing heterogeneous architecture (Fig. 1).

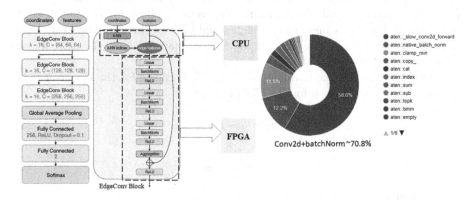

Fig. 1. ParticleNet Architecture and Time Analysis

2   Methodology

[7] proposed two hardware accelerator solutions including (a) **dedicated circuit designs**, often based on libraries of building blocks, and (b) **overlay-based co-designs** from different vendors and academic sources. [8] discussed the current design of hardware architectures is divided into two types:

(1) **the overlap pattern**, design a reusable execution engine, which offers flexibility by eliminating the need for hardware redesign when the model parameters change while the operators remain unchanged;
(2) **stream pattern**, each layer of the target model possesses independent computing units, similar to (a)dedicated circuit designs mentioned in [7].

In this work, We used the dedicated circuit designs with the Stream pattern. A given neural network, except for the parts executed on the CPU, is all implemented on the FPGA, and each layer has an independent execution unit. It often requires heavy net tuning in terms of quantization of weights, inputs. Therefore, before deploying neural networks, it is common to further optimize the models using techniques such as model pruning, model quantization, and graph fusion. In our research, we performed graph fusion and model quantization on the ParticleNet model to accelerate the model inference speed.

2.1   Operator Fusion

The fusion of Convolution (Conv) and Batch Normalization (BN) layers, is a technique used in deep learning to optimize the computational efficiency and memory usage of neural networks. In the inference mode, the fusion of Conv and BN layers eliminates the need for separate computations. The statistics (mean and variance) calculated during the training phase for BN are directly incorporated into the Conv layer's weights and biases. This way, the Conv layer can directly produce normalized output feature maps without the need for an additional BN layer.

In inference mode, the parameters $\gamma, \sigma, \epsilon, \beta$ are known for the BN layer. The parameters w, b are known for the CONV layer.

$$x_1 = w * x + b \tag{1}$$

$$x_2 = \gamma \cdot \frac{x_1 - \mu}{\sqrt{\sigma^2 + \epsilon}} + \beta \tag{2}$$

By combining (1) and (2), we derive

$$x_2 = \frac{\gamma * w}{\sqrt{\sigma^2 + \epsilon}} * x + \beta + \gamma * \frac{b - \mu}{\sqrt{\sigma^2 + \epsilon}} \tag{3}$$

$$\tilde{w} = \frac{\gamma * w}{\sqrt{\sigma^2 + \epsilon}}, \tilde{b} = \beta + \gamma * \frac{b - \mu}{\sqrt{\sigma^2 + \epsilon}}, x_2 = \tilde{w} * x + \tilde{b} \tag{4}$$

By integrating the Convolution (Conv) and Batch Normalization (BN) operations, it is possible to reduce computational overhead and accelerate inference speed.

2.2 Model Quantization

Generally speaking, the trained model parameters are float32. The network model parameter size of ParticleNet occupies 1.5MBytes. The on-chip resources of FPGA are very limited. If the network parameters of float32 are stored in the on-chip memory BRAM of FPGA, the on-chip memory will be insufficient. This requires quantification of model parameters. Model quantization is the process of converting float32 model parameters into low-precision fixed-point representations, thereby reducing the model's storage space and computational requirements. However, during the quantization process of deep neural network models, there is an inevitable loss of information learned during the previous training process, leading to a decrease in performance on the corresponding tasks. Therefore, given a specific quantization method and a quantized model, it is necessary to optimize the quantization parameters in the quantization function or the parameters in the original network model to some extent to restore the performance of the network model. Based on whether the quantized model is retrained, the current optimization methods can be classified into the following two categories: Quantization-Aware Training (QAT) and Post-Training Quantization. (PTQ) [5].

In general, the PTQ method is preferred for quantization, and the QAT method is only considered when the loss in precision becomes significant. Therefore, in this study, we prioritize the utilization of PTQ for quantization, which results in a precision loss of less than 1%. As a result, QAT quantization was not employed. The quantized results are presented in the following table (see Table 1):

Table 1. Table captions should be placed above the tables.

	Parameters (Bytes)	Data Type	Accuracy (%)
ParticleNet	**1.5MB**	Float32	93.90%
ParticleNet-Quantization	**466KB**	Int8	93.28%

The conversion formulas between fixed-point and floating-point numbers for quantization are as follows:

$$A_{float} = S(A_{int} - Z) \tag{5}$$

$$A_{int} = round(\frac{A_{float}}{S}) + Z \tag{6}$$

where A_{float} stands for the floating-point number, S is the scale constant, and A_{int} represents the fixed-point number. The constants S and Z are our quantization parameters. For 8-bit quantization, A_{int} is quantized as an 8-bit integer (for B-bit quantization, A_{int} is quantized as an B-bit integer). Some arrays, typically bias vectors, are quantized as 32-bit integers [9]. So, In this study, we adopted the PTQ method to quantize the weight parameters to int8 and the bias to int32. Denoting the weight matrix as $\mathbf{W}$ and the bias matrix as $\mathbf{B}$, the convolution operation is carried out as follows, Assuming $\otimes$ represents a convolution operation:

$$A_{n+1} = A_n \otimes W_n + B_n \tag{7}$$

because, we used the symmetric quantizer, which restricts zero-point to 0. With the symmetric quantizer, the conversion operations simplify to:

$$A_{float} = SA_{int} \tag{8}$$

$$A_{int} = round(\frac{A_{float}}{S}) \tag{9}$$

By expressing the variables $\mathbf{W}$, $\mathbf{B}$, and $\mathbf{A}$ in Equation (7) using quantized fixed-point numbers, we can derive Equation (10).

$$Aq_{n+1}S_{a_{n+1}} = Aq_n S_{a_n} \otimes Wq_n S_{w_n} + Bq_n S_{b_n} \tag{10}$$

the bias quantization scale S_b is as same as the product of the scales of the weights and of the input activations [9]. Specifically, $S_{b_n} = S_{a_n} S_{w_n}$.

$$Aq_{n+1} = (Aq_n \otimes Wq_n + Bq_n)\frac{S_{a_n} S_{w_n}}{S_{a_{n+1}}} \tag{11}$$

Following parameter quantization, Eq. (10) is replaced by Eq. (11) for computations. In Eq. (11), only the calculation when $M = \frac{S_{a_n} S_{w_n}}{S_{a_{n+1}}}$ involves floating-point operations. If $M = 2^{-n}$, the Eq. (11) can be fully converted to fixed-point calculations because, multiplication by 2^{-n} can be implemented with an efficient bit shift, albeit one that needs to have correct round-to-nearest behavior [9]. Power-of-two quantization, a special case of symmetric quantization, in which the scale factor is restricted to a power-of-two, $S = 2^{-k}$ [10]. We adopted Power-of-two quantization, enabling us to achieve full fixed-point computations through simple bit-shifting operations on hardware. The quantization process on the hardware is depicted in (Fig. 2) [10].

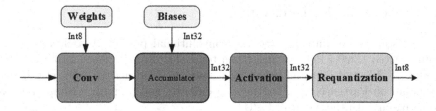

Fig. 2. Schematic overview of quantized forward pass for convolutional layer

3 HardWare Design

In this section, we present the overall architecture of ParticleNet on FPGA. We make it out of multikernels with the stream structure. The details are as follows.

3.1 Overall System Architecture

In this work, due to the complexity of ParticleNet, placing the dynamic graph on an FPGA poses certain challenges. As analyzed in Sect. 1, the convolutional module in ParticleNet is identified as the most time-consuming component. Therefore, we focus on offloading the most time-consuming module onto the FPGA,(see Fig. 3). In Fig. 3, we can be observed that the various Edconvblocks within ParticleNet have multiple interactions with external memory, which can potentially result in performance bottlenecks.

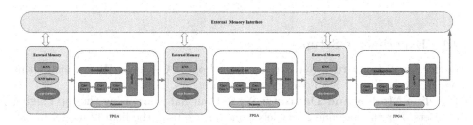

Fig. 3. Overall System Architecture for Software-Hardware Co-design

3.2 Implementation Design

The most time-consuming part of ParticleNet is distributed across three different EdconvBlocks. Since ParticleNet follows the structure of DGCNN, each EdconvBlock contains the algorithm for updating the graph. Therefore, we execute the graph update algorithm on the CPU, while the convolutional part within the EdconvBlock is offloaded to the FPGA for execution (see Fig. 3). In this work, we utilize the stream pattern, which involves allocating a separate compute unit for the convolutional operations in each EdconvBlock. As a result, we have three compute units. Each compute unit implements inter-layer pipelining and intra-layer parallelism for optimization, as illustrated in Fig. 4.

4   Results and Discussion

In this section, we implemented the convolutional part of ParticleNet on the Xilinx Alveo U200 FPGA, which has a total of 6840 DSP slices, 2160 BRAMs, 1182K LUTs, and 2364K FFs on this platform. We measure the overall performance on ParticleNet in Top tagging datasets and perform a comprehensive comparison with the result of CPU.

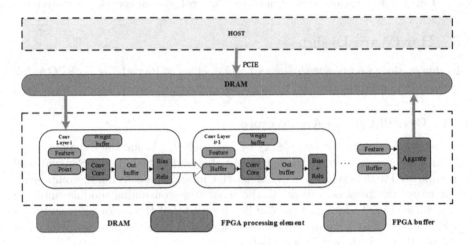

Fig. 4. Pipeline Architecture for Compute Unit

4.1   Experiments Setup

In this work, our ParticleNet is programmed on FPGA using HLS, which forms the HLS core library with configurable parameters.

Baseline Models and Datasets: We test our hardware using the top tagging datasets. For the baseline ParticleNet model, we used the float32 model implemented in PyTorch.

Hardware and Toolkits: The hardware design is synthesized and generated with Xilinx Vitis HLS 2021.1, and then we use Xilinx Vitis 2021.1 and Xilinx Vivado 2021.1 to synthesize and deploy the complete project. The target platform is Xilinx Alveo U200 FPGA Card to implement our ParticleNet to perform the ParticleNet model. Moreover, the host CPU controls the ParticleNet accelerator and initial DDRs via the PCIe interface. The built-in power report in Vivado gives power consumption numbers.

4.2   Results

Resource Utilization: In this work, he resource utilization of each compute unit of our hardware resources is presented in Table 2. we observe that the utilization of resources increases with the complexity of the network when employing a pipelined structure. This is because each convolution within the compute unit consumes resources, as it is a dedicated acceleration module.

Comparison with CPU: CPU BaseLine:Intel(R) Xeon(R) Silver 4116 CPU @ 2.10GHz. Using the previous Pytorch model is not optimized, instead of the onnx model. In our FPGA implementation, the recognition of a particular type of particle takes approximately 15 ms. On the other hand, using the PyTorch pt model, the recognition of the same type of particle takes around 35 ms. This FPGA implementation demonstrates a performance improvement achieves 2.3X improvement. The Power of hardware design (see Fig. 5)

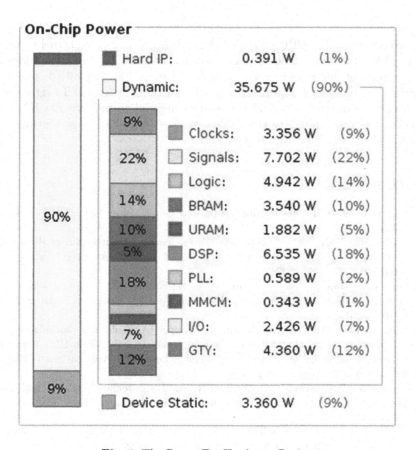

Fig. 5. The Power For Hardware Design

4.3   Discussion

Currently, we have quantized the most time-consuming part of ParticleNet and migrated it to execute on the FPGA, resulting in a performance improvement of

Table 2. Table captions should be placed above the tables.

Compute Unit		LUT	FF	BRAM	DSP
	Available	1182K	2364K	2160	6840
EdBlock-1	Used	39K	88K	74	904
	Utilized[%]	4.2	3.7	4.4	13.2
EdBlock-2	Used	77.7K	84K	148	1024
	Utilized[%]	8.4	3.5	8.8	15.0
EdBlock-3	Used	178.5K	160K	218	2320
	Utilized[%]	19.4	6.7	12.97	33.9

2.3x compared to the CPU. The FPGA implementation of the kernel consumes only 35 W of power, which is lower than that of the CPU and GPU. However, due to frequent interactions with the host, it cannot fully exploit the pipelining capabilities and can only utilize the pipeline within the kernel. To significantly enhance performance, it would be highly beneficial to migrate the KNN and EdgeFeature components to the FPGA. However, this task poses certain challenges and requires substantial research time.

References

1. Qu, H., Gouskos, L.: Jet tagging via particle clouds. Phys. Rev. D **101**(5), 056019 (2020)
2. Coelho, C.N., Jr., et al.: Automatic heterogeneous quantization of deep neural networks for low-latency inference on the edge for particle detectors. Nature Mach. Intell. **3**(8), 675–686 (2021)
3. Duarte, J., Han, S., Harris, P., Jindariani, S., Kreinar, E., Kreis, B., Wu, Z.: Fast inference of deep neural networks in FPGAs for particle physics. J. Instrum. **13**(07), P07027 (2018)
4. Que, Z., Loo, M., Fan, H., Pierini, M., Tapper, A., Luk, W.: Optimizing graph neural networks for jet tagging in particle physics on FPGAs. In 2022 32nd International Conference on Field-Programmable Logic and Applications (FPL), pp. 327–333. IEEE, August 2022
5. Nagel, M., Fournarakis, M., Amjad, R. A., Bondarenko, Y., Van Baalen, M., Blankevoort, T.: A white paper on neural network quantization. arXiv preprint arXiv:2106.08295 (2021)
6. Li, G., Liao, L., Lou, X., Shen, P., Song, W., Wang, S., Zhang, Z.: Classify the Higgs decays with the PFN and ParticleNet at electron-positron colliders. Chin. Phys. C **46**(11), 113001 (2022)
7. Plagwitz, P., Hannig, F., Ströbel, M., Strohmeyer, C., Teich, J.: A safari through FPGA-based neural network compilation and design automation flows. In 2021 IEEE 29th Annual International Symposium on Field-Programmable Custom Computing Machines (FCCM), pp. 10–19. IEEE, May 2021
8. Wang, T., et al.: Via: A novel vision-transformer accelerator based on fpga. IEEE Trans. Comput. Aided Des. Integr. Circuits Syst. **41**(11), 4088–4099 (2022)

9. Jacob, B., et al.: Quantization and training of neural networks for efficient integer-arithmetic-only inference. In Proceedings of the IEEE Conference on Computer Vision and Pattern Recognition, pp. 2704–2713 (2018)
10. Nagel, M., Fournarakis, M., Amjad, R.A., Bondarenko, Y., Van Baalen, M., Blankevoort, T.: A white paper on neural network quantization. arXiv preprint arXiv:2106.08295 (2021)
11. LNCS Homepage. http://www.springer.com/lncs. Accessed 4 Oct 2017

Application of Graph Neural Networks in Dark Photon Search with Visible Decays at Future Beam Dump Experiment

Zejia Lu[1,2,3], Xiang Chen[1,2,3], Jiahui Wu[1,2,3], Yulei Zhang[1,2,3(✉)], and Liang Li[1,2,3(✉)]

[1] School of Physics and Astronomy, Institute of Nuclear and Particle Physics, Shanghai Jiao Tong University, Shanghai, China
{avencast,liangliphy}@sjtu.edu.cn
[2] Shanghai Key Laboratory for Particle Physics and Cosmology, Shanghai, China
[3] Key Lab for Particle Physics, Astrophysics and Cosmology (MOE), Shanghai, China

Abstract. Beam dump experiments provide a distinctive opportunity to search for dark photons, which are compelling candidates for dark matter with low mass. In this study, we propose the application of Graph Neural Networks (GNN) in tracking reconstruction with beam dump experiments to obtain high resolution in both tracking and vertex reconstruction. Our findings demonstrate that in a typical 3-track scenario with the visible decay mode, the GNN approach significantly outperforms the traditional approach, improving the 3-track reconstruction efficiency by up to 88% in the low mass region. Furthermore, we show that improving the minimal vertex detection distance significantly impacts the signal sensitivity in dark photon searches with the visible decay mode. By reducing the minimal vertex distance from 5 mm to 0.1 mm, the exclusion upper limit on the dark photon mass ($m_{A'}$) can be improved by up to a factor of 3.

Keywords: GNN · Tracking Reconstruction · Deep Learning · Dark Photon Search · Fixed-Target Experiment

1 Introduction

Dark matter is one of the most intriguing mysterious that cannot be solved by the Standard Model. Numerous astrophysical and cosmological observations have provided strong evidence for the presence of dark matter, which is believed to make up about 85% of the matter in the universe. Despite its prevalence, the nature of dark matter particles remains elusive, and identifying their properties is a key challenge in particle physics research. Direct searches for Weakly Interacting Massive Particles (WIMPs) as one of most sought-after dark matter candidates have so far yielded null results [1]. This outcome has further motivated

C. Cruz et al. (Eds.): IC 2023, CCIS 2036, pp. 254–263, 2024.
https://doi.org/10.1007/978-981-97-0065-3_19

the exploration of alternative dark matter candidates, such as the dark photon, a hypothetical particle also known as a U(1) gauge boson of a hidden sector and is predicted to have a small mass and interact weakly with standard matter. Unlike WIMPs, the dark photon is predicted to have extremely weak interactions with standard matter, making it a challenging particle to detect using traditional experimental methods. Numerous experimental efforts are underway to search for dark photons in various decay channels and interaction modes [3–5].

Beam dump experiments provide a unique opportunity to explore dark photons as dark matter candidates due to their distinctive experimental setup. In a beam dump experiment, a high-energy particle beam is directed towards a target, where interactions between the beam particles and target nuclei can produce new particles, including dark photons, as illustrated in Fig. 1a. The generated dark photons can escape the target and travel through the experimental setup before decaying into visible particles that can be detected by the surrounding detectors. One of the key advantages of beam dump experiments is their ability to probe a wide range of dark photon masses and couplings. Unlike traditional direct detection experiments that are sensitive to a specific mass range, beam dump experiments have the potential to cover a broad spectrum of dark photon masses, including those that are challenging to access with other experimental methods. Furthermore, beam dump experiments are relatively cost-effective and can be conducted using existing accelerator facilities, making them attractive platforms for exploring new physics beyond the Standard Model.

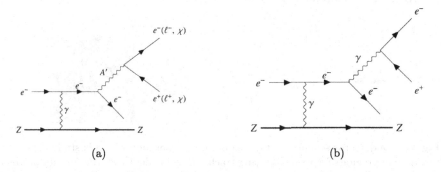

Fig. 1. Feynman diagram of (a) dark photon produced through bremsstrahlung and decaying into leptons pair or dark matter χ (b) QED radiative trident reaction, which serves as main background of dark photon visible decay [2]. A' is dark photon while γ is photon.

Dark photons couple with standard matter through the kinetic mixing term [2]

$$\mathcal{L}_{dark,\gamma} = -e\epsilon A'_{\mu} J^{\mu}_{em}, \qquad (1)$$

where ϵ is the coupling constant, A'_{μ} is the mediator field of the dark U(1) gauge group. The model has two free parameters, the coupling constant ϵ and the dark photon mass $m_{A'}$.

By comparing measured decay products with expected backgrounds, beam dump experiments can provide crucial insights into the nature of dark photons, either supporting their existence within specific parameter regions or setting stringent limits on these parameters. This process involves dark photon signature searches, background estimation, sensitivity projections, and statistical analyses. With advanced detector technologies and innovative analysis techniques like Graph Neural Networks, the sensitivity to dark photon signals can be significantly improved, further motivating the search for these elusive particles in beam dump experiments.

2   Experimental Setup and Simulation Framework

Several beam-dumped experiments have been proposed to search for the dark photon, such as LDMX [3], NA64 [4], and DarkSHINE experiments [5]. The experimental setup of our study is mainly based on the DarkSHINE experiment, an electron-on-fixed-target experiment using a single electron beam.

2.1   Experimental Setup

The overall setup of DarkSHINE is shown in Fig. 2. The beam is an 8 GeV electron beam with a high frequency. The whole detector is composed by several main sub-detectors systems: a tracking system, an electromagnetic calorimeter (ECAL), and a hadronic calorimeter (HCAL).

Fig. 2. The detector scheme for the beam-dump experiments in this study. There are two tracking regions for reconstructing track of the incident electron and its products after interacting with the nuclei in the target. ECAL and HCAL are behind the tracking region.

The tracking system is composed of silicon trackers and dipole magnets, serving distinct purposes within two separate tracking regions. The first region, known as the tagging tracker, is designated for the monitoring of incident electrons, whereas the second region, the recoil tracker, is used for tracking the recoil electrons and decay products. The module of each tracker has two silicon strips placed at a small angle for more precise position measurements of each energy hit. The tungsten target with $0.1X_0$ is placed between two tracking regions. The ECAL is designed to absorb the full energy of incident particles with good energy resolution, and the HCAL is placed after ECAL, serving for detecting hadronic backgrounds and capturing muons.

2.2 Event Simulation

We simulate all events with the detector setup mentioned above, utilizing specific software tools. The dark photon process is modeled using CalcHEP v3.8.10 [6] and executed via GEANT4 v10.6.10 [7]. For maintaining a good modeling of the dark photon process, we directly incorporate the particles' four-momentum truth information from the generator during the dark photon process simulation. Standard Model processes such as photon-nuclear interactions, electron-nuclear interactions, and photon decays into muon pairs, are simulated by GEANT4.

3 Analysis Strategy and Tracking Reconstruction

The main background of dark photon visible decay is the QED radiative trident reaction, as depicted in Fig. 1b. While several kinematic distinctions, such as the energy and exit angle of the recoil electron, exist, the key signature of visible decay is the displaced vertex [8]. As presented in Table 1, employing typical kinematics selections: (1) No. of track = 3, (2) $p(e^+) > 2$ GeV, (3) $p(\text{hard } e^-) <$ 6 GeV, (4) $\theta(\text{hard } e^-) > 0.1°$, the background rejection power can reach 0.2%, which is insufficient considering that the expected number of electrons on target reaches 3×10^{14} in DarkSHINE. Therefore, tracking and vertex reconstruction becomes vital in the quest for visible decay searching.

Table 1. The event count of the signal and inclusive background samples after kinematics selections.

Event type	total event count	cut1	cut2	cut3	cut4	efficiency
Visible decay	9918	5963	3889	3889	3436	**34.6%**
Inclusive background	1.84e7		3.33e5	5.18e4	5.18e4	3.42e4	**0.19%**

3.1 GNN-Based Tracking Reconstruction

We propose a novel approach to tracking reconstruction in high-energy physics, leveraging machine learning principles, particularly GNN.

Network Structure. We construct 2 GNN models based on transformer convolutions to process graph-structured data using PyTorch [11] and PyG [10]:

- **LinkNet** *edge classification task for track finding*: The LinkNet model is to
predict if an edge is a true particle trajectory passing through the two connected nodes. It employs a multi-layer perceptron (MLP) for both node and
edge feature embedding, with each MLP consisting of three layers. The input
dimensions are set to 6 for nodes (corresponding to spatial coordinates and
magnetic field components: x, y, z, Bx, By, Bz) and 3 for edges (representing
the relative polar coordinates: r, theta, phi). The hidden dimension across
the network is uniformly set to 128. Key to LinkNet's design is the use of
Transformer Convolutional layers, which combine the strengths of CNN and
transformer models. The network comprises six iterations, each with four
attention heads. The node features are updated during iteration using the
Transformer Convolutional layer, and the corresponding edge features are
updated by adding the new features on the surrounding nodes of each edge.
Layer Normalization and shortcut connections are integrated into each iteration.

For training, LinkNet is optimized using the Adam algorithm, motivated
by its adaptive learning rate capabilities. The initial learning rate is set at
0.00075, with a weight decay of 1.e-5. A tiered learning rate decay schedule
is implemented, reducing the learning rate by factors of 0.5, 0.2, and 0.1 at
epochs 30, 40, and 50, respectively. The model undergoes training for a total
of 55 epochs, a duration determined to be sufficient for convergence based on
preliminary experiments. The loss function used is the binary cross-entropy
with logits, suitable for the binary classification task intrinsic to track reconstruction – determining whether a given pair of hits belong to the same track.
The batch size is set to 64.

- **MomNet** *edge regression task for track fitting*: This network is specialized
in predicting the momentum associated with each edge, which represents a
potential particle trajectory in the detector. The MomNet architecture incorporates an MLP for embedding features of both nodes and edges. Each MLP
contains three layers, catering to the specific dimensions of the input features.
For nodes, the input dimension is 6, accounting for spatial coordinates and the
magnetic field components (x, y, z, Bx, By, Bz). For edges, the input dimension is set at 4, representing the relative polar coordinates (r, theta, phi) and
a link score from LinkNet that quantifies the likelihood of two nodes being
connected in a physical particle track. MomNet also uses Transformer Convolutional layers, which are adept at handling the complex relational dynamics
of the particle tracks. The network comprises 10 iterations, each with a significantly increased number of attention heads, set at 32. Each iteration of
the network updates the node features using the Transformer Convolutional
layer, while the corresponding edge features are refined by integrating the
updated node features surrounding each edge. MomNet is further enhanced
with layer normalization and shortcut connections in each iteration.

For the training process, MomNet utilizes the Adam optimizer, selected for its
adaptive learning rate properties. The initial learning rate is set at 0.00025,
with a weight decay of 1.e-5. The learning rate undergoes a tiered decay,
reducing by factors of 0.5 and 0.1 at epochs 40 and 50, respectively. The

total training duration is extended to 60 epochs, slightly longer than that of LinkNet, to ensure the model's convergence given its more complex task. The batch size is maintained at 64, consistent with the LinkNet configuration. For the MomNet model, a specialized loss function, the RelativeHuberLoss, is employed, which combines the MSE and L1 loss. The RelativeHuberLoss operates by first calculating the difference between the predicted values and the truth values. If the truth value is zero, which accounts for the case that this edge is false, and does not have true momentum along this edge, the absolute difference is used. In the other case for the truth edge, the relative difference is calculated.

Simulated Samples. For the training phase, we use 5 visible decay signal samples with masses of 20, 50, 100, 200, and 500 MeV, each with 1×10^6 electron-on-target (EOT) events. In these samples, single-track events are predominant in the tagging region and 3-track events in the recoil region. The input graph is built with digitized simulated hits, derived from a mean-shift algorithm-based clustering of original simulated hits. We utilize an orthogonal set of visible decay signal samples for performance evaluation, each with 1×10^6 EOT.

Single-Track Performance Evaluation. We evaluate the tracking performance by studying both single-track events (simple case) and 3-track events (more realistic case). The benchmark parameters are event reconstruction efficiency and track reconstruction resolution. The event reconstruction efficiency is defined as the percentage of the track events that are found and reconstructed by the reconstruction method. Track reconstruction resolution is defined as the full width at half maximum of the relative uncertainty distribution for the reconstructed track momentum. Owing to the variations in magnetic fields and tracker structures, we can assess the resolution of the reconstructed momentum in two distinct regions, as shown in Fig. 3. We have compared the GNN performance with the traditional tracking reconstruction method based on the Combinatorial Kalman Filter (CKF) method [5,9] using an orthogonal evaluation dataset with 50000 events. The GNN approach can reconstruct 98.9% single-track events, while the traditional approach gives 96.9%. The comparison is shown in Table 2. GNN gives better reconstructed momentum resolution than the CKF approach, improving the resolution from 1.5% to 0.6% in the tagging region, and 7.5% to 5.6% in the recoil region.

Multi-track Performance Evaluation. We evaluate the GNN performance to find 3-track events in the visible decay mode. The 3-track reconstruction efficiencies for 5 mass points are listed in Table 4. The low mass (long lifetime) region is the most sensitive region for beam dump experiments. GNN outperforms the CKF method by 30% to 88% in the region of interest (< 100 MeV) and

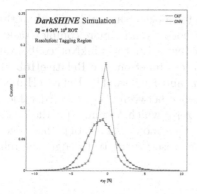
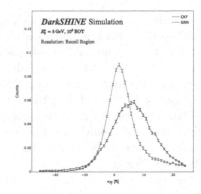

Fig. 3. The relative uncertainty of the reconstructed momentum in the tagging region (left) and recoil region (right).

Table 2. The reconstruction efficiency and momentum resolution comparison between the CKF and GNN methods.

Inclusive sample	CKF	GNN	Truth
Single-track Efficiency	96.9%	98.9%	99.9%
Resolution (Tagging)	1.5%	0.6%	-
Resolution (Recoil)	7.5%	5.6%	-

gives overall reconstruction efficiencies around 60%. The reconstructed momentum for each track is also shown in Fig. 4. There is a selection criterion requiring $P_i \geq 150$ MeV to ensure a good track quality in the tracker region. The application of GNN significantly improves the computational efficiency in terms of time per event, compared to the traditional CKF method, which is summarized in Table 3.

Future Prospects. The current study has demonstrated the effectiveness of GNNs for track fitting and finding tasks. These networks effectively leverage the natural graph structure inherent in detector data, creating a powerful framework for these complex tasks. Further research into vertex finding is also a promising direction. Several deep-learning methods have been proposed and used for vertex finding based on various experimental setups and constraints. Exploring more

Table 3. Comparative analysis of average computational time per event (in seconds) for the CKF method and the GNN method with and without pile-up.

Average Time [sec]	CKF	GNN
No pile-up	0.179	0.002
Pile-up $< \mu \approx 10 >$	300	0.010

Table 4. The 3-track reconstruction efficiency with respect to truth 3-track events.

$m_{A'}$[MeV]	20	50	100	200	500
CKF	31.8%	43.7%	51.5%	61.2%	68.4%
GNN	59.9%	61.9%	66.8%	65.2%	50.3%

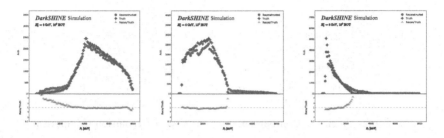

Fig. 4. The momentum distribution for reconstruction and truth for the leading track (left), subleading track (middle), and subsubleading track (right).

architectures and approaches becomes crucial. For instance, the use of Point Cloud Networks or Deep Sets could offer innovative strategies for addressing vertex-finding tasks, taking advantage of their unique capabilities for dealing with point-like structures or variable-sized sets of data.

3.2 Dark Photon Search in Visible Decay Mode

With the above improvements in both tracking and vertex reconstruction, the residual background level is expected to be 10 out of 3×10^{14} EOT. The 90% confidence level exclusion region for DarkSHINE experiment in the search for visible decays is calculated in Fig. 5. According to Sect. 3.1, the improved signal reconstruction efficiency for GNN method is expected to be around 60% in the most sensitive mass region (less than 100 MeV), while the efficiency for CKF method is between 30% and 50%. We have compared the two cases of signal efficiency being 30% and 60%, the result is shown in Fig. 5a. Furthermore, with improved track momentum resolution, GNN method is expected to provide better vertex reconstruction resolution, which is important for signal searching in the visible decay mode. We present three projected signal exclusion limits when the minimal vertex detection distance is set to be 0.1 mm, 1 mm, and 5 mm respectively, as shown in Fig. 5b. Taking $\epsilon^2 = 10^{-8}$ as an example, the upper limit on the dark photon mass $m_{A'}$ can be extended to 55 MeV, 95 MeV and 155 MeV with vertex resolutions of 5 mm, 1 mm and 0.1 mm respectively.

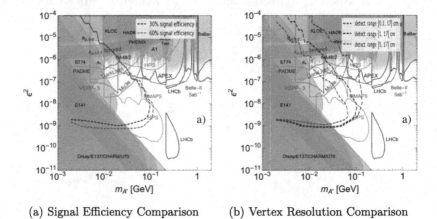

(a) Signal Efficiency Comparison   (b) Vertex Resolution Comparison

Fig. 5. 90% confidence level exclusion region for dark photon search in the visible decay mode for the DarkSHINE project when comparing different conditions: (a) signal efficiency comparison (b) vertex resolution comparison. Gray indicates regions that have been explored by other experiments [2] (Color figure online).

4   Conclusion and Outlook

We have demonstrated that in the single track and 3-tracks scenario, GNN approach outperforms the traditional approach based on Combinatorial Kalman Filter (CKF) and improves the 3-track event reconstruction efficiency by 88% in the most sensitive signal region. Furthermore, we have showed that improving the minimal vertex detection distance has significant impact on the signal sensitivity in dark photon searches in the visible decay mode. The exclusion upper limit on the dark photon mass $m_{A'}$ can be improved by up to a factor of 3 by reducing the minimal vertex distance from 5 mm to 0.1 mm.

Acknowledgements. This work was supported by the National Natural Science Foundation of China (Grant No. 11975153). We thank for the support from Key Laboratory for Particle Astrophysics and Cosmology (KLPPAC-MoE), Shanghai Key Laboratory for Particle Physics and Cosmology (SKLPPC).

References

1. Collaboration, P.X.: Limits on the luminance of dark matter from xenon recoil data. Nature **618**, 47–50 (2023). https://doi.org/10.1038/s41586-023-05982-0
2. Filippi, A., De Napoli, M.: Searching in the dark: the hunt for the dark photon. Rev. Phys. **5**, 100042 (2020)
3. Åkesson, T., et al.: [LDMX], Light Dark Matter eXperiment (LDMX), arXiv:1808.05219 [hep-ex]
4. Andreev, Y.M., et al.: [NA64], Search for Light Dark Matter with NA64 at CERN, arXiv:2307.02404 [hep-ex]

5. Chen, J., Chen, J.Y., Chen, J.F., et al.: Prospective study of light dark matter search with a newly proposed DarkSHINE experiment. Sci. China Phys. Mech. Astron. **66**, 211062 (2023). https://doi.org/10.1007/s11433-022-1983-8

6. Belyaev, A., Christensen, N.D., Pukhov, A.: CalcHEP 3.4 for collider physics within and beyond the Standard Model, Comput. Phys. Commun. **184**, 1729–1769 (2013) arXiv:1207.6082 [hep-ph]. https://doi.org/10.1016/j.cpc.2013.01.014

7. Agostinelli, S., et al.: [GEANT4], GEANT4-a simulation toolkit. Nucl. Instrum. Meth. A **506**, 250–303 (2003). https://doi.org/10.1016/S0168-9002(03)01368-8

8. Bjorken, J.D., Essig, R., Schuster, P., Toro, N.: New fixed-target experiments to search for dark gauge forces. Phys. Rev. D **80**(7), 075018 (2009)

9. Frühwirth, R.: Application of Kalman filtering to track and vertex fitting, Nuclear Instruments and Methods in Physics Research Section A: Accelerators, Spectrometers, Detectors and Associated Equipment, Volume 262, Issues 2–3, 1987, pp. 444–450, ISSN 0168–9002, https://doi.org/10.1016/0168-9002(87)90887-4

10. Fey, M., Lenssen, J.E.: Fast Graph Representation Learning with PyTorch Geometric. ArXiv. /abs/1903.02428 (2019)

11. Paszke, A., et al.: PyTorch: an imperative style, high-performance deep learning library. ArXiv, 2019, /abs/1912.01703. Accessed 15 Nov 2023

Neutrino Reconstruction in TRIDENT Based on Graph Neural Network

Cen Mo[1]([✉]), Fuyudi Zhang[2], and Liang Li[1]([✉])

[1] School of Physics and Astronomy, Shanghai Jiao Tong University, Shanghai, China
{mo_cen,liangliphy}@sjtu.edu.cn
[2] Tsung-Dao Lee Institute, Shanghai Jiao Tong University, Shanghai, China

Abstract. TRopIcal DEep-sea Neutrino Telescope (TRIDENT) is a next-generation neutrino telescope to be located in the South China Sea. With a large detector volume and the use of advanced hybrid digital optical modules (hDOMs), TRIDENT aims to discover multiple astrophysical neutrino sources and probe all-flavor neutrino physics. The reconstruction resolution of primary neutrinos is on the critical path to these scientific goals. We have developed a novel reconstruction method based on graph neural network (GNN) for TRIDENT. In this paper, we present the reconstruction performance of the GNN-based approach on both track- and shower-like neutrino events in TRIDENT.

Keywords: Neutrino telescopes · Reconstruction · Neural network

1 Introduction

In 2013, the first detection of astrophysical neutrinos was reported [1]. Unlike cosmic rays, high-energy neutrinos remain unaffected by galactic magnetic fields, preserving their trajectory and pointing directly back to their sources. This makes them ideal instruments for investigating the origins of high-energy cosmic rays.

The deep inelastic scattering (DIS) between high-energy neutrinos and nucleons in water is employed to detect astrophysical neutrinos. When ν_μ charged-current (CC) interactions occur, high-energy muons are generated and produce a kilometer-long track-like event topology. The track-like events are important in neutrino point-source searches, such as TXS 0506 [2] and NGC 1068 [3], due to their sub-degree level angular resolution. On the other hand, ν_e CC interactions and neutral-current (NC) interactions produce a cascade of secondary particles at the DIS vertex, resulting in the deposition of neutrino energy in a localized region and forming a shower-like topology. Despite their poor angular resolution, the distinctive event topology of shower-like events makes them easily distinguishable from atmospheric-neutrino background. As such, shower-like events play a critical role in the search for extended neutrino sources. For ν_τ CC interactions, a tau lepton is generated along with a hadronic cascade. The

C. Cruz et al. (Eds.): IC 2023, CCIS 2036, pp. 264–271, 2024.
https://doi.org/10.1007/978-981-97-0065-3_20

tau lepton travels some distance before decaying into a hadronic or electromagnetic cascade. If the ν_τ is sufficiently energetic, the two cascades resulting from the tau lepton's decay will be spatially separated, giving rise to a characteristic double cascade topology signature. In the case of lower energy ν_τ events, the identification of such events can be based on the presence of a double pulse in the readout waveform [4].

TRIDENT is a next-generation neutrino detector aiming to identify astrophysical neutrino sources with high precision. This telescope design incorporates hybrid digital modules (hDOMs) comprising multiple Photomultiplier Tubes (PMTs) and Silicon Photomultipliers (SiPMs). To achieve comprehensive neutrino detection capabilities, these hDOMs are strategically planned for deployment across a vast cubic kilometer region deep in the deep waters of the South China Sea.

To reconstruct the direction and energy of incoming neutrinos using information from Cherenkov photons, both machine learning-based and likelihood-based reconstruction methods have been widely used in neutrino telescopes. In IceCube, convolutional neural networks (CNNs) [5,6] and GNNs [7] have been assessed for their efficiency. KM3NeT employs likelihood methods for both ν_e and ν_μ in the reconstruction of direction and energy [8]. 3D CNNs are also implemented in KM3NeT/ORCA [9]. The likelihood method has relatively high reconstruction resolution but there is still room for improvement, especially in the case of ν_e events. The CNN approach faces challenges in handling sparse signals in TRIDENT which has a large detector volume.

In this study, we propose a novel reconstruction method based on GNN. We simulate ν_e CC and ν_μ CC events utilizing the preliminary full detector configuration of TRIDENT. Subsequently, a GNN architecture is designed and employed to facilitate the precise reconstruction of direction for the neutrino events.

2   Event Simulation

The comprehensive Monte Carlo simulations of neutrino events are executed in two steps.

In the initial step, the DIS processes are simulated in the CORSIKA8 framework [10]. To represent the TRIDENT detector region, a cylindrical volume is constructed, with a radius of 2500 m and a height of 1000 m, positioned at a depth of 2900 m below sea level. The PYTHIA8 program [11] is employed in CORSIKA8 to simulate the DIS processes. By employing different rules for ν_e and ν_μ neutrinos, accounting for their distinctive characteristics, the vertices are sampled accordingly. Given that the typical size of hadronic cascades is less than 50 m, to ensure an adequate number of Cherenkov photons for each event, the vertices of ν_e CC interactions are uniformly sampled within the detector region. Conversely, high-energy muons exhibit significant travel distances in sea water. As a result, the DIS vertices of ν_μ interactions are sampled over a larger region, the extent of which is contingent upon the energy of the muon involved. Particles decay and propagate through water until they reach the detector region.

Subsequently, the interactions of these particles and the response of detectors inside the telescope are further simulated using another dedicated program.

The detector response simulation is implemented with the Geant4 software framework [12,13]. Within a cylinder with a radius of 2000 m, a total of 1200 vertical strings are deployed in a Penrose tiling pattern, as depicted in Fig. 1. Each string comprises 20 hDOMs separated vertically by 30 m. During this process, the propagation and energy loss processes of particles are simulated. For electromagnetic cascades induced by high-energy electrons, a parameterized simulation method is employed to accelerate the simulation process, achieving a speed-up of approximately $\mathcal{O}(1000)$ times compared to traditional particle-by-particle simulations of the cascade. For the efficient handling of Cherenkov photons, all Cherenkov photons are propagated using the OptiX ray tracing framework [14] to utilize the acceleration of GPU. Finaly, the detector response to Cherenkov photons is fully simulated with Geant4.

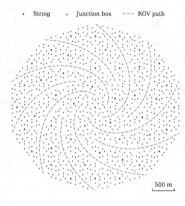

Fig. 1. Top view of TRIDENT detectors.

3   Network Architecture

In the context of neutrino telescopes, each recorded neutrino event can be intrinsically represented as a graph and can be reconstructed using GNN. For a given event, the triggered hDOMs serve as the nodes of the graph, forming an edge-less graph. The position (relative to the position of the initially triggered hDOM) and physics quantities of each hDOM comprises the coordinates and attributes of the corresponding node. To establish connections between nodes, edges are introduced such that each node is linked to its k nearest neighboring nodes. Here k is a user-defined hyperparameter. Additionally, the mean value of node attributes can serve as an indicator of the overall knowledge of a neutrino event. Thus, a neutrino event is noted as $G = \{pos_i, x_i, e_{ij}, u\}$, where

- pos_i and x_i represent the location and attributes of the i-th hDOM, respectively.
- e_{ij} represents the edge connecting the i-th and j-th hDOMs.
- u is a global attribute that describes the overall characteristics of the neutrino event.

The GNN architecture utilized in this study incorporates a fundamental building block known as the EdgeConv block, as illustrated in Fig. 2. This Edge-Conv block is adapted from the EdgeConv block employed in ParticleNet [15]. The EdgeConv block serves as a convolution-like operation. It commences by defining a latent vector for each edge e_{ij} as: $e_{ij} = \phi_\theta(u, x_i, x_j - x_i)$. Here, ϕ_θ denotes a multilayer perceptron (MLP) with trainable parameters θ. To obtain the latent vectors for the nodes, an aggregation operation is performed based on the connected edges, which is defined as: $x'_i = (\underset{j=1,\dots k}{Max} \{e_{ij}\} + x_i)$.

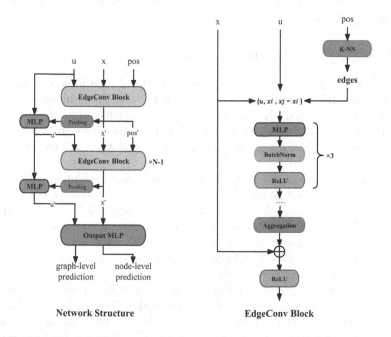

Network Structure   **EdgeConv Block**

Fig. 2. Architecture of GNN used in this study is shown on the left plot. The detailed structure of EdgeConv block is illustrated on the right plot.

The GNN architecture is built with several EdgeConv blocks. In each Edge-Conv block, the block updates the graph $G = \{pos_i, x_i, e_{ij}, u\}$ as follows:

$$x'_i = \text{EdgeConv}(G)$$
$$u' = \Phi_\Theta(u, \text{Global\_Average\_Pooling}(\{x'_i\}))$$

where Φ_Θ is another MLP with parameters Θ. By iteratively applying the Edge-Conv blocks, the GNN progressively enriches the input graph with higher-level information.

The final EdgeConv block is followed by an output MLP layer for reconstructing desired physical parameters. Depending on the context, the input to this layer can be the node attributes (x_i) for DOM-level reconstruction or the global attributes (u) for event-level reconstruction.

4   Results

The aforementioned GNN architecture is constructed with PyTorch Geometric [16,17] and is utilized to reconstruct the direction of ν_e with 100 TeV energy and ν_μ with energy ranges from 1 TeV to 1000 TeV. The ν_e samples are limited to a single energy level, as there is an insufficient number of samples in other energy ranges attributed to the slow speed of their simulation. In this section, we show the training methods and results.

4.1   Shower-Like Event Reconstruction

For the reconstruction of ν_e events, the attribute of each node is a histogram detailing the arrival times of photons at each hDOM. These histograms counts the number of received photons within every 5 ns time window. In a typical shower-like event, the majority of Cherenkov photons are received within 1000 ns. Therefore, the histograms are configured to split the 1000 ns interval into 200 time windows.

The GNN model used for shower-like event reconstruction consists of of 6 EdgeConv blocks and 2 layers of output MLP and it possesses 12,289,167 trainable paramters. The samples are divided into training (130k samples) and validation (10k samples) sets during the training session. The model is trained to directly predict the direction of ν_e, n_{ν_e}, with MSELoss as loss function. Subsequently, the trained models undergo testing using an additional 130k samples, yielding the results presented below.

The training result, shown in Fig. 3, demonstrates the angular error between the true ν_e direction and the reconstructed direction. The median angular error, as is represented by the red line, is about 1.3°. For comparison, the median angular error of 100 TeV ν_e events using the traditional likelihood method is found to be about 1.7° [8].

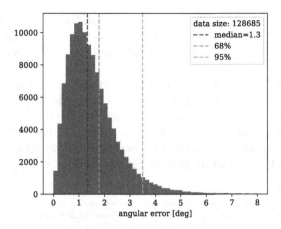

Fig. 3. The angular error for 100 TeV ν_e CC events. The red line represents the median angular error. The orange and blue lines exhibit the 68% and 90% quantiles. (Color figure online)

4.2   Track-Like Event Reconstruction

Muons in ν_μ CC events leave tracks within the telescope. The photons received by the hDOM with early arrival time are more likely to reach the hDOM with less scattering along the travelling path. Therefore the arrival time of the photons provides useful information when reconstructing the neutrino direction and it is taken as a node attribute. To obtain the information about the distance from the hDOM to muons, the number of photons received by each hDOM is also taken as a node attribute.

The GNN model used for track-like event reconstruction is made of 5 Edge-Conv blocks followed by 2 MLP layers (7,966,005 trainable parameters). The model is trained with MSELoss as loss function to predict the photon emission positions, r_i (as illustrated in Fig. 4), for all triggered hDOMs. Subsequently, the direction of the muon is then reconstructed using a linear fit on the r_i positions.

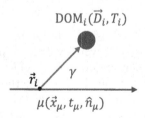

Fig. 4. Muon emits Cherenkov photon at r_i and triggers DOM_i.

In the low-energy range, graphs may have as few as 2 nodes, making it challenging to train the GNN. To address this, all training samples must consist of more than 7 nodes. The dataset is partitioned into training (210k samples) and validation (70k samples) sets. In the evaluation phase, the trained models are tested with an additional 220k samples, each comprising more than 2 nodes, to generate the results.

As the result, the distribution of angular error between the true ν_μ direction and the reconstructed direction is shown in Fig. 5. The red line represents the median angular error and the color bands exhibits the 68% and 90% quantiles. The model achieves an angular resolution at the 0.1° level for ν_μ events with sufficiently high energy. The angular resolution using the likelihood method also falls below 0.1° for sufficiently high energy events [8].

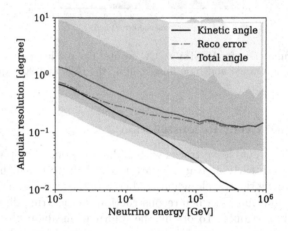

Fig. 5. The angular resolution of ν_μ CC events as a function of energy. The median angle between the reconstructed track and the true direction of μ and ν_μ is visualized by the green and red line, respectively. Color bands exhibits the 68% and 90% quantiles. Black line represents the median angle between direction of μ and ν_μ. (Color figure online)

5   Summary

In this paper, a GNN-based reconstruction method is proposed to reconstruct the direction of ν_e CC and ν_μ CC events with high precision in TRIDENT. For shower-like events, the median angular error achieved by this method is 1.3°, which significantly outperforms the likelihood method result by 75%. For track-like events, the median angular error reaches 0.1° when the neutrino energy is sufficiently high, which gives a comparable performance with the likelihood method.

For the next step, we plan to extend the GNN-based reconstruction method to reconstruct both the direction and energy of neutrino events in a wide kinematic range. We will also improve the robustness of the method against experimental uncertainties and noises.

Acknowledgements. We thank for the support from Key Laboratory for Particle Astrophysics and Cosmology (KLPPAC-MoE) and Shanghai Key Laboratory for Particle Physics and Cosmology (SKLPPC). This work was supported by the Oceanic Interdisciplinary Program of Shanghai Jiao Tong University (project number SL2022MS020).

References

1. Abbasi, R., et al.: Evidence for high-energy extraterrestrial neutrinos at the Ice-Cube detector. Science **342**(6161), 1242856 (2013)
2. Abbasi, R., et al.: Neutrino emission from the direction of the blazar TXS 0506+056 prior to the IceCube-170922a alert. Science **361**(6398), 147–151 (2018)
3. Abbasi, R., et al.: Evidence for neutrino emission from the nearby active galaxy NGC 1068. Science **378**(6619), 538–543 (2022)
4. Wille, L., Xu, D.: Astrophysical tau neutrino identification with IceCube waveforms (2019)
5. Abbasi, R., et al.: A convolutional neural network based cascade reconstruction for the IceCube neutrino observatory. J. Instrum. **16**(7), P07041 (2021)
6. Yu, F.J. Lazar, J., Argüelles, C.A.: Trigger-level event reconstruction for neutrino telescopes using sparse submanifold convolutional neural networks (2023). arXiv:2303.08812
7. Abbasi, R., et al.: Graph neural networks for low-energy event classification & reconstruction in IceCube. J. Instrum. **17**, P11003 (2022)
8. Melis, K., Heijboer, A., de Jong, M.: KM3NeT/ARCA event reconstruction algorithms, PoS. In: ICRC2017, p. 950 (2018)
9. Aiello, S., et al.: Event reconstruction for KM3net/ORCA using convolutional neural networks. J. Instrum. **15**, P10005–P10005 (2020)
10. Huege, T.: CORSIKA 8 - the next-generation air shower simulation framework. arXiv:2208.14240 (2022)
11. Bierlich, C., et al.: A comprehensive guide to the physics and usage of PYTHIA 8.3. arXiv:2203.11601 (2022)
12. Agostinelli, S., et al.: GEANT4 Collaboration. Nucl. Instrum. Meth. A **506**(3), 250–303 (2003)
13. Allison, J., et al.: GEANT4 Collaboration. IEEE Trans. Nuclear Sci. **53**(1), 270–278 (2006)
14. Blyth, S.: Opticks?: GPU optical photon simulation for particle physics using NVIDIA® OptiXTM. EPJ Web Conf. **214**, 02027 (2019)
15. Qu, H., Gouskos, L.: Jet tagging via particle clouds. Phys. Rev. D. **101**, 056019 (2020)
16. Paszke, A.: PyTorch: a imperative style, high-performance deep learning library. In: Advances in Neural Information Processing Systems, vol. 32, pp. 8024–8035, Curran Associates Inc. (2019)
17. Fey, M., Lenssen, J.E.: Fast graph representation learning with PyTorch Geometric. In: ICLR Workshop on Representation Learning on Graphs and Manifolds (2019)

Charged Particle Reconstruction for Future High Energy Colliders with Quantum Approximate Optimization Algorithm

Hideki Okawa[(✉)] [iD]

Institute of High Energy Physics, Chinese Academy of Sciences, Shijingshan 100049, Beijing, China
okawa@ihep.ac.cn

Abstract. Usage of cutting-edge artificial intelligence will be the baseline at future high energy colliders such as the High-Luminosity Large Hadron Collider, to cope with the enormously increasing demand of the computing resources. The rapid development of quantum machine learning could bring in further paradigm-shifting improvement to this challenge. One of the two highest CPU-consuming components, the charged particle reconstruction, the so-called track reconstruction, can be considered as a quadratic unconstrained binary optimization (QUBO) problem. The Quantum Approximate Optimization Algorithm (QAOA) is one of the most promising algorithms to solve such combinatorial problems and to seek for a quantum advantage in the era of the Noisy Intermediate-Scale Quantum computers. It is found that the QAOA shows promising performance and demonstrated itself as one of the candidates for the track reconstruction using quantum computers.

Keywords: Quantum Machine Learning · QUBO · QAOA · High Energy Physics · track reconstruction

1 Introduction

High energy physics aims to unveil the laws of the fundamental building blocks of matter, the elementary particles, and their interactions. High energy colliders have been one of the most promising approaches to discover new particles and deepen understanding of the underlying physics through precise measurements. After the revolutionary discovery of the Higgs boson in 2012 [9,23] at the ATLAS [8] and CMS [22] experiments at the Large Hadron Collider (LHC) [25], we are entering the precision measurement era of the Higgs sector, first of all, to be pursued at the High-Luminosity LHC (HL-LHC) [34], and to be followed by future colliders being proposed, such as the Circular Electron Positron Collider (CEPC) [17–20] to be hosted in China.

At the HL-LHC, we will enter the "Exa-byte" era, where the annual computing cost will increase by a factor of 10 to 20. Without various innovations,

C. Cruz et al. (Eds.): IC 2023, CCIS 2036, pp. 272–283, 2024.
https://doi.org/10.1007/978-981-97-0065-3_21

the experiment will not be able to operate. Usage of the Graphical Processing Units (GPUs) and other state-of-the-art artificial intelligence (AI) technologies such as deep learning will be the baseline at the HL-LHC. However, the emerging rapid development of quantum computing and implementation of machine learning techniques in such computers could bring in another leap. Two of the highly CPU consuming components at the LHC and HL-LHC are (1) the charged particle reconstruction, the so-called track reconstruction, for both in data and simulation and (2) simulation of electromagnetic and hadronic shower development in the calorimeter. Development of quantum machine learning techniques to overcome such challenges would not only be important for the HL-LHC, but also for a future Higgs factory CEPC and a next generation discovery machine the Super Proton-Proton Collider (SppC) [17,18] to be hosted in China as well as other such colliders under consideration in the world.

2   Track Reconstruction

Track reconstruction or tracking is the standard procedure in the collider experiments to identify charged particles traversing the detector and to measure their momenta. The curvature of particle trajectories (tracks) bent in a magnetic field will provide the momentum information. Tracks are reconstructed from hits in the silicon detectors, which have many irrelevant hits from secondary particles and detector noise, and require sophisticated algorithms. Tracking is one of the most crucial components of reconstruction in the collider experiments.

At the HL-LHC, additional proton-proton interactions per bunch crossing, the so-called pileup, becomes exceedingly high, and the CPU time required to run the track reconstruction explodes with pileup [10,21].

2.1   Current Classical Benchmarks

The Kalman Filter technique [30] has been often used as a standard algorithm to reconstruct the tracks. It is implemented in A Common Tracking Software (ACTS) [1], for example. Seeding from the inner detector layers, the tracks are extrapolated to predict the next hit and iterated to find the best quality combination. It is a well established procedure and has excellent performance but suffers from the computing time, especially when the track multiplicity per event becomes high.

Recently, usage of the graph neural network (GNN) is actively investigated at the LHC [35,37] as well as other collider experiment including the tau-lepton and charm-quark factory in China: Beijing Spectrometer (BES) III [14]. Hits in the silicon detectors can be regarded as "nodes" of the graphs, and segments reconstructed by connecting the hits can be considered as "edges". GNN-based algorithms provide compatible performance as the Kalman Filter, but the computing time scales approximately linearly instead of exponentially with the number of tracks. The GNN is thus considered to be one of the new standards in the era of the HL-LHC.

2.2   Quantum Approach

There have been several studies to run the track reconstruction with quantum computers. First of all, doublets are formed by connecting two hits in the silicon detectors. Then, triplets, segments with three silicon hits, are formed by connecting the doublets. Then, triplets are connected to reconstruct the tracks, by evaluating the consistency of the triplet momenta. Such procedure can be considered as a quadratic unconstrained binary optimization (QUBO) problem:

$$O(a, b, T) = \sum_{i=1}^{N} a_i T_i + \sum_{i=1}^{N} \sum_{j<i}^{N} b_{ij} T_i T_j, \tag{1}$$

where N is the number of triplets, T_i and T_j corresponds to the triplets and takes the value of either zero or one, a_i are the bias weights to evaluate the quality of the triplets, and b_{ij} are the coefficients that quantify the compatibility of two triplets ($b_{ij} = 0$ if no shared hit, $= 1$ if there is any conflict, and $= -S_{ij}$ if two hits are shared between the triplets). The coefficients $-S_{ij}$ quantify the consistency of the two triplet momenta by [12]:

$$S_{ij} = \frac{1 - \frac{1}{2}(|\delta(q/p_{Ti}, q/p_{Tj})| + max(\delta\theta_i, \delta\theta_j))}{(1 + H_i + H_j)^2}, \tag{2}$$

where δ is the difference between the curvature q/p_T or angle θ of the two triplets and H_i is the number of holes in the triplet.

The bias weights a_i have significant impact on the Hamiltonian energy landscape and thus on the track reconstruction and computation speed. They can be parameterized as:

$$a_i = \alpha \left(1 - e^{\frac{|d_0|}{\gamma}}\right) + \beta \left(1 - e^{\frac{|z_0|}{\lambda}}\right), \tag{3}$$

where d_0 and z_0 are the transverse and longitudinal displacements of the triplets from the primary vertex (the most significant proton-proton collision point in an event) and α, β, γ and λ are tunable parameters, which are taken to be 0.5, 0.2, 1.0 and 0.5, optimized in a previous study [39] for the same dataset (see Sect. 3).

The first quantum tracking studies [12,47] have been pursued with quantum annealing computers. The quantum annealer looks for the global minimum of a given function with quantum tunneling. It is a natural machine to solve QUBO problems by searching for the ground state of a Hamiltonian. As the size of the QUBO generally exceeds the number of available qubits in the current era of the Noisy Intermediate-Scale Quantum (NISQ) computers [42], the QUBO is split into subsets (sub-QUBOs) to search for the ground state. Performance evaluation between the D-Wave annealing machine and a classical emulation using a digital annealer is mentioned in Ref. [43]. The performance dependence against the size of the sub-QUBOs is evaluated in Ref. [44].

To exploit the quantum gate computers, a QUBO can be mapped to an Ising Hamiltonian and be solved using the Variational Quantum Eigensolver (VQE),

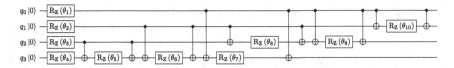

Fig. 1. Layout of the QAOA with the Quantum Alternative Operator Ansatz. The number of qubits is reduced to four for the demonstration purpose.

Quantum Approximate Optimization Algorithm (QAOA) [28], or other similar algorithms. Previous studies conducted at the LUXE experiment considered the TwoLocal ansatz with the R_Y gates and a circular CNOT entangling pattern [24, 31]. The LHCb Collaboration has also looked into the QUBO approach with the gate computer using the Harrow-Hassadim-Lloyd (HHL) algorithm [40].

Another distinctive approach is pursued with the hybrid quantum classical GNN, where some components of the classical GNN method [35, 37] are replaced by the Variational Quantum Layers. This hybrid model performs similarly to the classical approach, as also confirmed by Refs. [24, 31]. This hybrid quantum-classical GNN approach is out of the scope of this paper, and will not be mentioned further.

3   Datasets

In this study, the dataset from the TrackML Challenge is adopted [6,7]. It is an open source dataset representing the conditions for the ATLAS and CMS experiments at the HL-LHC. It assumes the particle multiplicities expected with the pileup condition of $\langle \mu \rangle$=200. The dataset is simplified by focusing on the barrel (central) region of the detector, thus removing the hits in the end-cap region, which is a common approach often adopted for tracking studies for the HL-LHC.

The QUBO matrix, namely the bias weights a_i and the compatibility coefficients b_{ij} as defined in Eq. 1 is extracted from the dataset using the hepqpr-qallase library [2]. The QUBO matrix is pretty sparse, as is the nature of the collider experiments and track reconstruction.

4   Methodology

The QAOA is considered in this study. It is a hybrid quantum-classical method, designed to solve combinatorial optimization problems. Even at the lowest circuit depth, the QAOA cannot efficiently be simulated by classical computers, but has non-trivial guarantees on the performance [28,29]. For cases where the minimum energy spectrum gap is very small, the computing time required in the quantum annealers is very long to remain adiabatic. The QAOA is known to outperform adiabatic quantum annealing by several orders of magnitude in such circumstances [45]. Thus, the QAOA is one of the most promising algorithm to seek for quantum advantage in the era of the NISQ computers.

Its libraries are implemented in pyqpanda-algorithm [3] by Origin Quantum (Chinese name: Benyuan). It adopts the Quantum Alternative Operator Ansatz [32]. The schematic layout of the quantum circuit is demonstrated in Fig. 1. For the actual hardware computation, the 6-qubit machine Wuyuan is used through the cloud service [4].

4.1   Optimization of QAOA Implementation

To optimize the conditions to run the QAOA, a 6×6 matrix is extracted from a TrackML QUBO to match the available number of qubits in Wuyuan. Performance is evaluated for two loss functions: CVaR [13] and Gibbs [38]; three optimizers which can handle bounds on the variables: L-BFGS-B [16,46], Sequential Least SQuares Programming (SLSQP) [36], and a Truncated Newtonian algorithm (TNC) [16]; and the number of layers of the QAOA.

Figure 2 shows the probability of finding the correct minimum energy with various loss functions, optimizers and the number of the QAOA layers for the quantum simulator and the actual quantum hardware. The probabilities tend to be low for the shallow QAOA, which is consistent with what is reported in Refs. [15,26,27,33]. Among the three optimizers, L-BFGS-B shows the best performance, followed by SLSQP. TNC shows largely degraded probabilities without much improvement against the number of layers, thus, is not presented in the figure. There is no significant difference between the CVaR and Gibbs loss functions, thus CVaR is adopted onward in this work. The real hardware shows compatible performance as the quantum simulator, and there is no sign of degradation even for deep-layer QAOAs up to 20 layers. This is in contrast to what is observed in Ref. [41] but consistent with Ref. [44]. This could be due to the difference in the QUBO considered, which is largely sparse in this paper as well as in Ref. [44].

It is worth emphasizing that in the actual implementation, a single QAOA job will run multiple shots and the combination with the highest probability will be selected. Thus, the accuracy of obtaining the correct answer is much higher than the probability itself, reaching 100% already at 5 layers as is presented in Fig. 3. In the following sections, seven QAOA layers are adopted, where the probability reaches the plateau and the accuracy is compatible with 100% within the statistical uncertainty.

4.2   Sub-QUBO Method

The number of triplet candidates define the number of qubits required. Obviously the quantum computing resources currently available in the NISQ era cannot cover the full QUBO for tracking. Thus, the QUBO is split into sub-QUBOs of size 6×6 to match the Wuyuan hardware.

There are various sub-QUBO algorithms proposed: qbsolv [5] (now in the dwave-hybrid library), for example. In this paper, a sub-QUBO method using multiple solution instances [11] is adopted. This method has a strong theoretical justification, whereas other existing approaches are heuristic and lack in

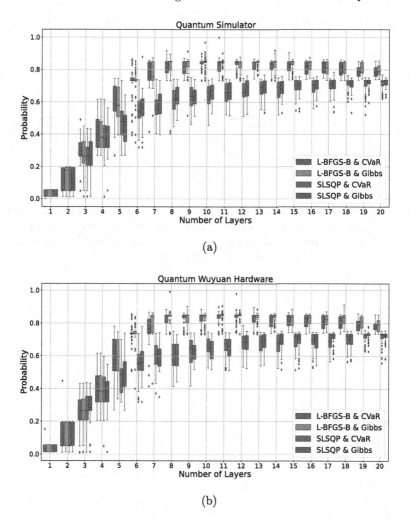

Fig. 2. Probability of finding the correct combination with the L-BFGS-B or SLSQP optimizers and the CVaR or Gibbs loss functions, presented against the number of QAOA layers for the quantum simulator (a) and Wuyuan hardware (b).

such a foundation [11]. In this multiple solution instance method, three parameters (N_I, N_E, N_S) are considered. First of all, N_I quasi-optimal solutions are extracted from the full-QUBO classically. Then N_S solution instances from N_I are randomly selected. The method focuses on a particular binary variable T_i (see Eq. 1), rank them in accordance to how much they vary over N_S solution instances. Highly varying T_i will be included in the sub-QUBO model. The pick-up process of N_S solutions from the QAOA is repeated N_E times and N_E sub-QUBO models are considered. Finally, a pool of N_I solutions is returned and the best solution is chosen.

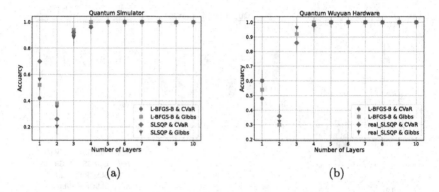

Fig. 3. Accuracy of finding the correct combination in a single job with the L-BFGS-B or SLSQP optimizers and the CVaR or Gibbs loss functions, presented against the number of QAOA layers for the quantum simulator (a) and real hardware (b). The statistical uncertainty is presented in the error bars.

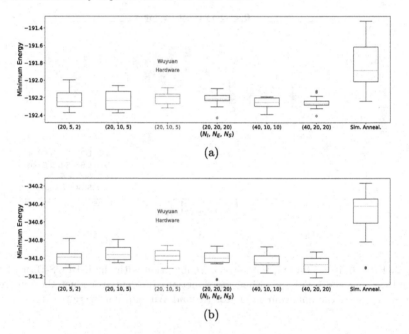

Fig. 4. Minimum energy estimated by the multiple solution instance method for various (N_I, N_E, N_S) parameters with the quantum simulator and Wuyuan hardware compared to a classical simulated thermal annealing. Two examples with the full QUBO size of 778×778 (a) and 1431×1431 (b) are presented.

Figure 4 shows the presumed minimum energy found by the multiple solution instance method for various sets of (N_I, N_E, N_S) parameters with the quantum simulator compared to a simple classical optimizer with the simulated

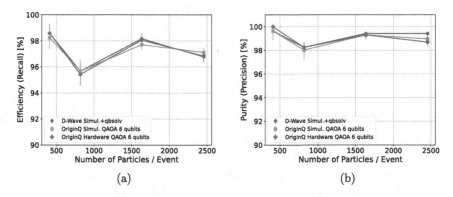

Fig. 5. Efficiency (a) and purity (b) as a function of particle multiplicity utilizing the QAOA simulator, Wuyuan hardware or the D-Wave simulator with qbsolv.

thermal annealing. Those two examples have the full QUBO size of 778×778 and 1431×1431 respectively, and correspond to the first two points later presented in Fig. 5 in Sect. 5. The output from the real quantum hardware Wuyuan is also presented for one set of the (N_I, N_E, N_S) parameters. It is clearly observed that this sub-QUBO method using the QAOA succeeds in obtaining lower energy than the classical simulated annealing. There is no obvious dependence in performance on the three parameters. These two features of the method are consistent with what have been reported in the original proposal of this sub-QUBO method [11]. As the computing time increases with the size of the parameters, $(N_I, N_E, N_S) = (20, 10, 5)$ is adopted for the final evaluation on the tracking performance, which will be summarized in the next section. It is also promising to see that there is no degradation in performance with the actual quantum hardware despite the presence of noise.

5   Results and Discussions

Several events are selected containing a few thousands particles and noise. The track reconstruction is pursued by running the sub-QUBO method with the three parameters (N_I, N_E, N_S) defined in the previous section. The QAOA utilizes seven layers and the CVaR loss function, and split into sub-QUBOs with the size of six qubits.

Performance of the track reconstruction is evaluated in terms of efficiency (recall) and purity (precision). They are defined as the following:

$$\text{Efficiency} = \frac{TP}{TP + FN} = \frac{\#\text{ of matched reconstructed doublets}}{\#\text{ of true doublets}}, \quad (4)$$

$$\text{Purity} = \frac{TP}{TP + FP} = \frac{\#\text{ of matched reconstructed doublets}}{\#\text{ of all reconstructed doublets}}, \quad (5)$$

where TP is the true positives, FN the false negatives, and FP the false positives. TP corresponds to the number of reconstructed doublets matching to

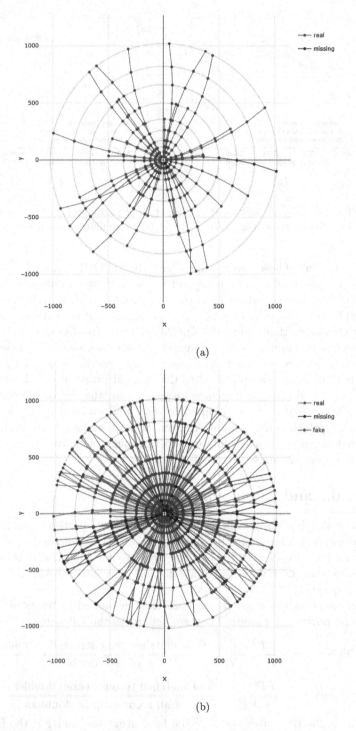

Fig. 6. Event displays of tracks reconstructed with the QAOA utilizing the Wuyuan hardware. They are generated with hepqpr-qallase library [2].

the correct (true) doublets. FN is the number of true doublets that are not reconstructed, thus $TP + FN$ is simply the number of true doublets. FP is the number of reconstructed doublets that do not match to the true doublets. In the high energy physics terminology, they are called the "fake" doublets.

Figure 5 shows the track efficiency and purity using the QAOA with the quantum simulator or Wuyuan hardware. The performance is compared to the D-Wave simulator approach implemented in Ref. [12]. The D-Wave simulator with qbsolv and NEAL show almost indistinguishable results, so only the qbsolv values are shown in the figure. The track reconstruction performance with QAOA is compatible with the D-Wave annealing approach, and there is no sign of degradation in the actual hardware. Figure 6 presents two event displays from the lowest and highest particle density events considered in Fig. 5.

This work demonstrates that the QAOA is a promising candidate to be considered for the track reconstruction at the future colliders. With utilizing a robust sub-QUBO method and the QAOA, high performance can be achieved even with a small size of qubits. The compatible performance obtained with the real quantum hardware further supports its potential usability.

It is not yet at the stage to evaluate the quantum advantage, since the QAOA with six qubits can run quickly on the quantum simulator as well with compatible performance. The non-trivial impact is to be investigated with higher qubit conditions, which would be beyond the reach of the quantum simulator and is left for future studies.

Acknowledgements. The author would like to thank Federico Meloni and David Spataro for discussions regarding the quantum tracking and Andreas Salzburger for his suggestion on the TrackML dataset. The author would also like to thank Ziwei Cui and Lei Li from Origin Quantum (Benyuan) for various feedback. The author is supported by NSFC under contract No. 12075060. This work is benefited by the libraries and quantum computing resources provided by Origin Quantum.

References

1. https://acts.readthedocs.io/en/latest/
2. https://github.com/derlin/hepqpr-qallse
3. https://pyqpanda-algorithm-tutorial.readthedocs.io/en/latest/
4. https://qcloud.originqc.com.cn/
5. https://docs.ocean.dwavesys.com/projects/qbsolv/en/latest/index.html
6. Amrouche, S., et al.: The tracking machine learning challenge: accuracy phase. In: Escalera, S., Herbrich, R. (eds.) The NeurIPS 2018 Competition. TSSCML, pp. 231–264. Springer, Cham (2020). https://doi.org/10.1007/978-3-030-29135-8_9
7. Amrouche, S., et al.: The tracking machine learning challenge: throughput phase. Comput. Softw. Big Sci. **7**(1), 1 (2023). https://doi.org/10.1007/s41781-023-00094-w
8. ATLAS Collaboration: The ATLAS experiment at the CERN large hadron collider. JINST **3**, S08003 (2008). https://doi.org/10.1088/1748-0221/3/08/S08003
9. ATLAS Collaboration: Observation of a new particle in the search for the standard model higgs boson with the ATLAS detector at the LHC. Phys. Lett. **B 716**, 1–29 (2012). https://doi.org/10.1016/j.physletb.2012.08.020

10. ATLAS Collaboration: Fast track reconstruction for HL-LHC. ATL-PHYS-PUB-2019-041 (2019). https://cds.cern.ch/record/2693670

11. Atobe, Y., Tawada, M., Togawa, N.: Hybrid annealing method based on subQUBO model extraction with multiple solution instances. IEEE Trans. Comp. **71**(10), 2606 (2022)

12. Bapst, F., et al.: A pattern recognition algorithm for quantum annealers. Comput. Softw. Big Sci. **4**(1), 1–7 (2019). https://doi.org/10.1007/s41781-019-0032-5

13. Barkoutsos, P.K., Nannicini, G., Robert, A., Tavernelli, I., Woerner, S.: Improving variational quantum optimization using CVaR. Quantum **4**, 256 (2020). https://doi.org/10.22331/q-2020-04-20-256

14. BES IIII Collaboration: Design and construction of the BESIII detector. Nucl. Instrum. Meth. A **614**, 345–399 (2010). https://doi.org/10.1016/j.nima.2009.12.050

15. Bravyi, S., Kliesch, A., Koenig, R., Tang, E.: Obstacles to variational quantum optimization from symmetry protection. Phys. Rev. Lett. **125**, 260505 (2020). https://doi.org/10.1103/PhysRevLett.125.260505

16. Byrd, R.H., Lu, P., Nocedal, J., Zhu, C.: A limited memory algorithm for bound constrained optimization. SIAM J. Sci. Comput. **16**(5), 1190–1208 (1995)

17. CEPC-SPPC Study Group: CEPC-SPPC Preliminary Conceptual Design Report. 1. Physics and Detector. IHEP-CEPC-DR-2015-01, IHEP-TH-2015-01, IHEP-EP-2015-01 (2015)

18. CEPC-SPPC Study Group: CEPC-SPPC Preliminary Conceptual Design Report. 2. Accelerator. IHEP-CEPC-DR-2015-01, IHEP-AC-2015-01 (2015)

19. CEPC Study Group: CEPC Conceptual Design Report: Volume 1 - Accelerator. IHEP-CEPC-DR-2018-01, IHEP-AC-2018-01 (2018)

20. CEPC Study Group: CEPC Conceptual Design Report: Volume 2 - Physics & Detector. IHEP-CEPC-DR-2018-02, IHEP-EP-2018-01, IHEP-TH-2018-01 (2018)

21. Cerati, G.B.: Tracking and vertexing algorithms at high pileup. Conference Report CMS-CR-2014-345 (2014). https://cds.cern.ch/record/1966040

22. CMS Collaboration: The CMS experiment at the CERN LHC. JINST **3**, S08004 (2008). https://doi.org/10.1088/1748-0221/3/08/S08004

23. CMS Collaboration: Observation of a new boson at a Mass of 125 GeV with the CMS experiment at the LHC. Phys. Lett. B **716**, 30–61 (2012). https://doi.org/10.1016/j.physletb.2012.08.021

24. Crippa, A., et al.: Quantum algorithms for charged particle track reconstruction in the LUXE experiment. DESY-23-045, MIT-CTP/5481. arXiv:2304.01690 (2023)

25. Evans, L., Bryant, P.: LHC machine. JINST **3**, S08001 (2008). https://doi.org/10.1088/1748-0221/3/08/S08001

26. Farhi, E., Gamarnik, D., Gutmann, S.: The quantum approximate optimization algorithm needs to see the whole graph: a typical case. arXiv:2004.09002 (2020)

27. Farhi, E., Gamarnik, D., Gutmann, S.: The quantum approximate optimization algorithm needs to see the whole graph: worst case examples. arXiv:2005.08747 (2020)

28. Farhi, E., Goldstone, J., Gutmann, S.: A quantum approximate optimization algorithm. arXiv:1411.4028 (2014)

29. Farhi, E., Goldstone, J., Gutmann, S.: A quantum approximate optimization algorithm applied to a bounded occurrence constraint problem. arXiv:1412.6062 (2015)

30. Fruhwirth, R.: Application of Kalman filtering to track and vertex fitting. Nucl. Instrum. Meth. A **262**, 444–450 (1987). https://doi.org/10.1016/0168-9002(87)90887-4

31. Funcke, L., et al.: Studying quantum algorithms for particle track reconstruction in the LUXE experiment. J. Phys: Conf. Ser. **2438**(1), 012127 (2023). https://doi.org/10.1088/1742-6596/2438/1/012127
32. Hadfield, S., Wang, Z., O'Gorman, B., Rieffel, E.G., Venturelli, D., Biswas, R.: From the quantum approximate optimization algorithm to a quantum alternating operator ansatz. Algorithms **12**(2) (2019). https://doi.org/10.3390/a12020034
33. Hastings, M.B.: Classical and quantum bounded depth approximation algorithms. arXiv:1905.07047 (2019)
34. Béjar Alonso, I., et al. (Eds.): High-Luminosity Large Hadron Collider (HL-LHC): Technical design report. CERN Yellow Reports: Monographs, CERN, Geneva (2020). https://doi.org/10.23731/CYRM-2020-0010, https://cds.cern.ch/record/2749422
35. Ju, X., et al.: Graph neural networks for particle reconstruction in high energy physics detectors. In: 33rd Annual Conference on Neural Information Processing Systems (2020)
36. Kraft, D.: A software package for sequential quadratic programming. Tech. Rep. DFVLR-FB, 88–28, DLR German Aerospace Center - Institute for Flight Mechanics, Koln, Germany (1988)
37. Lazar, A., et al.: Accelerating the Inference of the Exa. TrkX Pipeline. J. Phys: Conf. Ser. **2438**(1), 012008 (2023). https://doi.org/10.1088/1742-6596/2438/1/012008
38. Li, L., Fan, M., Coram, M., Riley, P., Leichenauer, S.: Quantum optimization with a novel Gibbs objective function and ansatz architecture search. Phys. Rev. Res. **2**, 023074 (2020). https://doi.org/10.1103/PhysRevResearch.2.023074
39. Linder, L.: Using a quantum annealer for particle tracking at LHC, Master Thesis at EPFL (2019)
40. Nicotra, D., et al.: A quantum algorithm for track reconstruction in the LHCb vertex detector (2023)
41. Pellow-Jarman, A., McFarthing, S., Sinayskiy, I., Pillay, A., Petruccione, F.: QAOA Performance in noisy devices: the effect of classical optimizers and ansatz depth. arXiv:2307.10149 (2023)
42. Preskill, J.: Quantum computing in the NISQ era and beyond. Quantum **2**, 79 (2018). https://doi.org/10.22331/q-2018-08-06-79
43. Saito, M., et al.: Quantum annealing algorithms for track pattern recognition. EPJ Web Conf. **245**, 10006 (2020). https://doi.org/10.1051/epjconf/202024510006
44. Schwägerl, T., et al.: Particle track reconstruction with noisy intermediate-scale quantum computers. arXiv:2303.13249 (2023)
45. Zhou, L., Wang, S.T., Choi, S., Pichler, H., Lukin, M.D.: Quantum approximate optimization algorithm: performance, mechanism, and implementation on near-term devices. Phys. Rev. X **10**, 021067 (2020). https://doi.org/10.1103/PhysRevX.10.021067
46. Zhu, C., Byrd, R.H., Lu, P., Nocedal, J.: Algorithm 778: L-BFGS-B: Fortran subroutines for large-scale bound-constrained optimization. ACM Trans. Math. Softw. **23**(4), 550–560 (1997)
47. Zlokapa, A., et al.: Charged particle tracking with quantum annealing-inspired optimization. Q. Mach. Intell. **3**, 27 (2021). https://doi.org/10.1007/s42484-021-00054-w

AI for Law

A Levy Scheme
for User-Generated-Content Platforms
and Its Implication for Generative
AI Providers

Weijie Huang👤 and Xi Chen(✉)👤

Law School, Shenzhen University, Shenzhen 518065, Guangdong, China
chenxi8601@126.com

Abstract. The democratization of the technology to re-create content
and make content publicly available online has spurred the wave of UGC
(user-generated-content). Nevertheless, UGC has faced a serious dilemma
under current copyright law. Copyright owners can hardly gain remuneration for the works on which UGC is based. UGC creators are at
risk of receiving cease-and-desist letters or even being sued, although
they have not gained profit from UGC creation. UGC platforms have
substantially profited from UGC but can get exempted from copyright
infringement liability. By exploring the intermediary-oriented approach,
this paper imports a non-commercial UGC creation levy on UGC platforms. Under the proposed scheme, copyright owners can gain remuneration while non-commercial UGC creators are free to create UGC
based on copyrighted works. The proposed scheme also provides some
insight for how copyright law addresses the burgeoning AIGC (artificial
intelligence-generated-content).

Keywords: Copyright law · User-generated-content · Safe harbor
doctrine · Intermediary-oriented approach · Non-commercial UGC
creation levy · Generative AI

1 Introduction

The widely available digital tools that allow ordinary users to engage in cultural creation, the hyper-connectivity of the Internet and the increasing amount
of spare time have resulted in what Clay Shirky [21] called "cognitive surplus".
This in turn has given rise to an unprecedented wave of user-generated-content
(UGC). There are two categories of UGC. One type of UGC, named user-
authored-content, is independently created by the creator, rather than based
on other people's works. The other type, named user-derived content, is based
on the transformation or combination of pre-existing copyrighted works, such
as samplings, fan fiction, and remixes. This paper mainly discusses user-derived

Supported by the Department of Education of Guangdong Province.

content. Although UGC is normally created for non-commercial purposes, UGC platforms have greatly profited from it. TikTok, a UGC platform for creating and sharing short videos, was launched in China in 2016. It rapidly expanded to overseas markets, ranking as the most frequently downloaded app. Tik Tok was valued over US\$75 billion in 2022. Many other platforms, such as Instagram, Snapchat, Vimeo, and Pinterest, have sprung up within the last decade.

The wave of UGC creation has had a significant social and cultural impact, such as increased user autonomy and participation [7], the democratization of access to the media and accentuated cultural and social fragmentation [23]. UGC has also triggered fierce challenges to current legal mechanisms, especially copyright law [16]. The basic principle of copyright law is that if you want to use a work created by others, you should ask the copyright owner for permission, unless your use comes within the umbrella of "fair use" or "compulsory license" [11]. The exclusive-right approach intends to prevent free riding to provide an economic incentive for authors to create copyrighted works. Nevertheless, most UGC creators, who create UGC just for fun rather than for profit, have not acquired authorization from copyright owners and thereby constitute infringement. However, it seems unfair to punish the mass of non-commercial UGC creators [13].

Our primary aim is to propose a non-commercial levy scheme to address the challenges mentioned above. This paper proceeds as follows: Sect. 2 analyzes how the current copyright law leads to the dilemma of UGC creation because of the safe harbor doctrine. Section 3 discusses the historical experience of copyright law in regulating intermediaries, followed by an explanation of how the intermediary-oriented approach has promoted the production of copyrighted works on one hand and protected the freedom of end consumers to use copyrighted works on the other hand. Drawing on the historical intermediary-oriented approach, Sect. 4 introduces a non-commercial levy scheme which requires UGC platforms to remunerate copyright owners and exempts non-commercial UGC creators. Section 5 examines the justification of the proposed levy scheme. Section 6 concludes this paper with an outlook for future work.

2   The UGC Dilemma Due to the Lack of Regulation on Intermediaries

With the increasing popularity of UGC based on the power of peer distribution in the network, UGC has begun to attract copyright owners' attention. For example, a US blogger, Andy Baio, received a cease-and-desist letter requiring him to pay US\$ 175,000 in settlement because he posted an eight-bit tribute to Miles Davis' "Kind Of Blue". The chilling effect of copyright law has had an even more negative impact on average, non-commercial UGC creators than on creators of popular UGC. Ordinary users, who do not have experience in dealing with case-and-desist letters, are more likely to pay the required money to settle down the issue rather than taking the litigation cost. Guilda Rostama [17] called the legal uncertainty of UGC creation "the source of a great deal of frustration among members of the public". Many scholarly writings have addressed the UGC dilemma either by accommodating UGC creation to fair use, or by suggesting a compulsory licensing scheme for UGC creation [8]. However, regardless of the feasibility of these proposals, these proposals regulate the relationship between UGC creators and copyright owners, but ignore another important party in UGC creation, namely, UGC platforms.

UGC platforms, UGC creators, and copyright owners formulate the triangular relationship of UGC creation. It is UGC platforms, not UGC creators, that have gained huge profit from UGC. Nicholas Carr [3] analogised this phenomenon to digital sharecropping: "the distribution of production into the hands of the many and the concentration of the economic rewards into the hands of the few". The UGC creators who have not profited from UGC surely would not pay for the copyrighted works they use, which in turn prevents copyright owners from acquiring compensation.

The dilemma that UGC is exploited for free while copyright owners cannot get compensation stems from the lack of regulations on UGC platforms. Under the extant copyright law, UGC platforms can be protected by the safe harbor doctrine. The safe harbor doctrine can exempt UGC platforms from liability for copyright infringement constituted by platform users, as long as the UGC platform takes down the UGC upon receipt of a notice alleging that the UGC constitutes copyright infringement. The statutory protection for UGC platforms has forced copyright owners to take enforcement actions against individual users. However, this strategy is also doomed to fail due to the diffusion of the end users and the minimal damage each user creates [14]. Most importantly, imposing liability on non-commercial users can hardly gain sympathy from courts. The difficulty of obtaining ex-post remuneration from either UGC platforms or UGC creators has driven copyright owners to adopt ex-ante measures and extra-legal privileges: blocking access to copyrighted material or removing UGC through takedown notices. Under the current copyright rule, the only winner is UGC platforms because they profit from UGC and can be exempt from liability. UGC creators are at risk of being sued or having their UGC taken down. Copyright owners cannot obtain compensation for their works, and thus may set technical measures to hinder the public's access to intellectual products.

Copyright law has been deemed the son of technology [6]. Nevertheless, the preceding discussion shows that the current copyright regime has not adapted well to the wave of UGC creation. To prevent technology's child from turning against its mother, the copyright rules for UGC should be reframed. Actually, copyright law has long-established tradition of regulating intermediaries, which in turn preserved the end users' freedom to take advantage of the latest technology to consume copyrighted works.

3   The Intermediary-Oriented Approach Underlying Copyright Law History

Copyright law has been underpinned by an intermediary-oriented approach which, by balancing the interests of intermediaries, secures end users' access to and use of copyrighted works. We identified two types of important intermediaries in copyright law: producers and distributors.

3.1   Producers and Distributors in Copyright Law

Producers as Copyright Owners. Producers are those who make large-scale productions of copies of copyrighted works and who own the copyright of these works. Examples of producers include publishing houses, record labels and film studios. Because of producers' capacity of self-financing cultural production, their cross-subsidization strategy by investing in diversified copyrighted works, and their ability to achieve economies of scale, granting copyright to producers, in turn, allows end users to have easy access to a diverse range of cultural products at a reasonably low price [24].

User-Distributors Governing by Compulsory License. Distributors are the providers of devices and service through which users can consume copyrighted works in a new channel or medium. Until the first half of the 20th century, distributors were direct users of copyrighted works. We name them user-distributors, such as manufacturers of phonograph records that mechanically reproduced and publicly performed musical compositions.

In addition to the regular exclusive proprietary regimes, user-distributors were also governed by compulsory licensing schemes as a way to balance copyright owners'/producers' incentives in cultural production and distributors' incentives in developing new distribution technology [22]. For example, compulsory licensing mechanism has been formulated for mechanical reproduction of musical compositions. Copyright owners of musical compositions had the right to license a record manufacturer to produce a record. However, once a composition was recorded, it would be subject to a compulsory license. That is, any other record manufacturer can record the musical composition at a fixed price without permission from the copyright owner. End users, in turn, can enjoy a wide range of copyrighted works through the latest technology at a low price.

Facilitator-Distributors Governing by Levy Schemes. With technical advances in such areas as photocopying and digital recording, the ability to reproduce works was transferred from professional distributors to a large number of individual users. In this scenario, distributors, such as the manufacturers of photocopiers, video cassette recorders, and digital audio recorders, did not directly use the copyrighted works but provided devices to facilitate the use of the works. These devices were designed primarily for or were closely associated with the purpose of enabling users' reproduction of copyrighted works [10]. The providers of these devices are referred to as facilitator-distributors in this paper.

Though facilitator-distributors did not use copyrighted works themselves, they facilitated and profited from end users' use of the copyrighted works in the new distribution channel. Therefore, copyright law still imposed liability on facilitator-distributors. Facilitator-distributors were accommodated by levy schemes. Facilitator-distributors pay levies to copyright owners/producers, thereby exempting non-commercial end users exempted from copyright infringement liability. For example, under private copying levies, individual users can make private photocopies at no cost, but the providers of the photocopying devices are required to remunerate the copyright owners. The primary reason for imposing levies on private copying is the significant revenue losses incurred by large-scale private photocopying. Although each user only had a modest influence on the market for the copyrighted works, a colossal number of users would substantially harm the copyright owners' interests.

Above all, copyright law has adopted an intermediary-oriented approach, through a series of non-proprietary regimes such as compulsory license and levies, to enable end users to freely obtain and use a variety of copyrighted works via the latest technology. Jonathan Barnett [1] explained this as follows: "The intermediary's private interest in influencing the state to expand copyright to cover a novel production or distribution technology and the public's interest in producing and distributing content through that technology at the lowest cost possible would go hand in hand". The intermediary-oriented approach indicates how copyright law has addressed a large scale of non-commercial use when new technology decentralizes the capacity to use copyrighted works.

3.2 The Safe Harbor Doctrine Exempting the Liability of Intermediaries

However, early in the Internet age, the safe harbor rule was introduced into copyright law, interrupting the long-established intermediary-oriented approach. The primary reason to establish the safe harbor rule is to provide breathing space for online platforms to promote the internet economy. The safe harbor doctrine also gained justification in the passive role online platforms played as mere conduits to distribute content at the early Internet age [2].

The shift from an intermediary-oriented approach to the safe harbor rule is the fundamental reason for the dilemma we meet in the UGC age. Copyright owners can hardly get a share from the new revenue stream brought about by UGC. UGC creators are at risk of being sued. UGC platforms earn plenty of money through UGC but do not need to pay for either UGC creators or copyright owners [18]. Since UGC platforms have departed from the passive distributor assumption of the safe harbor doctrine and played an active role in facilitating the use of copyrighted works, it is time to return to the intermediary-oriented tradition.

4   Imposing Levies on UGC Platforms

Since UGC platforms facilitate UGC creation based on pre-existing works, UGC platforms have played as facilitator-distributors of pre-existing works contained in UGC. Based on UGC platforms' active role in facilitating and profiting from the use of copyrighted works contained in UGC, this paper argues that the intermediary-oriented approach should be applied to UGC platforms to solve the current UGC dilemma. Drawing upon historical experience, a levy scheme may be imposed on UGC platforms to allow users to use copyrighted works to create non-commercial UGC. We name it the "non-commercial UGC creation levy".

4.1   Levy Rather Than Compulsory License

A report commissioned by the UK Intellectual Property Office found no substantial difference between the compulsory licensing systems and the levy schemes, claiming that they were just different approaches adopted by different jurisdictions [10]. The EU member states have preferred levy schemes and the US has preferred compulsory licensing. However, compulsory license is directly imposed on users, while levies are imposed on the providers of the device that facilitate end users' use of copyrighted works. In this regard, compulsory licensing is designed for professional users, namely user-distributors. Levy schemes, in contrast, intend to promise the freedom of a significant number of end users to use copyrighted works by collecting levies from intermediaries, namely facilitator-distributors. Furthermore, under compulsory licensing, license fees are counted based on the specific number of times and the way how users used the copyrighted works. While levy schemes focus more on the profits the facilitator-distributors made from end users' use of copyrighted works regardless of how end users used the works. Considering the large scale of UGC creators and UGC platforms' active role in facilitating UGC creation, it is more appropriate to impose a levy scheme on UGC platforms than to impose a compulsory license on UGC creators.

4.2   Levying UGC Platforms with Non-commercial UGC Creation Exempted

We propose a non-commercial UGC creation levy scheme. There are four parties under the proposed scheme: UGC creators, copyright owners, UGC platforms, and collective management organizations (CMOs). UGC creators make certain use of copyrighted works to create UGC and upload it to UGC platforms. UGC platforms profit from UGC and pay levies to CMOs. Copyright owners receive levies from CMOs, and UGC creators can be exempted from copyright infringement liability (see Fig. 1).

Under the non-commercial UGC creation levy, levies should be imposed on UGC platforms that have substantially profited from UGC creation. Small platforms that have not substantially benefitted from UGC creation can be exempted from the levy scheme. Copyright royalty judges or similar authorities could set some tests to decide if the value of a UGC platform has been substantially enhanced by facilitating UGC creation, and if a UGC platform should be subject to the proposed levy scheme. The levies should be allocated to copyright owners according to the popularity of their copyrighted works.

Non-commercial UGC creators can use copyrighted works to create UGC for free according to the proposed levy scheme. We formulate three thresholds for a

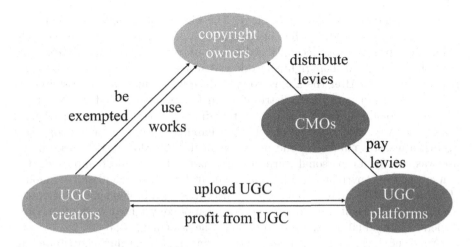

Fig. 1. The framework of the non-commercial UGC creation levy scheme

leviable non-commercial UGC: (i) the revenues the UGC creator captured should not exceed a predetermined threshold; (ii) the traffic the UGC attracted should not exceed a predetermined upper limit; and (iii) if the revenues or the traffic generated by the UGC exceeded the threshold, the content of UGC identical to the content of a pre-existing work should be lower than a predetermined percentage. The threshold for the revenue, traffic and overlapping rate would be predetermined by an authoritative third-party. The revenue condition and the traffic condition ensure that the UGC does not engender economic significance. The condition on substantial reproduction ensures that even if the UGC has gained significant revenues or popularity, the revenues or popularity does not directly come from the use of the copyrighted works. The three conditions draw a clear line for leviable UGC to maintain a balance between encouraging UGC creation and protecting the copyright owners' exploitation of copyrighted works.

5 The Justification of the Non-commercial UGC Creation Levy

5.1 Tailored to the Democratization of Re-Creating Content and Making Content Available to the Public

The proposed levy scheme covers a wider scope of leviable use than the existing levy scheme. The current levy scheme is strictly restricted to private, non-commercial copying. Under the proposed scheme, the leviable UGC creation not only reproduces but re-creates the copyrighted work. Moreover, UGC is publicly available for everyone connected to the Internet. Though most previous research has not explained the justifications for extending the levy scheme beyond its original scope, extending levy schemes to non-commercial UGC creation corresponds to the democratization of the capacity to re-create and publicly communicate content engendered by new technology. The history of levies shows that they were introduced to address the democratization of technology. Before the technology to reproduce was decentralized to end users, it was the distributors who controlled the use of copyrighted work and whose use was governed by compulsory licensing. After the ability to reproduce was democratized, distributors transformed from being user-distributors to facilitator-distributors. Nevertheless, even though the facilitator-distributors did not directly use the copyrighted works, they were still attached to copyright liability because enforcing copyright against a massive number of end users was inefficient. Moreover, the end users' use was mostly for personal purposes and had only a modest impact on the market for the copyrighted works. This made it difficult to calculate the amount of license fees. It is more reasonable to collect fees from professional distributors who profit from facilitating users' reproduction of copyrighted works than from dispersed individual users. Therefore, levy schemes emerged [4].

Analogue technology such as photocopying and recording contributed to decentralizing the ability to reproduce works, and thus the pre-Internet facilitator-distributors such as the manufacturers of photocopiers and recorders were subject to private copying levies. However, the UGC age has led to the democratization of not only the capacity to reproduce, but also the capacity to re-create and publish copyrighted works. Accordingly, the levy scheme should cover UGC creation and online distribution.

In the pre-Internet age, or what Lawrence [12] called the "read/only (RO) age", end users could only re-create text-based works, such as quotations. The re-creative use of works conveyed through other media could only be conducted by professional users. In the UGC age, or the "read/write (RW) age", with the advancement in the tools for amateurs' (re-)creation, "quoting" sounds, videos and images has become much easier. Compared to traditional facilitator-distributors that facilitate end users' reproduction, UGC platforms which facilitate end users' re-creation are even more suited to levy schemes because, among other factors, UGC creators make productive rather than consumptive use of the copyrighted works. Further, due to the peer-to-peer online distribution mode, the UGC age has also brought about the decentralization of the capacity to make copyrighted works existing in the form of UGC available to the public. Therefore, the levy scheme in the UGC age should be extended to not only re-creating copyrighted works but also making the works publicly available.

5.2 Fair Remuneration to Copyright Owners

The remuneration scheme is also another advantage of the levy system, especially compared to the fair use doctrine. The fair use doctrine has been inclined towards polarization, like an on/off switch. It has either provided copyright owners with overwhelming control over the use, or entitled users to use the work without remunerating the copyright owner. The polarizing binary choice has threatened to sacrifice copyright owners' interests, and led copyright industries to lobby to narrow the scope of fair use, which in turn casts a dark shadow on non-commercial end users [5].

Even if an individual non-commercial UGC creation does not intervene in the normal exploitation of the copyrighted work and can be exempted by fair use, a large-scale of UGC will lead to a loss of revenue for the copyright owner. In order to prevent copyright owners from opposing this revolutionary technology democratization due to their complete loss of interests in copyrighted works, it is still necessary to provide copyright owners with reasonable remuneration.

5.3 Privacy Concern

Protecting privacy is another advantage of the levy scheme. Unlike the compulsory licensing mechanisms that require payment from direct users, levy schemes collect fees from third parties, thereby alleviating the concern that copyright owners can obtain detailed information on a specific use.

In the 1964 landmark case of Personalausweise, based on privacy rights, Germany crafted the first levy scheme in the world [20]. The plaintiff, a German collection society called GEMA, required audio recording device manufacturers to disclose information on their purchasers so that GEMA could determine whether these purchasers had obtained a license for home taping. The German Supreme Court rejected GEMA's request, arguing that although the audio recording manufacturers had contributorily prejudiced the interests of copyright owners, requiring the disclosure of individual users' proprietary information would encroach on their right to the inviolability of their home and their privacy. A levy scheme based on the sale and importation of audio recording devices was thus introduced as a way to reward copyright owners and meanwhile respect end users' fundamental right to privacy.

5.4 Cultural Promotion

The goal of promoting cultural diversity has also played a crucial role in the formulation of the levy scheme. Most EU countries require a certain percentage of the levy funds to be set aside for specified social and cultural purposes. For instance, France allocates 25% of its levy revenues to public institutions for cultural purposes, such as funding performances and festivals. Despite the political controversies over the way levies have been allocated, "levies are at least a reasonable policy option" according to the French economist Fabrice Rochelandet [9].

Moreover, a levy scheme can promote culture diversity by cross-subsidizing. Some countries have allocated additional shares of their levies to certain types of works and others have allocated a specified percentage to new authors. In an economic sense, a levy scheme also narrows the gap between the revenue received by the most popular and least popular works. Because private copying increases with a work's popularity, the income of popular works would be reduced more than the less popular works under a levy scheme.

6   Concluding Remarks

Copyright law has an ancient tradition of levying facilitator-distributors to allow non-commercial use of copyrighted works by a mass of end users to allow non-commercial home copying, and levying various kinds of reprographic and recording devices to allow private copying. The essence of these levy schemes was to require third parties to pay copyright owners and to exempt large numbers of end users. By emphasizing the role of UGC platforms in facilitating the distribution of copyrighted works on which UGC is based, this paper acknowledges that UGC platforms should be regarded as a type of facilitator-distributor and a non-commercial UGC levy scheme should be imported. In this way, countless number of users can use copyrighted works to create UGC for free, and copyright owners can gain necessary incentive for creating copyrighted works.

Future work includes figuring out how to place the levy scheme into national copyright laws, and how to deal with the relationship between the levy scheme and fair use and compulsory licenses. Hopefully the proposed non-commercial UGC creation levy scheme will provide some inspiration for how copyright law can respond to the recent boom of AIGC (Artificial Intelligence-Generated-Content). The groundbreaking generative AI providers has empowered amateur users to create highly creative content. Although it is still controversial whether AIGC can be protected by copyright, legal proceedings have been initiated on whether AIGC's use of copyrighted works constitutes copyright infringement. For example, Getty Images sued Stability AI, creators of the popular AI art tool Stable Diffusion, over alleged copyright violation in January 2023. A month later, three artists brought a class-action lawsuit against Stability AI, asserting that Stability AI made unlawful copies of their copyrighted images. Although currently copyright owners are suing generative AI providers rather than AI tool users, when the law makes it clear that these deep-pocketed AI providers do not need to bear the responsibility, the copyright owner will inevitably turn to target the users. According to the principle of "substantial non-infringing uses" under Sony case, generative AI providers may indeed not be liable for copyright infringement [15]. Therefore, it is urgent to find a way to make a balance between promoting the development of AI and preserving copyright owners' interests. Applying a levy scheme to non-commercial AIGC creation is an option to unleash creativity in the AIGC age without compromising the incentives of copyright owners.

Nevertheless, generative AI providers is different from UGC platforms in the way to utilize copyrighted works. UGC platforms are facilitator-distributors that promote the distribution of copyrighted works contained in UGC. But generative AI providers are not merely facilitator-distributors, because they not only provider a platform for users to upload AIGC, but generate AIGC themselves. The users only initiate the generating process by entering keywords. On the other hand, generative AI providers are not user-distributors either because AI's use of copyrighted works is not similar to a human users' use of copyrighted works. The overall logic of AI generating content is to mark the relevant works with necessary parameters, and carry out statistics and analysis of mathematical nature through parameter marking, but the relevant works have not actually become part of the model [19]. Whether and how AI's utilization of works will affect the application of the levy system to generative AI providers is a topic for future research.

Acknowledgement. This work was supported by the Department of Education of Guangdong Province in China under Grant No. 2023GXJK463.

References

1. Barnett, J.M.: Copyright without creators. Rev. Law Econ. **9**(3), 389–438 (2014)
2. Bostoen, F.: Online platforms and pricing: adapting abuse of dominance assessments to the economic reality of free products. Comput. Law Secur. Rev. **35**(3), 263–280 (2019)
3. Carr, N.: Sharecropping the long tail. Rough Type **19** (2006)
4. Cohen, J.E., Loren, L.P., Okediji, R.L., O'Rourke, M.A.: Copyright in a Global Information Economy. Aspen Publishing, Boston (2019)
5. Fisher, W.W., III.: Promises To Keep: Technology, Law, and the Future of Entertainment. Stanford University Press, California (2004)
6. Goldstein, P.: Copyright's Highway: From Gutenberg to the Celestial Jukebox. Stanford University Press, California (2003)
7. Gotlieb, M.R., Sarge, M.A.: Civic learning and self-determination: A model of user-generated content and civic readiness among actualizing citizens. Commun. Theory **31**(1), 127–149 (2021)
8. Hetcher, S.: User-generated content and the future of copyright: part one-investiture of ownership. Vand. J. Ent. Tech. L. **10**, 863 (2007)
9. Katzenbach, C., Ulbricht, L.: Algorithmic governance. Internet Policy Rev. **8**(4), 1–18 (2019)
10. Kretschmer, M.: Private copying and fair compensation: an empirical study of copyright levies in Europe. Intellectual Property Office Research Paper (2011/9) (2011)
11. Lemley, M.A., Merges, R.P., Balganesh, S.: Intellectual property in the new technological age: 2020. Volume I: Perspectives, Trade Secrets and Patents (2020)
12. Lessig, L.: Remix: Making art and Commerce Thrive in the Hybrid Economy. Bloomsbury Academic, London (2008)
13. Li, Y., Huang, W.: Taking users' rights seriously: proposed UGC solutions for spurring creativity in the internet age. Queen Mary J. Intellect. Property **9**(1), 61–91 (2019)
14. Lunney Jr, G.S.: The death of copyright: digital technology, private copying, and the digital millennium copyright act. Virginia Law Rev. 813–920 (2001)
15. Murray, M.D.: Generative and AI authored artworks and copyright law. Hastings Comm. Ent. LJ **45**, 27 (2023)
16. Quintais, J.P., De Gregorio, G., Magalhães, J.C.: How platforms govern users' copyright-protected content: exploring the power of private ordering and its implications. Comput. Law Secur. Rev. **48**, 105792 (2023)
17. Rostama, G.: Remix culture and amateur creativity: a copyright dilemma. WIPO Magaz. **3**, 22–25 (2015)
18. Salar, N.: The Canadian UGC exception: an attempt to revolutionise the fair use defence for user generated content? J. Intellect. Property Rights (JIPR) **28**(2), 107–113 (2023)

19. Samuelson, P.: Generative AI meets copyright. Science **381**(6654), 158–161 (2023)
20. Senftleben, M.: User-generated content-towards a new use privilege in EU copyright law. Research Handbook on Intellectual Property and Digital Technologies, Tanya Aplin, ed., Cheltenham: Edward Elgar Publishing (2019)
21. Shirky, C.: Cognitive surplus: creativity and generosity in a connected age. Penguin UK (2010)
22. Vickery, G., Wunsch-Vincent, S.: Participative web and user-created content: web 2.0 wikis and social networking. Organization for Economic Cooperation and Development (OECD) (2007)
23. Wagner, K.B.: Tiktok and its mediatic split: the promotion of ecumenical user-generated content alongside sinocentric media globalization. Media, Cult. Soc. **45**(2), 323–337 (2023)
24. Williams, J.L.: Automation is not "hacking": why courts must reject attempts to use the CFAA as an anti-competitive sword. BUJ Sci. Tech. L. **24**, 416 (2018)

Moving Beyond Text: Multi-modal Expansion of the Toulmin Model for Enhanced AI Legal Reasoning

Jiaxing Li[✉]

College of Philosophy, Nankai University, Tianjin 300350, China
1120200792@mail.nankai.edu.cn

Abstract. In the burgeoning field of artificial intelligence's application in law, enhancing the precision of legal reasoning is a pressing challenge. The Toulmin model, a fundamental framework in AI for legal problem-solving, primarily focuses on textual data, yet it faces limitations in the context of today's multi-modal digital environment. This paper aims to expand the scope of the Toulmin model to incorporate multi-modal consideration. Based on the multi-modal argumentation theory, the expansion of the Toulmin model will focus on multi-modal data sources, such as text, images, audio, and video, multi-modal warrant, and multi-modal rebuttals, so that enriching AI's legal argumentation capabilities. Concluding with its theoretical and practical implications, this paper sets a direction for future research in AI and law, highlighting the critical role of multi-modal concern in advancing AI's ability to handle complex legal scenarios with enhanced efficacy and accuracy.

Keywords: multi-modal · the Toulmin model · AI · legal reasoning

1 Introduction

Legal reasoning is the cornerstone of the jurisprudential process and constitutes the methods and processes by which lawyers and judges interpret and apply legal texts in specific situations. Broadly speaking, legal reasoning encompasses a range of activities including the interpretation of statutes and precedents, the synthesis of relevant case law, the analysis of factual evidence, and the construction of conclusions that argue for a particular law. However, the advent of artificial intelligence (AI) has initiated a transformative shift, automating and augmenting aspects of legal reasoning. Over the past decade, the integration of AI in legal sectors has notably deepened, revolutionizing traditional practices in legal research, case analysis, and predictive jurisprudence.

AI's application in law, despite its rapid maturation, faces critical challenges in ensuring accuracy and reliability in legal reasoning. Traditional AI models, predominantly text-centric, have shown a lack of nuanced understanding necessary for deciphering the complexities of legal debate. This deficiency is particularly evident in handling multi-modal data and generating interpretable, legally

C. Cruz et al. (Eds.): IC 2023, CCIS 2036, pp. 299–308, 2024.
https://doi.org/10.1007/978-981-97-0065-3_23

coherent reasoning outcomes, which fail to capture the nuances present in other forms of data such as images, audio, and video. Legal reasoning is a complex and multidimensional process, involving the interpretation and application of a multitude of legal texts, cases, and rules. In traditional practice, lawyers and judges engage in comprehensive judgment and argumentation through a deep understanding of facts and legal rules. While AI excels in processing large-scale data and identifying patterns, its text-centric approach faces difficulties in comprehending the complexities of legal contexts, handling multi-modal data, and generating interpretable and legally coherent reasoning outcomes. Furthermore, the intricate network of legal arguments often relies on unspoken premises, cultural differences, and a deep understanding of social norms, which purely textual models might overlook.

The Toulmin Model, a renowned framework for understanding argument structures, emerges as a potent tool in legal AI. Developed by philosopher Stephen Toulmin in 1958, the model provides a methodical approach for dissecting arguments into constituent parts such as data, claim, warrant, backing, qualifier, and rebuttal. [1] Its structured approach to argumentation provides potential benefits for AI applications in legal reasoning, serving as a basis for AI systems to analyze and construct legal arguments. However, the traditional structure of the Toulmin Model, primarily designed for textual data, requires adaptation to meet the demands of contemporary legal practices. As legal arguments increasingly incorporate multi-modal data, there is a pressing need to expand the scope of the Toulmin Model. This expansion involves integrating capabilities for analyzing and interpreting visual and auditory data, thereby enabling AI systems to process a wider range of evidence types. The adaptation of the Toulmin Model for multi-modal data is not just an enhancement but a necessary evolution to ensure that legal AI systems can comprehend and interpret the full spectrum of evidence in legal scenarios.

The primary objective of this study is to investigate how multi-modal argumentation can augment AI's engagement of legal reasoning. Multi-modal argumentation extends beyond traditional textual interpretations, encompassing all forms of representation in a cultural context. In legal AI, this implies analyzing not only written documents but also other forms of evidence such as courtroom recordings and video testimonies, thus offering a more comprehensive understanding of legal cases. Through a theoretical lens of Toulmin's model and multi-modal argumentation theory, the paper extends Toulmin's model in the AI domain and emphasizes the value of multi-modal perspective in enhancing the depth and breadth of analysis capabilities.

2 The Toulmin Model and Legal Reasoning

2.1 The Introduction of the Toulmin Model

The Toulmin Model, a unique theoretical framework for argumentation, is developed by the British philosopher Stephen Toulmin in 1958, providing an effective tool for analyzing legal arguments. In The Uses of Argument, Toulmin distinguishes analytic arguments and substantive arguments and asserts that arguments that appear in different fields and disciplinary areas are mostly substantive arguments rather than analytic arguments. After critiquing the inability of traditional formal logic to accurately analyze everyday arguments and analogizing with jurisprudential argumentation, he outlines the different elements of argumentation and constructs a framework – the Layout of Arguments, which is later commonly referred to as the Toulmin Model. According to Toulmin, every argument is composed of a claim(C), the reason for that claim(D), and the warrant (W) that licenses the step from this reason to that claim, backing (B)which stand "behind warrants", rebuttal (R), and modal qualifiers(Q). And the framework appears as follows:

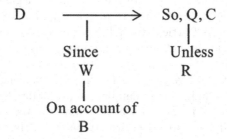

Fig. 1. The Toulmin model on *The Uses of Argument* (2003, 97)

In the context of legal AI, the Toulmin Model has been instrumental, guiding the development of algorithms for legal reasoning and argumentation. The adaptability of the Toulmin Model in AI is evident in its diverse applications, from clinical decision-making extensions by Fox and Modgil(2006) [2] to its use in Online Dispute Resolution systems (Zeleznikow 2006) [3]. These applications demonstrate the model's utility beyond traditional text-based legal reasoning, accommodating a more dynamic and context-sensitive approach to legal decision-making. Besides, this structure is pivotal in designing AI algorithms that simulate human-like legal reasoning. By breaking down complex legal arguments into their constituent elements, AI systems can better assess the validity of arguments and simulate legal reasoning processes [4,5]. Its application in AI-powered legal research tools aids in organizing and analyzing legal arguments, thereby enhancing the efficiency and accuracy of legal research and case analysis [6] (Fig. 1).

2.2 Challenges Faced in the Context of Multi-modal Digital Data

Despite its widespread recognition, the Toulmin Model encounters critique for its inability to sufficiently dissect complex discursive patterns in legal settings. It effectively manages'easy cases' that rely on clear deductive reasoning, with minimal disputes overrule interpretation or fact classification [7]. However, the model's limitations become apparent in'hard cases', where external justification is essential, prompting a detailed scrutiny of the premises and expanding the scope of decision-making beyond basic deductive logic.

Critics, such as Gasper and George (1998), argue that the model forces arguments into a'single-claim, single-ground' mode, which may not reflect the multifaceted nature of practical legal reasoning. This critique highlights a potential limitation of the model in adequately capturing the nuances and complexities inherent in legal discourse [8]. Traditionally, the model has been text-centric, but it struggles to accommodate the broader spectrum of evidence in contemporary legal contexts, which often includes visual and auditory data along with written documents.

This limitation poses a significant challenge in legal AI, as the model must evolve to incorporate and interpret diverse data types accurately. This shortfall is particularly striking in'hard cases'. In these scenarios, legal decisions go beyond simple deductive logic, necessitating a thorough examination of the law's'working logic', as Toulmin himself articulated [1]. Such cases demand an exploration of legal reasoning that transcends mere rule application, revealing the depth and intricacy of the law.

In the context of AI-driven legal reasoning, the challenge is even more pronounced. The model's limitation in handling multi-modal digital data is a significant hurdle. AI systems, in their endeavor to navigate the intricate landscape of legal reasoning, must not only structure arguments within the Toulmin framework but also accurately interpret and align diverse types of data-textual, numerical, visual, and auditory. The evolution of the Toulmin Model is thus imperative to effectively incorporate and interpret these varied data types, reflecting the true complexity and multi-dimensionality of modern legal discourse.

3   Expansion of the Toulmin Model

3.1   The Theoretical Basis for Multi-modal Integration

Building on the identified challenges of the Toulmin Model in AI-driven legal reasoning, recent developments in multi-modal argumentation theory present a promising avenue for advancement[1]. This emerging theory posits that arguments, including counterarguments, are more effectively presented, and analyzed through various modes beyond text. [9,10] Our research aims to integrate this theory with AI to thoroughly explore the capabilities of multi-modal data-encompassing audio, video, and text-in capturing the intricacies of legal argumentation. This exploration suggests a necessary extension of theoretical frameworks such as the Toulmin model to include multi-modal data, thereby broadening AI's ability to process and interpret complex legal scenarios.

---

[1] Multi-modal argumentation theory broadly refers to multimodal, the mode of arguing, like visual, auditory elements raised by Leo Groarke, as well as the multi-modal, the mode of rationality, developed by Michael Gilbert. This paper focuses on both aspects to illustrate "multi-modal".

The rapid digitization within the legal sector has ushered in a plethora of data types, including images, audio recordings, and videos, each contributing unique perspectives essential for understanding legal cases. However, there is a bias against translating multimodal elements into textual expressions, the consideration of multi-modal elements in practical arguments is necessary. Multi-modal data sources bring critical nuances, such as tone, inflection, and visual context, which are indispensable for an exhaustive analysis of legal cases-challenges that AI systems limited to textual data alone cannot address. The integration of these diverse data forms into legal AI systems marks a pivotal theoretical and practical leap, equipping them to conduct thorough analyses of cases by considering all forms of available evidence.

Pioneered by Michael A. Gilbert, multi-modal argumentation theory critiques the traditional focus on verbal reasoning, advocating for the inclusion of non-logical elements like emotions and intuition in the argumentation process. Logical mode is the most important mode of traditional argumentation, which focuses on the relationship between reasoning and argumentation, people through rigorous reasoning and argumentation to reach a conclusion, to ensure the rationality and effectiveness of the argument. Emotional mode emphasizes the role of feelings and emotions in the argumentation process. The emotional mode examines how emotions affect people's beliefs, attitudes, and decisions, and explores the ways in which emotions are expressed and influenced in argumentation, and sometimes the message that is intended to be conveyed is in the emotions rather than in the words. The visceral mode puts the focus on limb perception and limb experience. Humans perceive the world through their bodies, and body language and gestures often convey messages. Gestures, gestures, and facial expressions also influence perceptions and attitudes. The kisceral mode emphasizes the importance of intuition in the argumentation process. It includes forms of argumentation that involve intuition, mysticism, premonition, religion, the occult, and non-sensory forms of knowledge and persuasion in general [9]. An example is a person who refuses to take a picture at night because he senses evil in the place.

This theory, which highlights cognitive aspects and diverse pathways of arguments, becomes increasingly relevant in legal settings where both emotional and moral factors significantly influence decision-making. Recognizing the value of non-textual data allows AI systems to replicate human reasoning more closely, encompassing a full spectrum of evidence [11]. This is particularly crucial in areas like family law, where emotional and cultural considerations play a key role, especially in cases touching on minority rights and cultural sensitivities [9,12].

In legal reasoning, multi-modal argumentation offers a more holistic view of cases. Gilbert's theory, which embraces logical, emotional, visceral, and kisceral modes, is essential in trials where emotional and moral considerations are as important as logical arguments. Analyzing non-verbal cues in courtroom videos enables AI systems to detect emotional and contextual subtleties not captured in written records, leading to more accurate and equitable decisions. Therefore, expanding the Toulmin Model to include multi-modal elements is crucial for addressing the complexities of modern legal practices. This expansion enhances the model's ability to analyze diverse data types, interpreting non-textual evidence and its various roles in legal argumentation. An adaptable Toulmin Model is necessary to fully capture and analyze the intricate nature of legal arguments in today's digital landscape, ensuring that AI systems can effectively assist legal professionals in their decision-making.

3.2   The Multi-modal Expansion of the Toulmin Model

The Toulmin Model's significance in legal argumentation stems from its emphasis on the role of contextuality and field-dependency in arguments, aligning closely with the defeasibility inherent in legal reasoning. Argumentation schemes, akin to classical logic's inference rules, are adopted in AI primarily because the critical questions associated with them align with the defeasible environments. Therefore, in expanding the Toulmin Model, this paper focuses on elements related to defeasibility. According to Bex and Verheij's [13] recharacterization of the Toulmin model of argumentation based on rebuttal, its defeasibility is manifested through rebuttals that weaken the strength of the warrant, thus rendering it defeasible. Moreover, in practical arguments, the expansion of the types of rebuttals dictates the expansion of data, warrant, backing, and claim. Based on the emphasis on domain-dependence and substantive validity of Toulmin's model, it expands the scope of rationality in traditional formal logic to encompass rationality under a broader multidimensional synthesis related to the domain. Thus, this model in open to multimodal expansion. In the following part, we will now focus on the multi-modal expansion of data, warrant, and rebuttal to enhance the application of the Toulmin Model in the domains of Artificial Intelligence and law.

Expanding Data: Integrating Various Data Types. In the context of modern legal reasoning and the Toulmin Model, expanding the definition of 'Data' to include non-textual elements is crucial. This expansion reflects the necessity to adapt to the increasingly multi-modal nature of evidence in legal proceedings. Traditional textual data, while fundamental, no longer suffice in isolation due to the richness and complexity of modern legal cases. The visual presentation of evidence can profoundly influence legal decisions. In criminal cases, for example, real-time visual accounts from CCTV footage or bodycam videos provide compelling evidence, often more directly than written statements. Audio recordings can reveal tone, inflection, and emotion, adding layers of interpretation to verbal testimony. Similarly, images and videos can capture spatial and temporal details crucial for constructing a comprehensive narrative of events.

The integration of multi-modal data into the Toulmin Model is not merely a matter of including additional types of evidence but also involves understanding the unique contributions each mode brings to legal reasoning. This incorporation of multi-modal data enriches the analytical process, providing a comprehensive understanding of cases and aligning with the realism advocated in legal theory.

Expanding Warrant: Considering Multi-modal Modes. The expansion of the "Warrant" component within the Toulmin Model to incorporate multi-modal argumentation theory signifies a pivotal shift in understanding and applying this model, especially in legal contexts. This shift involves a nuanced appreciation of the various dimensions of rationality of argumentation: logical, emotional, visceral, and kisceral modes. Such an expansion aligns the model more closely with the complexities of human reasoning and decision-making, reflecting the real-world dynamics of arguments.

The incorporation of multi-modal dimensions into the Toulmin Model's "Warrant" component enriches the model's applicability and effectiveness in legal argumentation. By acknowledging and integrating these diverse modes of reasoning, the model becomes a more versatile and comprehensive tool, capable of capturing the multifaceted nature of arguments. This expanded approach not only enables a deeper understanding of legal arguments but also ensures that the Toulmin Model remains a relevant and effective tool in the evolving landscape of legal analysis and reasoning.

Expanding Rebuttal: Enhancing the Dynamism of Legal Arguments. The role of rebuttals in legal argumentation is pivotal, as they challenge the warrants and directly reflect the effectiveness of argumentation. Stephen Toulmin, emphasizing that a robust argument withstands criticism, highlights the importance of rebuttals in his model. This notion is further developed by scholars like Verheij, who evolved the concept of rebuttal in the Toulmin Model, emphasizing its role in formal argument evaluation [14].

Defeasibility, as discussed by Pollock (2010) [15] and Prakken (1997,2010) [16,17]refers to the idea that conclusions can be revised or rejected based on new evidence or rebuttal, a concept particularly relevant in legal contexts where new evidence can significantly alter the course of an argument. Verheij [18] contributes to this understanding by developing a theory of evaluation in dialectical interpretation, where the evaluation status of statements in an argument is determined by the assumptions of the argument. This approach highlights the dynamic nature of argumentation and the significance of considering the audience and the field in constructing arguments.

In the context of modern legal practice, rebuttals are not merely defensive mechanisms but crucial tools for comprehensive defense. With the increasing diversity of legal evidence, including digital media, audio, and visual elements, the nature of counterarguments and rebuttals must adapt. For instance, the analysis of non-verbal cues in witness testimonies, such as body language and facial expressions, can provide substantial grounds for rebuttal. This recognition of multi-modal data enhances legal argumentation by utilizing diverse evidence forms, such as discrepancies between verbal testimony and video evidence or inconsistencies in body language, to fortify legal arguments.

The expansion of the "Rebuttal" component in the Toulmin Model, especially when integrated with AI in legal reasoning, offers a dynamic and comprehensive approach to legal argumentation. It not only accommodates the growing complexity and multi-modality of legal evidence but also enhances the capacity of legal practitioners and AI systems to anticipate and address potential counterarguments effectively. As AI continues to evolve in the legal domain, its role in identifying and assessing rebuttals will be increasingly vital, contributing to developing robust and resilient legal arguments. This evolution in the Toulmin Model ensures that legal reasoning remains adaptable and responsive to the multifaceted nature of evidence in the digital era.

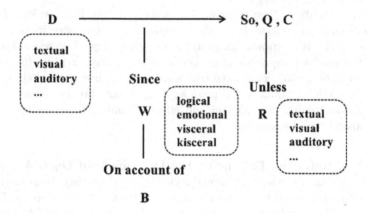

Fig. 2. The expanded Toulmin model

Incorporating multimodal aspects into data, warrant, and rebuttal, the expanded Toulmin model, shown in the Fig. 2, lays a solid theoretical foundation for AI systems to analyze and evaluate legal reasoning. It becomes especially pertinent for complex legal arguments, where the consideration of multimodal rebuttals prompts a re-evaluation of the data and warrant's effectiveness, offering new perspectives for assessing legal reasoning. For instance, a legal claim like "Mary slipped in a mall, and according to law A, the mall has a duty to ensure customer safety; therefore, the mall should compensate her." The primary analysis based on the Toulmin model would be D: Mary slipped in a mall; C: the mall should compensate Mary; W: the mall has a duty to ensure customer safety, if the customer was hurt, the mall should give compensation to the customer; B: the law A.

When analyzing this legal reasoning, there may be additional multimodal elements in the facts, so this reasoning is properly analyzed through the extended Toulmin model. The inference should consider other elements related to it, incorporating CCTV footage, audio and the like. By using multimodal rebuttal as a lens, various kinds of rebuttals, situations in which an argument or reasoning can be invalidated, are explored, including visual rebuttal and auditory rebuttal evidence. For example, while viewing the surveillance video if it is found that Mary was distracted by her cell phone, thus causing her to fail to notice the signs placed at the mall, thus entering the danger zone, and falling and injuring herself. This new information in this case challenges the initial claim that the mall was not primarily responsible. However, again through another visual presentation and emotional consideration, for example, her injuries are serious and given the emotional modality, the mall should provide compensatory care. Combined with specific weights, a final decision is given against this legal reasoning.

4   Conclusion: The Future of Legal Argumentation in the Age of AI

In summary, the expanded Toulmin model not only aids in the evaluation of legal arguments but also provides valuable insights for the construction of AI legal systems and machine learning algorithms. The integration of multi-modal data into this model reflects the evolving nature of legal evidence and argumentation in the digital age. By integrating multimodal data, these systems can process and analyze legal cases more comprehensively, considering a wider range of evidence and perspectives.

However, this advancement also brings forth challenges that must be addressed. The ethical implications of using AI in legal contexts, the accuracy of AI systems in interpreting multi-modal data, and the need for continuous technological innovation are all critical considerations for the future development of legal AI. One of the primary challenges in integrating multi-modal data into AI systems lies in the complexity of processing and analyzing diverse data types. AI systems must be equipped with advanced algorithms capable of handling not just textual data but also visual and auditory information. This necessitates substantial computational power and sophisticated machine learning techniques. Current Convolutional Neural Networks (CNNs) for image and video processing and Transformer models for Natural Language Processing (NLP) can provide the technical basis for the execution of this process. However, the seamless integration of these different types of data, especially in the context of complex arguments, remains an area of development. Artificial intelligence systems still face significant challenges in data fusion, contextual understanding, and emotion processing. Integrating various forms of data requires the development of AI models that can accurately interpret this information within the legal framework, placing it in the appropriate context. Moreover, ethical and privacy concerns also arise when using multi-modal data in legal AI systems. The ethical use of data, especially in sensitive legal cases, is paramount to ensure that AI systems align with legal principles and ethical standards.

In conclusion, the extended Toulmin Model, enriched with the capabilities of AI to process and analyze multi-modal data, represents a significant stride forward in legal reasoning. This approach not only aligns with the digital transformation of the legal sector but also enhances the efficacy and depth of legal argumentation. As AI technology progresses, its role in shaping robust, accurate, and fair legal arguments is poised to become increasingly integral, offering a glimpse into the future of legal practice in the digital era.

308   J. Li

References

1. Toulmin, S.E.: The Uses of Argument, 2nd edn. Cambridge University Press, Cambridge (2003)
2. Fox, J., Modgil, S.: From arguments to decisions. Extending the Toulmin view. In: D. Hitchcock, B. Verheij (eds.), Arguing on the Toulmin model. New Essays in Argument Analysis and Evaluation, pp. 273–287. Springer, Netherland (2006)
3. Zeleznikow, J.: Using Toulmin argumentation to support dispute settlement in discretionary domains. In: D. Hitchcock, B. Verheij (eds.), Arguing on the Toulmin model. New Essays in Argument Analysis and Evaluation, pp. 289–302. Springer, Netherland (2006)
4. Prakken, H.: Artificial Intelligence & Law, Logic and Argument Schemes. In: Hitchcock, D., Verheij, B. (eds.), Arguing on the Toulmin Model. New Essays in Argument Analysis and Evaluation, pp. 231–246. Springer, Netherland (2006)
5. Bex, F.J., van Koppen, P., Prakken, H.: Verheij, B: a hybrid formal theory of arguments, stories and criminal evidence. Artif. Intell. Law 18(2), 123–152 (2010)
6. Tans, O.: The fluidity of warrants: using the toulmin model to analyse practical discourse. In: Hitchcock, D., Verheij, B. (eds.), Arguing on the Toulmin model. New Essays in Argument Analysis and Evaluation, pp. 219–230. Springer, Netherland (2006)
7. MacCormick, N.: Legal Reasoning and Legal Theory. Oxford University Press, Oxford (1978)
8. Gasper, D.R., George, R.V.: Analyzing argumentation in planning and public policy: assessing, improving, and transcending the Toulmin model. Environ. Plann. B. Plann. Des. 25, 367–390 (1998)
9. Gilbert, M.: Coalescent Argumentation. Lawrence Erlbaum, Mahwah (1997)
10. Groarke, L.: Going multimodal: what is a mode of arguing and why does it matter? Argumentation 28(4), 133–155 (2015)
11. Roque, G.: Should visual arguments be propositional in order to be arguments? Argumentation 29(2), 177–195 (2015)
12. Gilbert, M.A.: Multi-modal argumentation. Philos. Soc. Sci. 24(2), 159–177 (1994)
13. Bex, F., Verheij, B.: Solving a murder case by asking critical questions: an approach to fact-finding in terms of argumentation and story schemes. Argumentation 26(3), 325–353 (2012)
14. Verheij, B.: Evaluating arguments based on Toulmin's scheme. In: D. Hitchcock, B. Verheij (eds.), Arguing on the Toulmin model. New Essays in Argument Analysis and Evaluation, pp. 181–202. Springer, Netherland (2006)
15. Pollock, J.L.: Defeasible reasoning and degrees of justification. Argument Comput. 1(1), 7–22 (2010)
16. Prakken, H.: Logical Tools for Modelling Legal Argument. A Study of Defeasible Reasoning in Law. Kluwer, Mahwah (1997)
17. Prakken, H.: An abstract framework for argumentation with structured arguments. Argument Comput. 1(1), 93–124 (2010)
18. Verheij, B.: Dialectical argumentation with argumentation schemes: an approach to legal logic. Artif. Intell. Law 11(1–2), 167–195 (2003)

The Worldwide Contradiction of the GAI Regulatory Theory Paradigm and China's Response: Focus on the Theories of Normative Models and Regulatory Systems

Laitan Ren[✉] and Jingjing Wu

The Law School of Hainan University, Hainan 570208, China
18959289980@163.com

Abstract. In the period when mankind is entering the intelligent era, Generative Artificial Intelligence (GAI), as a leapfrog achievement of Artificial Intelligence (AI) development, has become the core technology of intelligent industrial transformation. However, as one of the emerging industries in the international community, GAI's normative model and management system theory as a "pre-consideration" have defects, resulting in regulatory gaps or weak sectors in its transformation practice, which makes some entities in the GAI industry abuse GAI resources. In order to avoid this problem, all countries have adopted different single planning governance paradigms, but the applicable single planning governance paradigm followed by each country has its own limitations. This paper discusses the network gauge theory system under system optimization. Therefore, it is necessary to analyze the essential causes of this worldwide problem, clarify the specific response limitations of countries to follow the applicable single planning governance paradigm theory, and the implementation dilemma under different attempts to explore and update the regulatory system theory. In addition, it tries to explain and complete China's theoretical arrangement and solution exploration on the dual dimension of regulation and management, explore the theory of network normative mode from the perspective of system optimization, explore the balance setting theory of governance-type supervision from the principle of two-law inverse, solve the two major imprisons of international GAI governance with dynamic integration methods.

Keywords: Generative Artificial Intelligence · Regulatory Theory Paradigm · Standardize Models · Management Systems · Conflicts and Resolution

1 Introduction

In the World Economic Forum's *Top 10 Emerging Technologies Report 2023* [18], Generative Artificial Intelligence (GAI) has been called a revolutionary technology that can expand the boundaries of human endeavour. As a phased development result of Artificial Intelligence (AI), GAI has become an important force to promote human beings into the intelligent era and plays a pivotal role in industrial transformation. The global industry has gradually increased its investment in GAI, and the advent of ChatGPT marks a new era for the GAI industry. However, in contrast to the booming GAI industry, there is a paradigm of regulatory theory with lag,[1] and the overall supervision of the global GAI industry is scattered. On the one hand, countries around the world have not yet formulated special norms, there are still a large number of gaps or overlapping restrictions on the GAI industry. On the other hand, due to the scattered implementation of the normative body and the undetermined exploration of the management system, the GAI industry disorderly development.

At present, GAI, as a major emerging industry, in the case of deficiencies in the above-mentioned normative system and management system theory, the regulation and setting of institutions arranged in accordance with such theoretical guidelines will lead to weak supervision, which may make many entities participating in the development of this industry, such as technology providers, service providers, service users, regulators will not afraid of supervision and strive to maximize their own interests. They will compete for and abuse GAI resources. And it is clear that if the causes of this dilemma are not solved in time, the dilemma will continue for a long time, endangering the healthy development of the GAI industry.

In order to ensure the development of GAI, the perspective of theoretical law focuses on the GAI industry standard model and management system theory that need to be discussed in advance. Pay attention to the current situation of international development of its reality transformation, and analyze the essential reasons for its predicament, so as to help the road of theoretical improvement.

2   Sporadic Governance: Attribution of Worldwide Problems and Dilemmas

As an emerging field, GAI, due to imperfect theory in the regulatory process, has become a worldwide problem.

Countries are actively exploring feasible regulatory and translating theories into practical arrangements. However, with the continuous expansion of the application field and application depth of GAI, the existing normative model and management system theory are increasingly no longer applicable.

From the perspective of the normative model, the EU restricts GAI by dividing risk levels, but this move often has the problem of insufficient flexibility in actual application, and the United States chooses to restrict GAI by new application and new norm, but in fact, it is difficult to follow up the amendments

---

[1] The regulatory theory paradigm referred to in this paper includes two items: normative model theory and management system theory, that is, from the binary development of "norms" and "management of implementing norms".

or regulations in a timely manner. For example, in the continuous interaction of GAI, there are Scitech ethics problems such as algorithm discrimination and value bias, and China formulates norms according to the prior "Subject-oriented" theory, but in practical applications, there are often unclear topics of normative responsibility.

From the perspective of management systems, many European companies oppose the relevant arrangements of the EU, believing that such a premature, overly strong and centralized regulatory system theory and guidance will stifle the opportunity to reenter the technological frontier. Conversely, more than 350 AI experts, including Sam Altman, the father of ChatGPT, and Kevin Scott, CTO of Microsoft, issued a statement at the Center for AI Safety calling for intervention and increased regulation of the industry as a whole [17]. That is, it is believed that the realistic setting guided by the theory of supervision that is too late, lax, and scattered will form a huge disaster.

Under the current situation of frequent problems in the above-mentioned normative mode and management system, many participants in the operation of the GAI industry have abused GAI resources at will under the pretext of the current unregulated and incompletely solved problems. These behaviors undoubtedly reflect the reality that the GAI industry is in danger in the transformation of reality due to imperfect guiding theory.

The implementation of laws is often reflected in reality, and it can be seen that either due to the omission of normative models in various countries around the world, or the disorderly implementation of the implementation of normative bodies, resulting in the scattered dispersion of governance forces. Therefore, the current situation should be explored from the perspectives of normative model and management system theory.

2.1 The Limitations of the Single Planning Governance Paradigm

The main paradigms of AI governance in the current international community can be summarized as "The risk governance paradigm theory" with the EU as the typical one, "The applied governance paradigm theory" with the United States as the typical theory, and "The subject governance paradigm theory" with China as the typical (Table 1). However, the mainstream governance model theory is mainly oriented to the traditional AI model, and it may be difficult to fully adapt it in the current iterative update of GAI [22], The limitations of the regulatory lag and effectiveness generated by the theory after the transformation of reality may emerge gradually, and it is difficult to fully standard.

Table 1. Main paradigm theories of AI governance in the international community and corresponding countries (organizations).

Canonical model theory	The main body	The name of the main norm of the theoretical guide	Specification brief
Risk governance paradigm theory	European Union	*Artificial Intelligence Act*	The Act classifies AI systems into four levels: minimum, limited, high and unacceptable risk, and differentiates the supervision of each level [26].
("Risk" benchmark)	Germany	*Recommendations for Data and Algorithms*	The core idea of the Proposal is to establish a five-level risk rating system for the use of data by digital service enterprises, and adopt different regulatory measures for enterprises with different risk types [7]
	Canada	*Digital Charter Enforcement Act 2022 (DCIA 2022)*	The Artificial Intelligence and Data Law of DCIA Part III 2022 requires measures to identify, assess and mitigate harm to use and issue explanations for high-impact systems [3]
	Russia	*Concept of Relationship Regulation and Development in the Field of Artificial Intelligence and Robotics by 2024*	The third part of the General Provisions of the Concept, "Principles for regulating the relationship between the field of artificial intelligence and robotics", states that regulation applies a risk-oriented model [2]
Apply governance paradigm theory ("Scene" benchmark)	United States	1. *Algorithmic Accountability Act 2022* 2. *Deep Counterfeiting Liability Act* 3. *Autopilot Act* 4. *Ethical Use of Facial Recognition Act*	In the field of artificial intelligence, the United States has not introduced unified, universal, and comprehensive legislation,[a] but has promoted special legislation for specific application fields such as algorithm recommendation and deep synthesis.[b]
Subjective governance paradigm theory ("Principal" benchmark)	China	1. *Measures for the Administration of Internet Information Services* 2. *Provisions on the Administration of Internet Information Service Algorithm Recommendations* 3. *Provisions on the Administration of Deep Synthesis of Internet Information Services*	It can be seen from relevant laws, regulations and normative documents that most of China's provisions on AI-related elements are regulated by "subject responsibility".[c]
	Japan	1. *Guidelines for Artificial Intelligence Development (Draft)* 2. *Artificial Intelligence Utilization Pointer* 3. *Regulatory Guidelines for Implementing AI Principles Version 1.1*	Japan's idea of AI regulatory model is that the government sets goals, enterprises independently regulate,- and multi-subject participation, and builds a regulatory system with a soft law paradigm based on this [24].

[a] The more comprehensive *GAINS Act* and the *Blueprint for an AI Bill of Right* are not universal (which is required by the U.S. DOC and the FTC), and the latter white paper form is not legally binding, so it is difficult to become a strong guiding norm.

[b] For example, the *Algorithmic Accountability Act of 2022* in the area of "algorithm recommendations" and the *Deepfakes Accountability Act* on "deep synthesis".

[c] For example, Article 7 of the *Provisions on the Administration of Algorithm-generated Recommendations for Internet Information Services* clarifies the "responsibility of algorithm security subjects" and relates them throughout. Article 9 of the *Provisions on the Administration of Deep Synthesis of Internet Information Services* clarifies the responsibility of information security entities and relates them throughout.

Regarding the theoretical dimension of risk governance paradigm, the EU has established an AI governance model of risk benchmark in *The AI Act*, which divides AI into "Minimal or no risk", "Limited risk", "High risk", "Unacceptable risk", and supervises at different levels. However, this theoretical guidance model may have the disadvantages of singularity and staticism for highly dynamic GAI. Specifically, the possibility of partial generalization is relatively large, and it can be seen from the specific norms of the European Union, Germany, Canada and other countries (organizations) that most of them only clarify the scenarios and main responsibilities of high-risk or high-impact systems, while the relevant provisions under other risk levels are often generalized. Coupled with the diversity of GAI's application scenarios and the continuous innovation and expansion of the industry, it is difficult for the existing specifications to provide a sufficient reference basis, and it is impossible to summarize all the situations to be seen in an exemplary manner. Second, the dynamic response is poor, for the continuous development of GAI with versatility, it may be suitable for the entertainment field, may also be applicable to the military field, it has the characteristics of penetrating application from low risk to high risk scenarios, then the static risk induction example like the EU loses its significance in special circumstances.[2]

In terms of "The theoretical dimension of applied governance paradigm", the practical change of the application model theory based on the United States is to make special legislative regulations for different scenarios of AI application. However, with the application of pre-training, deep learning, plug-ins, etc., the alignment ability of GAI (alignment) has gradually improved, the application scenarios have exploded, and the "GAI+" effect has become prominent [15].[3] In addition, ChatGPT can also interact with more than 5,000 third-party plug-ins [4], in the situation of rapid growth of application scenarios caused by model tandem, it is impossible to form a new regulation after a scene, and the supervision model by scenario may face inefficiency.

As far as "The theoretical direction of the subject governance paradigm" is concerned, China often stipulates different subject responsibilities for different subjects in the governance of the field of artificial intelligence, which seems to be able to give due obligations to all participants under the same participation content, but in practice it also lacks theoretical inscription and flexibility. Specifically, it is difficult to provide focused instructions under multi-risk concurrency. For example, although the *Provisions On The Administrations Of Internet Information Service* stipulates that service providers have the responsibility of algorithm entities and provides specific requirements, there is no further indication on how to deal with risks such as algorithm operation collapse, scientific and technological ethics chaos, personal information data leakage, and security emergency incidents that may often occur in GAI. Second, the application of subject transformation is not considered, for example, in the whole application process of GAI, there are three major subjects: technology providers, service providers and service users. Technology providers and service providers generally need to fulfill the responsibility of ensuring the normal operation of algorithms and training algorithms. However, for service users, during the interaction with deep learning GAI applications, the information, data, and dialogues provided by them may

---

[2] For example, according to the risk classification method of the AI Act, chatbots belong to limited risk scenarios, and ChatGPT belongs to this category in situations such as private entertainment conversations; However, if the generation of false information or reactionary information affects public order, it can no longer be included in the limited risk and only given a low degree of supervision.

[3] Alignment refers to guiding an AI system progressively to meet the designer's intended usage goals.

314 L. Ren and J. Wu

feed the model, promote the evolution of the model [12]. In this regard, service users may also have duty attributes similar to those of technology providers, so it may be difficult to accurately define the responsibility of the subject in a fixed definition.

2.2 The Dilemma of Theoretical Implementation in the Attempt to Update the Regulatory System

As a disruptive innovation technology, GAI will force governance and regulatory reform [23], the traditional regulatory system theory will face a huge impact, and the control of theoretical explanations such as regulatory timing and regulatory methods may be inadequate, so that there will be a lot of abuse in the regulatory "Vacuum period" and "No man's region". Therefore, it is particularly important to analyze the sources of the implementation dilemma of the current regulatory theory system in order to solve the follow-up problems.

GAI is a new proposition for technology, governance and supervision should also be redone with a new answer sheet. How to smoothly and efficiently in the process of answering questions has become the focus of the international community. The British philosopher of technology David Collingridge, D. pointed out in his book *The social control of technology* that if you are worried about the adverse consequences of a technology and control it too early, then the technology will be difficult to use explosively; on the contrary, if it is controlled too late, it may go out of control, and then solving the problem will cost a lot of money, or even not solve [20]. The hesitation of controlling the new technology is "Collingridge's Dilemma". In order to solve this dilemma, countries have made different theoretical attempts on GAI supervision: for example, the United States adopts an "Encouragement regulatory theory" model for GAI, delegating the industry to self-discipline, while the EU adopts a "Conservative regulatory theory" to form a unified management [8]. However, the regulatory model tried by many countries still faces contradictions, "Encouraged regulation" is conducive to GAI R&D and development, but it is prone to the risk of insufficient supervision and development going astray, resulting in regulatory dilemmas; "Conservative regulatory theory" is conducive to overall guidance and strengthening protection, but it will also lead to poor industry enthusiasm and make supervision use-less. Therefore, a mere model of regulation theory of conservative or r encouraged is not a good approach.

Considering that the risks involved in the use of GAI will involve multiple fields, the international community usually adopts two regulatory system settings to seek regulatory settings covering the full risk spectrum (Table 2). One is " Multi-party regulatory system", for example, China's regulatory authorities for GAI cover many national agencies such as the Cyberspace Administration of China, the Ministry of Industry and Information Technology, and the Ministry of Science and Technology. The other is "Centralized regulatory system", which is to bring the responsibility related to the whole process of GAI supervision into a single department (agency) for supervision, such as the Telecom Regulatory Authority of India (Trai) proposed the establishment of the Indian Artificial

Table 2. The main regulatory systems and corresponding countries (organizations) of the international community on generative AI governance [1].

Regulatory system Canonical model theory	The main body	Regulatory authorities give a brief description of examples
Multi-party regulatory system	China	For example, the Cyberspace Administration of China (Network Management Technology Bureau of the Central Cyberspace Administration of China) is responsible for algorithm filing and evaluation. The Telecommunications Departments and Public Security Departments are responsible for relevant supervision and management work
	United States	The competent authorities in each field supervise the GAI within their own areas of responsibility
	Canada	For example, the Canadian Public Prosecutor's Office, the Office of the Privacy Commissioner and other departments are responsible for AI-related enforcement in their respective fields
	United Kingdom	For example, the UK Information Commissioner's Office, as a data protection authority, supervises the data processing of AI-related subjects. The Competition and Markets Bureau supervises unfair competition and harm to consumers, etc.
	Australia	For example, the Information Commissioner's Office is responsible for enforcing privacy laws to investigate AI-related privacy violations. The Department of Industry, Science and Resources is often involved in the development of AI policy documents
	Japan	For example, Japan's Personal Information Protection Commission regulates the processing of data by AI. The Ministry of Economy, Trade and Industry is often involved in the formulation of policies and guidelines for AI governance, etc.
	Singapore	For example, the Singapore Personal Data Protection Commission will supervise the use and processing of AI data from the level of data protection
Centralized regulatory system	European Union	There is currently no independent regulatory body for AI in the EU, and AI-related enforcement in their respective fields such as the European Data Protection Board (EDPB) and national market surveillance authorities are responsible for AI-related enforcement. However, the latest AI Bill also proposes to set up an "AI Office" at the EU level as a dedicated independent regulatory body
	India	The Telecom Regulatory Authority of India has proposed the establishment of the Artificial Intelligence and Data Authority of India as an independent regulator

Intelligence and Data Authority (AIDAI) as a dedicated and independent regulator [19].

However, both regulatory models have some flaws. For the "Multi-party regulatory system", the participation of too many entities in the governance of GAI may lead to a situation where regulatory entities blame each other in some complex situations, large cost inputs, and small potential benefits. Conversely, in the application scenario where the interests are significant, there may be a situation of competition for supervision. At the same time, this system may also lead to legislation that cannot give detailed guidelines to regulators, or specific rules that conflict with each other, or produce unequal distribution of governance resources. For the "Centralized supervision system", it may be able to solve the shortcomings of multi-entity supervision to a certain extent, but first of all, with the continuous expansion of GAI application scenarios, if there is a special regulatory body, it need to continuously expand the corresponding regulatory authorities and even restructure them, which will face great challenges in terms of cost and utility. Secondly, if the above problems can really be solved and form an independent regulatory body that is perfectly inclusive in all fields, it will also face a paradigm shift in the governance model - will the power overlap between the respective institutions of multiple supervision and the independent institutions with centralized supervision?

3   Dynamic Integration: China's Response and Development Direction

In response to the evolving social situation of GAI and to consider the balance between the efficient development of new technologies and orderly supervision, seven ministries and commissions in China issued the *Interim Measures for the Management of Generative Artificial Intelligence Services* (hereinafter referred to as the *Interim Measures*) on July 10, 2023. It is necessary to discuss it again, so that it is possible to improve China's GAI regulatory system and provide lessons for the world.

3.1   Mesh Specification Under System Optimization

The Marxist system view holds that the system is a whole composed of different elements based on certain relationships, and there are internal logical connections within or between various fields that are almost systematic form [13]. In this sense, the legal norm itself is a large system, and the normative governance paradigm is its subsystem. Furthermore, there are coupling connections between different governance paradigm systems, and coupling the system may achieve the purpose of optimizing the system.

The three specific governance paradigm systems, "Risk Governance Paradigm", "Application Governance Paradigm" and "Subject Governance Paradigm", all have certain shortcomings, but integrating each system can effectively avoid the existence of shortcomings. Therefore, in implementing the previous model of taking entity responsibility as the criterion, Article 3 of the

Interim Measures proposes that the supervision of GAI's should also be classified and graded, reflecting the mesh normative model that balances subjects, scenarios (classification) and risks (grade), and effectively combines the advantages of the three mainstream governance paradigms.

However, as a preliminary legal norm, the *Interim Measures* have certain omissions or ambiguities in their arrangements. In this regard, The author mainly discusses the dynamic integration level of the governance paradigm, and will not discuss the specific clause setting in addition.

First of all, although the "Subject Governance Paradigm" is a continuation of the previous governance model in the *Interim Measures*, in order to prevent the paradigm shift, it is not appropriate for the subject to be classified as a service provider and service user. It can be seen from the specific provisions of the *Interim Measures* that service suppliers bear the dual obligations of technology providers and service providers within this regulation. However, the author believes that , GAI, as a highly technical new technology, in addition to the convenience and safety of model retraining, the rest of the content involving primary technology, including initial model provision, technical problem solving, technical operation and maintenance, etc., should be implemented by more professional technology providers. This arrangement can also enable GAI service providers to not bear excessive technical burdens, and at the same time can invest more experience in "services" to optimize the application experience. Therefore, for the future application governance of GAI, it is more appropriate to divide the subject into three major categories: technology provider, service provider and service user to regulate its behavior, and when the risk of subject conversion occurs, the obligation of the target conversion subject should also be constrained by the pre-conversion entity.

Secondly, as far as the "classification and grade" provisions that integrate the "Application governance paradigm" and the "Risk governance paradigm" are concerned, there is no more expansion, and its abstraction makes it difficult to develop and apply the method in the initial governance of GAI. At the same time, in order to minimize the cost of each management need to determine which type and level of GAI services are located in different institutions, a more stable formal appearance marked with its type and level can play a role similar to qualification description. However, the provisions should not be too narrow, otherwise it will fall into the short-comings of the original risk and application governance paradigm when applied alone, which is too partial and inflexible.

In view of this, reference may be made to articles 17[4] and 23 paragraph 1[5] of the *Interim Measures*, this qualification description is provided by filing

---

[4] Article 17 provides: "Where generative AI services with public opinion attributes or social mobilization capabilities are provided, security assessments shall be carried out in accordance with relevant national provisions, and the procedures for filing algorithms, modifications, and cancellation filings shall be performed in accordance with the *Provisions on the Administration of Internet Information Services* Algorithm Recommendations".

[5] Article 23 stipulates: "Laws and administrative regulations provide that the provision of generative AI services shall obtain relevant administrative licenses, and providers shall obtain licenses in accordance with law".

or licensing. However, the author believes that due to the diversity of GAI use scenarios and risks, only "Generative AI service providers with public opinion attributes or social mobilization capabilities" should not be allowed to file as the *Interim Measures* do, and such services may require a more binding model other than filing to mitigate the risk of large social impact that they may generate. Therefore, the author believes that adopting the two-level constraint model of filing and licensing to structure the negative list licensing system may have a good application here.

Specifically, the first step is to draw on the risk classification standards of the EU and Germany, and consider that in order to better improve the efficient use of GAI in different scenarios, a more detailed split method is adopted to classify the risk level into "Minor risk", "Limited risk", "High risk", "Major risk" and "Unacceptable risk". The second step is to consider the integration of scenario, and divide it into five levels of "Low impact", "General impact", "Large impact", "Extensive impact" and "Unacceptable impact" according to the audience degree and consequence scope of different application scenarios, and form a distribution based on the risk level (Fig. 1). The third step is to prohibit GAI applications from providing services that include "unacceptable risk" or "unacceptable impact" (Fig. 1 shaded background section). GAI applications (Fig. 1 strip background part), should be included in the negative list and obtain administrative permission before providing services. For situations other than the above (blank background in Fig. 1), if GAI has a small possibility of damage in such cases, service filing is fine. In addition, the license and filing certificate will be announced in a conspicuous place each time the service is provided, which can also provide obvious scenario risk judgment for service users to form good psychological expectations.

At the same time, it is considered that the specific provisions of China's *Measures for the Administration of Internet Information Services and Provisions on Security Assessment of Internet Information Services with Public Opinion Attributes or Social Mobilization Capabilities* have relevant provisions on "Review and consent" and "Security assessment" under GAI special applications.[6] Therefore, in the field of special scenarios, it is necessary to implement

---

[6] For example, Article 5 of the *Measures for the Administration of Internet Information Services* stipulates: "Where Internet information services such as news, publishing, education, health care, pharmaceuticals and medical devices are subject to the review and consent of the relevant competent departments in accordance with laws, administrative regulations and relevant national provisions, they shall obtain the review and consent of the relevant competent departments in accordance with law before applying for a business license or performing filing formalities."

The Article 3 of the Provisions on the *Security Assessment for Internet Information Services with Characteristics of Public Opinions or Capable of Social Mobilization* provides: "In any of the following circumstances, Internet information service providers shall carry out security assessments on their own in accordance with these Provisions, and be responsible for the assessment results......".

the two pre-procedures of "Review and consent" and "Security assessment" before proceeding with the "permission-filing" process (Fig. 2), so as to meet the expectations of special operation of macro supervision and special operation of micro self-inspection, and also have a good normative reference role for the external intervention and self-regulation of the international GAI.[7]

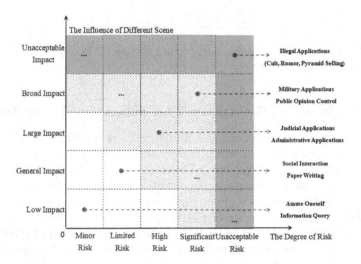

Fig. 1. Illustration of the negative list system based on GAI's application of "Classification and grading" governance.[7]

---

[7] The specific situations in the figure are only examples (e.g., "broad impact" and "significant risk" application scenarios, "political applications", etc.), and the non-examples are not non-existent but omitted (e.g., "widespread impact" and "unacceptable risk" scenarios include "causing large-scale data breaches"). At the same time, in order to prevent the regulations from being too loose, based on the specific provisions in the international community, for example, the European Union classifies chatbots as limited risks, but their risks may be smaller when personal entertainment chats for entertainment, and the risks may be greater when inserted into social software to chat, so the former is classified as "minor risk" and "low impact" situations, and the latter is classified as "limited risk" and "general impact" situations. At the same time, in order to prevent excessive restrictions, the scene is not too limited, such as "social daily life" including transportation booking, restaurant reservation, social chat, etc.; "Information query" includes weather inquiry, date query, etc.

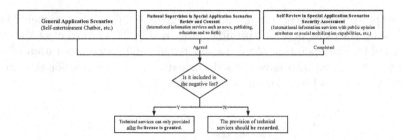

Fig. 2. Illustration of the negative list system based on GAI's application of "Classification and grading" governance.

3.2 The Balance of Governance-Type Supervision Under Antinomy

The German philosopher Immanuel Kant proposed the "Antinomy" principle, refers to the phenomenon in which two theories of the same object or problem are independent but contradictory. Under the general topic of GAI supervision, in order to solve the two specific disputes between the "Vacuum period" of supervision and "No man's region", there are "the emergence of new technologies should be immediately supervised to prevent disorder - the emergence of new technologies should be suspended to make them develop efficiently", and "multiple entities give full play to their respective advantages to supervise in an all-round way - centralizing regulatory rights helps to regulate more efficiently". It can be seen from this that finding the balance point under the Antinomy is an effective attempt to couple the stable order and efficient promotion to achieve the supervision behavior of compliance.

In the early stage of the development of new technologies and new formats such as "Internet+", China proposed to minimize prior access and strengthen supervision during and after the event, and advocated the concept of "Inclusive and prudent supervision" [8]. This regulatory concept has been continued and included in Article 3 of the *Interim Measures*,[8] which has become a general legal norm in the field of GAI to resolve the contradiction between the two pairs of regulators. Among them, "Inclusive" refers to the attitude of tolerance for new business formats that are greater than the unknown and do not touch the bottom line of safety. "Prudence" includes setting a observation period when the development trend of a new business format is unknown and cracking down on illegal acts without discrimination [10]. At the same time, it is like the "Subsumtion" method of testing the compliance of Internet platforms [21]. To test whether GAI's regulatory conduct is compliant, it can also be applied to "Inclusiveness", that is, to determine whether the regulatory behavior complies with the legal

---

[8] Article 3 stipulates: "The State adheres to the principle of attaching equal importance to development and security, promoting innovation and governing according to law, takes effective measures to encourage the innovative development of generative AI, and implements inclusive and prudent and categorical supervision of generative AI services".

norm of "Inclusiveness and Prudence". However, "Inclusiveness and Prudence" is not further specifically explained in the *Interim Measures*, and Chap. 4 of the *Interim Measures* stipulates that the establishment of regulatory entities is led by the competent authorities, and the "Prudent" part may have been satisfied, but there is still a lack of "Inclusiveness" for industry entities to become regulators to self-regulate themselves and enhance development enthusiasm. Therefore, in order to better improve the specific norms of "inclusiveness and prudence" and clarify their application logic, it is necessary to re-analyze and supplement this summary legal norm to form a more perfect "Inclusive" comparative standard to test the compliance of supervision behavior.

As mentioned above, "Inclusiveness and Prudence" should focus on how to solve "Period" and "Regional" problems in a targeted manner, and enhance "Inclusiveness". Some scholars have proposed that "Governance-based regulation" with three core functions of "Open and collaborative supervision rights", "Diversified and integrated supervision methods" and "Compatible and adaptive regulatory measures" [25] may solve the above problems well. Drawing on its core advantages, this paper proposes a regulatory pyramid model (Fig. 3), and explains why it is suitable to solve problems and how to operate it.

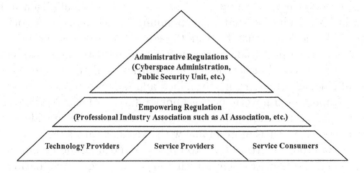

Fig. 3. Schematic diagram of the regulatory pyramid model.

As mentioned above, the problem of "Period" mainly lies in the control of the timing and strictness of supervision, and there are drawbacks in the strict intervention of the competent authorities in the early stage of GAI development while technology providers, service providers, and users develop under constraints, or the mode selection in which GAI is self-adaption in the early stage of development and the competent authorities intervene later. In other words, whenever the two main groups of competent authorities, technology providers, service providers, and users fluctuate greatly, it is difficult to find a balance point of Antinomy. Therefore, the regulatory pyramid allows administrative authorities and industry entities to participate in GAI supervision at the same time, and adds neutral supervision of professional, popular and manageable professional industry associations between the two, effectively balancing the excessive expansion of one of them. Realizing that administrators and industry players have participated in supervision from beginning to end and developed in a relatively stable state can effectively solve the problem of inappropriate regulatory timing and unsuitable supervision.

Second, how does the regulatory pyramid solve the problem of "Region"? As mentioned above, the problem of "Region" mainly lies in the exploration of the arrangement and division of labor of regulatory entities. Under the existing model, the redundancy and fragmentation of the "Multi-party regulatory system", the paradigm shift risk of the centralized regulatory system and the high restructuring costs make them unable to become an adapted GAI regulatory system. In the regulatory pyramid, by empowering some non-decisive administrative powers to professional industry associations to make them coordinate supervision, one can centralize part of the regulatory power to a certain extent, and to a certain extent, it can replace multiple administrative subjects with the same regulatory power, and prevent regulatory decentralization and power conflicts. Second, since there are already many areas in the existing IP system to explore attempts to coordinate regulation between competent authorities and professional trade associations (e.g. geographical indications), there is a deep practical experience that can also reduce the risk of a paradigm shift. Also, there are many professional industry associations related to GAI (such as the Internet Society of China, the Shanghai Artificial Intelligence Technology Association)[9]. The international community, such as the "Association for Generative AI", no longer has to bear the cost of a large restructuring. Third, because of its strong professionalism, professional industry associations can use their good control capabilities, design capabilities, and training capabilities to explore and efficiently implement diversified and integrated regulatory methods, such as developing GAI governance technology (govern tech) to explore AI self-governance, and promoting ethical algorithm design to achieve wide application when the output content is more in line with excellent values. Explore the use of "Regulatory Sandbox" to train GAI to optimize and upgrade technical models. Fourth, professional trade associations help promote regulatory harmonization. Professional industry associations or industry alliances formed can provide GAI industry players with more detailed and targeted regulations by issuing self-binding conventions and norms,[10] that is, their "Downward benefits". At the

---

[9] Although this article establishes industry associations as intermediate first-level regulatory bodies, in fact, scientific research institutions, universities, etc. can also cooperate with the strong subject of supervision. However, this paper believes that the selected middle-level regulatory body should not only connect with the masses, but also form good communication with the government. Therefore, industry associations, social groups with easy contact with the government and a practical bias, have become the best choice.

[10] Practices such as Sina, Tencent, and other 18 well-known Internet companies jointly initiated the establishment of an alliance and released the *Regulation and Competition Norms for the Use of Online Copyright and Data Information in China*, an industry standards, that is, through professional industry associations or industry alliances formed by them, and through the issuance of self-discipline conventions and norms, to provide more detailed regulations for relevant industries.

same time, it can also report the regulatory data it directly contacts to the competent authorities, and directly communicate with the administrators, and adjust the mode of regulatory behavior, that is, its "Upward benefit". Through the information connection of the two benefits of upward and downward, the flexible adjustment and unification of regulatory measures can be realized.

It can be seen that unlike traditional regulatory methods, the regulatory pyramid generally focuses on listing administrative entities as regulators, and it also includes GAI technology providers, service providers, and service users in three types of industries to achieve self-regulation. This kind of industry self-discipline method can not only cooperate with the competent authorities to improve regulatory efficiency, reduce regulatory costs, and achieve detailed supervision, but also enable the industry to maximize initiative and achieve independent control and development.

4   Postscript: GAI, Is It a New Era or a Phantom?

Countries are now actively exploring the setting and improvement of GAI industry normative models and management systems, but if they expand their horizons, they will find that the focus on GAI only exists in parts of North America, Europe, and East Asia, and even some of these countries (regions) have not yet made any attempts, but only indicate that there will be no mandatory supervision planning in the future [6]. It can be seen that the current attention of the GAI industry is still difficult to reach the global level. In this process, there will be many reports indicating that people are watching the GAI industry, focusing on whether it is a "bubble" or a "long-term trend", will it become a "phantom" like the "Metaverse" [11]?

In fact, whether from the perspective of technical mechanism or policy trends, GAI will open a new era of new applications of AI. From a technical point of view, the "Metaverse" and GAI are not mutually exclusive, but complement each other. Technology is the underlying architecture of the "Metaverse", and when many technologies are still in the conceptual stage, the systematic project of the "Metaverse" will face the barrel effect and is difficult to achieve widely [9], so the "Metaverse" has been outraged due to the lack of hardware and too hasty ideas. As one of the three major infrastructures of the metaverse [6], GAI can provide important technical support for the "Metaverse" in the virtual layer, virtual connection layer, and reality layer [16]. That is, it is even a technological milestone that can drive the "Metaverse" to catch fire again. At the same time, GAI is a phased achievement of the gradual and stable development of AI, which has been transformed into reality and has a sustainable development of technical heritage. From the perspective of policy, the introduction of policies in many countries in the world at the GAI stage reflects an explosive update concern. Taking China as an example, in the context of data being listed as the fifth major element of society, the data element market has opened, and the value of elements can only be realized after combining with AI to generate intelligence [5]. In this context of the dual focus of data and intelligence, China has successively issued the *Opinions of the CPC Central Committee and the State Council on Building a Data Infrastructure System to Better Play the Role of Data Elements* (hereinafter referred to as *Data Twenty*), *Interim Measures for the Management of Generative Artificial Intelligence Services, Practice Guidelines for Cybersecurity Standards - Generative Artificial Intelligence Service Content Identification Method, Information Security Technology - Data Security Risk Assessment Method,* and a series of other opinions, regulations, standards, guidelines and other policy documents, it shows that the country's development of the GAI industry will be implemented from conceptualization to culture to ensure and then sustainable development of technology and reality.

Looking at the new era opened by the GAI, it seems that some areas of problems have not been dealt with beyond the existing regulations [14]. It is hoped that in the future, countries around the world will explore the development theory of GAI industry, believe that China can bring new experience focusing on the development of GAI industry for international reference, look forward to the new experience of continuous iterative upgrading of GAI technology, and firmly believe in the new blue ocean brought by the new era of GAI.

References

1. AnJie Broad Releases Global Generative Artificial Intelligence Regulatory Research Report. https://mp.weixin.qq.com/s/SMJMbq1BRyqxyIJ_nnZiWw, 30 Aug 2023
2. Approval of the Development Concept for Adjusting the Relationship between Artificial Intelligence and Robot Technology before 2024. https://www.zakonrf.info/rasporiazhenie-pravitelstvo-rf-2129-r-19082020/. 30 Aug 2023
3. Canada: Digital Charter Implementation Act 2022 - What you need to know. https://www.dataguidance.com/opinion/canada-digital-charter-implementation-act-2022-what. 30 Aug 2023
4. ChatGPT major update: the ability to connect to acquire new knowledge and interact with 5000+ applications. https://www.thepaper.cn/newsDetail_forward_22428596. 30 Aug 2023
5. CTO of Daily Interaction: In the era of GAI, big models promote the commercialization of data element, https://mbd.baidu.com/newspage/data/landingsuper?pageType=1& context=%7B%22nid%22%3A%22news_9067945949199531083%22,%22ssid%22%3A%- 22e6ec3739%22%7D. 30 Aug 2030
6. GAI technology and the metaverse. https://caifuhao.eastmoney.com/news/20221218095132-727531050. 30 Aug 2023
7. Germany's data ethics commission releases 75 recommendations with EU-wide application in mind. https://algorithmwatch.org/en/germanys-data-ethics-commission-releases-75-recommendations-with-eu-wide-application-in-mind/. 30 Aug 2023
8. In the face of the GAI boom, what is the regulatory attitude of various governments? https://mbd.baidu.com/newspage/data/landingsuper?pageType=1&context=%7B%22nid%22%3A%22news_9458492380571645530%22,%22ssid%22%3A%22e6ec3739%22%7D. 30 Aug 2023

9. KPMG China: The metaverse needs to break through the technology bottle-neck. https://mq.mbd.baidu.com/r/15enmKjndcI?f=cp&u=18fec69e8d4d7ea1. 30 Aug 2023

10. Li Keqiang explained in detail why the new business format is "inclusive and prudent" supervision? https://www.gov.cn/guowuyuan/2018-09/12/content_5321209.htm. 30 Aug 2023

11. Metaverse team layoffs on a large scale, will ChatGPT be the next meta-verse? https://m.163.com/dy/article/ICU2FTLM0518SE41.html?spss=adap_pc. 30 Aug 2023

12. Regulating AI in Europe: Four Problems and Four Solutions. https://www.adalove-laceinstitute.org/wp-content/uploads/2022/03/Expert-opinion-Lilian-Edwards-Regulating-AI-in-Europe. 30 Aug 2023

13. Several principles of the Marxist system view. http://www.xinhuanet.com/politics/2021-05/11/c_1127430925.htm. 30 Aug 2023

14. Some thoughts on the legal regulation of generative artificial intelligence. https://mp.weixin.qq.com/s/fGYeySaP5UtG-XtU6YSAuA. 30 Aug 2023

15. Tencent Releases GAI Development Trend Report: Embracing the Next Era of Artificial Intelligence. https://mp.weixin.qq.com/s/9AjTpyL4HmQ6BDhWIDbD0A

16. The relationship between the metaverse and GAI. https://mp.weixin.qq.com/s/QBzu-53Aw05lKmqY7LaFcJQ. 30 Aug 2023

17. "The threat is comparable to nuclear war", more than 350 experts and executives jointly warned! https://mbd.baidu.com/newspage/data/landingsuper?pageType=1&context=. 30 Aug 2023

18. Top 10 Emerging Technologies of 2023. https://www.weforum.org/reports/top-10-emerging-technologies-of-2023/digest. 30 Aug 2023

19. Trai issues recommendations on AI, stresses 'urgent' need for regulatory framework. http://www.livemint.com/technology/tech-news/trai-issues-recommendations-on-ai-says-regulatory-framework-for-development-of-responsible-ai-urgently-needed-11689859911432.html. 30 Aug 2023

20. Collingridge, D.: The social control of technology (1982)

21. Enming, X., Laitan, R.: The "phased" obligation normative construction of internet platform private information processing: Taking the deconstruction of positive legal order as the dimension. Shanghai Legal Studies 5(14), 93 (2023)

22. Hacker, P., Engel, A., Mauer, M.: Regulating chatgpt and other large generative ai models. In: Proceedings of the 2023 ACM Conference on Fairness, Accountability, and Transparency, pp. 1112–1123 (2023)

23. Wenxuan, B.: Risk regulation dilemma of generative artificial intelligence and its solution: from the perspective of chatgpt regulation. J. Comparative Law 2023(3), 170 (2022)

24. Xiangli, L., Hongjun, X.: Ethical regulation of artificial intelligence in the soft law paradigm: an analysis of the Japanese institution. Contemporary Econ. Japan 42(4), 28 (2023)

25. Xin, Z.: Algorithmic governance challenges and governance regulation for genera-tive ai. Modern Law Sci. 45(3), 117–122 (2023)

26. Xiong, Z., Zheng, L., Hui, Z.: The regulatory path of artificial intelligence in the eu and its enlightenment to china: taking the artificial intelligence act as the analysis object. E-Government 65(9), 63–72 (2022)

Intelligent Forecasting of Trademark Registration Appeal with TF-IDF and XGBoost

Qun Wang[1]([envelope]) (iD), ShuHao Qian[2], JiaHuan Yan[3](iD), Hao Wang[4](iD),
and XiaoTao Guo[5]([envelope]) (iD)

[1] Shanghai Maritime University, HaiGang Avenue, Shanghai 201306, China
202230311326@stu.shmtu.edu.cn
[2] Shenzhen Research Institute of Big Data, Room 201, Daoyuan Building,
2001 Longxiang Road, Longgang District, Shenzhen 518115, China
[3] East China University of Political Science and Law, Wanhangdu Road,
Shanghai 200042, China
[4] Tongji University, Siping Road, Shanghai 200092, China
[5] Hangzhou Dianzi University, 2 Xiasha, Jianggan District, Hangzhou 310018,
Zhejiang, China

Abstract. Against the backdrop of rapid advancements in information technology, predictive algorithms are increasingly being integrated into various industries and domains. Despite the global prominence of this trend, the application of such algorithms within the niche of trademark law, particularly in China, has not yet been explored or developed extensively. Given the escalating volume of trademark-related disputes and the strain on the existing administrative review systems tasked with managing such caseloads, the incorporation of predictive algorithms to enhance the efficiency of processing these non-litigious administrative actions is paramount. This study innovates by synthesizing the TF-IDF algorithm with the XGBoost model to develop a first predictive model for trademark rejection appeals. The model demonstrates remarkable performance with an accuracy rate of 68%, marking a significant academic contribution by filling a research void and proving its practical worth. From the perspective of trademark applicants, the model offers data-driven decision support that mitigates time and financial costs. For administrative and review bodies, it promises to reduce systemic costs associated with handling trademark rejection appeal cases, thereby optimizing efficiency. The model's codebase is made available to the public, accessible at: https://github.com/ValeriaWong/Trademark_Appeals.

Keywords: Trademark · Trademark Registration Appeal ·
Forecasting Model · XGBoost

1 Introduction

With the continuous deepening of China's trademark and brand strategy, China's trademark and brand construction has been strengthened, and the number of effective trademark registrations has ranked first in the world for 16 consecutive years. [4] However, as the number of trademark applications continues to expand, how to reduce the cost of time and money required for trademark applications and save administrative resources for trademark application review has become an increasingly prominent issue. According to China's "Trademark Law", the legal process for successfully applying for a registered trademark takes as short as 12 months and as long as 30 months. The examination period before making a decision on preliminary examination and publication is 9 months, and the publication period is 3 months. If an opposition is filed during the publication period, the Trademark Office will initiate opposition proceedings for a further 12 months after the expiration of the publication period, and even under special circumstances, this period can be extended to 18 months [1]. Although China's trademark examination department has reduced the average examination cycle of trademark registration to less than 8 months, its time cost still cannot be taken lightly [2]. In addition to the time cost, the monetary cost that a registered trademark applicant spends in order to file a trademark application should not be ignored. In today's increasingly competitive business environment, it has become imperative to capture business opportunities by increasing the success rate of trademark applications to minimize time costs.

According to the provisions of China's Trademark Law concerning the statutory process of registered trademark application, the applicant for registered trademark shall submit an application for registration to the Trademark Office, and it shall make administrative decisions such as preliminary examination and approval announcement, decision on granting registration, decision on rejecting the application, and decision on not registering the application after examination. The administrative decision documents of the above decisions are an objective reflection of the original appearance of the trademark application examination practice in China, and also an important basis for the application of text mining technology for artificial intelligence prediction. The decision documents of the Trademark Office provide a valuable data source for this study, which makes it possible to predict the trial results of trademark applications through machine learning models.

This study uses machine learning and natural language processing techniques to predict the outcome of the Trademark Office's adjudication of registered trademark applications, mainly using prediction algorithms to analyze China's trademark refusal re-examination processing data. Predictive algorithm [3] is a kind of algorithm based on decision-making algorithm to analyze historical data, summarize the historical features of the data, and predict the possible results in the future on the basis of this algorithm. In this study, we use the predictive algorithm to mine the text of a large number of trademark office decisions, and then construct a predictive algorithm model that can predict the outcome of a particular trademark application after inputting the reasons for the trademark application, trademark information, and classification of the registered trademark applicant. The innovation of this research lies in the fact that it is the first time in the world to explore the outcome prediction model of trademark refusal review, which fills the research gaps in this field and demonstrates the cutting-edge of the research. Meanwhile, in terms of application value and accuracy, the prediction algorithm constructed in this study has a high accuracy of predicting the outcome of trademark refusal review, with a prediction accuracy as high as 68%, which demonstrates its great potential in practical application [4].

From the perspective of importance, the construction of this trademark refusal review prediction model can provide a research basis for the research work on the task of predicting refusal review scenarios in the field of trademark non-prosecution, and the subsequent research work can be more improved on this basis. From the perspective of necessity, the construction of this prediction model has two aspects of necessity:

First, for trademark applicants, in the traditional situation, trademark registration applicants and agents can only manually check the key information in the trademark application guide and the administrative decision documents made by the Trademark Office to speculate on the accurate trademark examination rules, in order to improve the probability of successful application. This process is inefficient, heavily dependent on experience, and the coverage is not comprehensive enough, and it is easy to miss some important key information. After receiving the notice of refusal, when deciding whether to file a refusal review, if the applicant for trademark registration is able to predict the result of the administrative decision of the Trademark Office on the application for registration of a trademark that he/she is going to submit by means of an algorithmic model, the cost of the applicant for registration of a trademark will undoubtedly be greatly reduced. The development of algorithmic predictive models can significantly reduce the time and financial costs involved by providing data-driven predictions of Trademark Office decisions. Such models will mitigate the business risks associated with long-term uncertainty about the status of a trademark by, for example, predicting success rates and providing applicants and organizations with strategic advice on whether or not to file a reexamination of a refusal. Applicants can refine their applications or reexamination requests based on model predictions to maximize compliance with statutory requirements and improve chances of approval. In addition, when selecting a trademark agency, applicants often consider the agency's track record of successful filings. The construction of predictive models with good generalizability can provide trademark applicants with data on the accuracy of their trademark applications, thereby guiding applicants to choose agencies with higher success rates and increasing their likelihood of obtaining trademark registrations.

Second, for the staff of the Trademark Review Board, in the face of the reality that China has limited resources for trademark adjudication and a large number of trademark refusal review cases, the prediction model can provide valuable insights while ensuring the fairness of the review. By analyzing the features of the predicted outcome of a trademark review, the staff of the Trademark

Review Commission can obtain more perspectives to help them make decisions. Such model predictions can be used as a complementary tool to improve the efficiency of trademark reexamination reviews when the circumstances of the case to be handled are simple or similar to those of cases that have been handled in the past. At the same time, for the administrative authorities responsible for trademark registration, the implementation of such predictive models can reduce the influx of unqualified applications, promote more efficient allocation of resources and streamline the examination process [6]. This practice not only optimizes the allocation of resources for trademark evaluation, but also speeds up the examination process, benefiting the administrative authorities, the judicial system and the applicants.

To summarize, the algorithmic model for predicting trademark evaluation results based on a large number of administrative decisions made by the Trademark Office can not only save the time cost of registered trademark applicants, but also save national administrative resources, which is beneficial to the country and the people.

2   Related Work

Previous studies have focused on traditional analysis of legal instruments and empirical judgments, lacking quantitative and quantitative analysis methods.

In recent years, there has been a gradual emergence of research on the use of text mining technology to predict the outcome of legal decisions and the outcome of administrative decisions by mining the text content of legal documents, administrative decision documents and so on. For example, some people have extracted the reasons for appeal and relevant influencing factors of criminal procedure cases through text mining of second-instance judgment documents, and the text mining results can provide more decision-making references for the court's first-instance trial of criminal cases, intelligent sentencing, and so on [5]. However, the relevant research results in the field of registered trademark applications are still relatively small. in the five years between 2013 and 2017, China filed 15,846,000 trademark applications, accounting for 58.2% of the total number of applications globally, and there were 5,748,000 trademark registrations filed in 2017 alone, which is the world's largest [4]. The administrative resources required for such a huge volume of trademark applications as well as the time and money costs of registered trademark applicants will also be huge, so exploring how to apply emerging artificial intelligence technologies such as machine learning and natural language processing to the field of registered trademark applications is of great significance in helping registered trademark applicants to reduce the costs of time and money, to reduce the number of low-quality and low-efficiency trademark applications, and to strengthen the development strategy of branding and innovation-driven development.

Currently, most of the engineering research around the trademark field focuses on natural scene recognition of trademarks [16,20–23], feature extraction and similarity retrieval of graphical trademarks [16,23–25], and trademark similarity analysis [17]. In recent years, there have also been some tasks of analyzing legal documents and predicting results for trademarks. However, once it comes to the research of using text mining techniques to predict the results of legal judgments and administrative decisions by mining the text contents of legal documents and administrative decision documents, we have only found some models or systems for analyzing and predicting documents in opposition cases, such as Zhang, H et al. [14] used SVM and K-nearest neighbor methods to use the data features output from single preprocessed approximation judgments and comprehensive preprocessed approximation judgments as the input for secondary feature filtering, and the input features were included in the training set along with relevant features initialized by semantic similarity, which was suitable for approximation judgments of the data of similar goods in class 34 of the Trademark Notice of Initial Publication to make decision references for the need to file oppositions. Onomatics' intelligent trademark analysis system, [26], has been evaluated in a large number of actual trademark opposition cases, including more than 30,000 from the USPTO TTAB and more than 20,000 from the OHIM Opposition Division, and achieved an accuracy rate of 79.0%. cases, achieving an accuracy rate of 79.9% and a recall rate of 94.9%.

Beyond that, work potentially related to trademark rejection appeal cases is still at the stage of extracting key features. A. S. Li [15] et al. use algorithms such as TF-IDF and LDA to extract key features based on ontology patterns for 4,835 U.S. trademark law cases litigated in U.S. district and federal courts, including the cases' facts, issues in dispute, judgment results, and applicable rules and laws. The TF-IDF was used for text mining to discover key features of the litigated cases. However, we did not find any analytical or predictive models for domestic and foreign trademark rejection appeal cases.

In recent years, the rapid development of machine learning and natural language processing techniques has led to the attention of researchers on their wide range of applications in many fields. XGBoost (Extreme Gradient Boosting Tree), as an efficient end-to-end tree boosting system, has been significantly applied in many important fields such as financial stock prediction [11], medical prediction [12], and energy prediction [13], due to its outstanding performance and the interpretability of the decision tree-based algorithm. XGBoost's success is attributed to its unique algorithm design, such as Clever penalization of trees, Newton Boosting, and Parallel tree structure boosting for sparse data. The success of XGBoost is attributed to its unique algorithm design such as Clever penalization of trees, Newton Boosting, and Parallel tree structure boosting to optimize the processing of sparse data, etc. [7–10]. These features have enabled XGBoost to demonstrate excellent predictive power and interpretability in a wide range of real-world problems.

Although XGBoost has shown its powerful prediction ability in many fields, its application research in the legal field, especially in trademark law, still needs to be deepened. To the best of our knowledge, the TF-IDF and XGBoost prediction model proposed in this paper is the first current model for predicting the outcome of trademark refusal review cases.

3   Predictive Modeling

3.1   Data Soures and Model Construction Ideas

Under the existing trademark dispute resolution mechanism, if a prior right holder believes that a trademark applied for or registered by another person infringes on his/her rights, he/she needs to file an opposition or invalidation claim with the trademark administrative authority, and if the claim is found to be valid, the trademark applied for or registered by the other person is ruled not to be registered or invalidated. In this study, we mainly refer to the data from the Trademark Office of the State Intellectual Property Office (SIPO), which provides comprehensive and authoritative information on trademarks, including information on trademark applications and decisions on refusal review. This website is an authoritative source in the field of trademarks in China, which not only provides professional search of trademarks, but also the "Trademark Review and Adjudication Decision/Decision Instruments" section can also search for non-litigation instruments such as refusal review and declaration of invalidity in the next period of time. After information collection and extraction, a total of 18,459 samples of trademark refusal review documents were collected, and the main fields included trademark name, content of the document, name of the cited trademark, result of refusal review, basis of the decision, rejected goods, passed goods, applicant, authorized agent, and the main reasons of the applicant's review, etc. In the data cleaning process, we have collected a total of 18,459 samples of trademark refusal review documents. In data cleansing, we deleted fields with a missing rate of more than 90% and filled in the remaining missing values to ensure data quality. Based on the domain knowledge, the similarity between the trademark name and the cited trademark name was calculated using simple cosine similarity after constructing the word vector space using TF-IDF.

In this paper, mutual information scores are used to assess the correlation between feature values and target variables, and to improve model performance and reduce the risk of overfitting by calculating the mutual information between features and target variables. Mutual information, denoted as I(X;Y), is a non-negative value used in information theory to measure the statistical correlation between two random variables X and Y. It is used to measure the correlation between two random variables. [28] It quantifies the amount of information obtained from one random variable about another. The concept of mutual information is based on Shannon entropy and can be defined as (1):

$$I\left(X;Y\right) = \sum_{x \in X} \sum_{y \in Y} p(x,y) \log \frac{p(x,y)}{p(x)p(y)} \tag{1}$$

where p(x, y) is the joint probability distribution function of X and Y, and p(x) and p(y) are the marginal probability distribution functions of X and Y respectively. Taking the trademark refusal review documents as an example, suppose there are a series of trademark names and the refusal review results of trademark documents corresponding to them. By calculating the mutual information between the trademark name and the result of the review, we can quantitatively determine that there is a statistical dependence between the two. That is to say, the information of trademark name can reduce the uncertainty about the result of validation and increase the certainty of the prediction of the result.

Mutual information analysis can identify the degree of statistical correlation between different features and the target value, and thus provide a basis for feature selection and effect enhancement of the model. Based on the mutual information of the features, we can select the most relevant features to construct the model and eliminate the redundant and irrelevant features. Based on the results of mutual information analysis, this study will select the key features and verify the effects of different features on the model performance.

Table 1. Mutual Information Analysis Score Statistics

Name: MI Scores	dtype: float64
brandName	1.47672
applicant	1.2034
cited_brand_name	0.380868
authorized_agent	0.287432
main_reason_for_review_by_applicant	0.062146
similarity_score	0.004975

As can be seen from Table 1, the trademark name and the identity of the applicant are the two factors most strongly correlated with the result of trademark refusal review. Correspondingly, in the process of trademark application and reexamination, the choice of trademark name as well as the applicant's main business scope and brand building will have an important impact on the result of rejection and reexamination. Other features such as the cited trademark name and the commissioned agent also have a certain impact, while the applicant's main reason for reexamination and similarity scores have a lesser impact. These mutual information score data can provide valuable guidance on feature selection for building prediction models.

In view of this, based on the data derived from mutual information scoring, this paper predicts and analyzes the refusal reexamination by using the trademark similarity rule as the judgment criterion and combining the TF-IDF algorithm and XGBoost model. We hope that through this method, we can more effectively handle and resolve trademark confirmation disputes, and further optimize and improve the existing trademark registration and protection mechanism.

3.2   TF-IDF Algorithm

When processing the text data of trademark instruments, the text is first processed by cutting and breaking the key words. Subsequently, the TF-IDF (Term Frequency-Inverse Document Frequency) algorithm [27] is utilized to measure the relative importance of individual words in the text. The core of the TF-IDF algorithm lies in the fact that it assigns weights to a word by the frequency of its occurrence in the text (TF) and by the frequency of the word across the corpus (IDF) of the entire corpus to give it weight. Here, TF refers to the frequency of occurrence of the word in a single document, while IDF represents the rarity of the word in the entire document set and is used to measure the amount of unique information it provides. Combining these two metrics, the TF-IDF algorithm reveals not only the importance of the word in a single document, but also its rarity and informative properties in the entire corpus.

The specific steps to calculate TF-IDF are as follows: To calculate the word frequency (TF), this study adopts the primitive counting method, i.e., the number of occurrences of a word in a document is divided by the total number of occurrences of all words in that document:

$$TF_{ij} = \frac{n_{ij}}{\sum_k n_{ik}} \tag{2}$$

where n_{ij} is the number of occurrences of the word t_i in the document d_j, and $\sum_k n_{ik}$ is the sum of all occurrences of the word in the document d_j.

Calculate the inverse document frequency (IDF), which is the logarithm of the total number of documents in an anthology divided by the number of documents containing the word t_i, this study adopts the smoothed inverse document frequency for smoothing to prevent division by zero, which ensures that the IDF value remains well-defined even when a word appears in every document.

$$IDF_i = \log \frac{1 + |D|}{1 + |\{j : t_i \in d_j\}|} + 1 \tag{3}$$

where $|D|$ is the total number of documents in the anthology; $|\{j : t_i \in d_j\}|$ is the number of documents containing the word t_i; and, the numerator and denominator are added with a term for smoothing to prevent division by zero. The final 1 is added to ensure that each word in the document set has at least one non-zero IDF value, even if it occurs in every document in the corpus.

Combining TF and IDF yields the TF-IDF value, which represents the importance of the word in the document:

$$TFIDF_{ij} = TF_{ij} \times IDF_i \tag{4}$$

For example, if the word "distinctive feature" appears 5 times in a document, and the total number of occurrences of all the words in that document is 200, then its TF value is 0.025 (5 divided by 200). If 10 documents in a set of 100 documents contain the word "distinctive feature", their IDF values are $\log(\frac{1+100}{1+10}) + 1$ or $\log(\frac{101}{11}) + 1$. Therefore, the TF-IDF value of the "distinctive feature" in this document is $0.025 \times (\log(\frac{101}{11}) + 1)$. This TF-IDF value is used to determine the weight of the words, which can be further used for document classification or regression analysis. In summary, TF-IDF plays a key role in therejection appeal prediction model of trademark instruments by converting text into numerical features recognizable by machine learning models. In this study, we selected the top 100 terms in terms of TF-IDF weights as numerical representations of the instruments. Each legal instrument is represented by a 100-dimensional vector, and each dimension corresponds to the TF-IDF weight of a vocabulary, providing support for the model to capture the core information of the instrument. In this way, we take advantage of the frequency of vocabulary occurrences in documents and their distributional properties across the corpus to highlight those words that are highly discriminative to the content of the instrument, which in turn enhances the accuracy and efficacy of the model predictions.

3.3 XGBoost Model

XGBoost (Extreme Gradient Boosting Decision Tree) is an open source machine learning project being developed by Tianqi Chen [10] and others. It is essentially a variant of Gradient Boosting Decision Tree (GBDT) [10], which adopts the Boosting idea. The basic idea is to stack base classifiers step by step, and each layer will give higher weight to the previous layer of base classifiers misclassified samples during the training process [29]. In the scenario of trademark refusal review, the algorithm can automatically extract important features and construct an efficient prediction model by learning historical data. By training and tuning the model, we can achieve accurate prediction of trademark refusal reexamination results.

Specifically, we take as input the dataset of non-prosecution documents, which contains 18,459 samples and 18 features, where

$$D = \left\{ (m_i, n_i) \mid m_i \in R^{18}, n_i \in R \right\}$$

, and

$$m_i = \{ m_{i1}, m_{i2}, \ldots, m_{i18} \mid i = 1, 2, \ldots, 18459 \}$$

The main task of the XGBoost model is to build q trees so that the predicted values $\hat{n}_i^{(q)}$ up to the qth tree, i.e., we will get an XGBoost model containing q trees for predicting the outcome of the trademark refusal review [29]. The predicted values of the model are calculated as follows:

$$\hat{n}_i^{(0)} = 0 \tag{5}$$

$$\hat{n}_i^{(1)} = f_1(m_i) = \hat{n}_i^{(0)} + f_1(m_i) \tag{6}$$

$$\hat{n}_i^{(2)} = f_1(m_i) + f_2(m_i) = \hat{n}_i^{(1)} + f_2(m_i) \tag{7}$$

$$\vdots \tag{8}$$

$$\hat{n}_i^{(q)} = \sum_{k=1}^{q} f_k(m_i) = \hat{n}_i^{(q-1)} + f_q(m_i) \qquad (9)$$

These formulas describe the contribution of each tree to the predicted value, while in each iteration of the gradient boosting decision tree algorithm, a weak classifier $f_k(m_i)$, i.e., DT [29], is generated, and the predicted value of this iteration, $\hat{n}_i^{(q)}$, is the sum of the predicted value of the previous iteration, $\hat{n}_i^{(q-1)}$, and the classification result of the current round, $f_q(m_i)$, where we have defined an objective function:

$$\min L^{(q)}(n, \hat{n}^{(q)}) = \min \left(\sum_{i=1}^{18459} l\left(n_i, \hat{n}_i^{(q)}\right) + \sum_{k=1}^{q} \Omega(f_k) \right) \qquad (10)$$

The function contains a loss function and a regularization term. The loss function $l(n_i, \hat{n}_i^{(t)})$ is used to measure the error between the true value and the predicted value, while the regularization term $\sum_{k=1}^{q} \Omega(f_k)$ is used to control the complexity of the model and prevent the occurrence of the overfitting phenomenon. In this way, we can obtain a predictive model of trademark refusal review that is both accurate and stable, providing valuable decision support for practical applications (Fig. 1).

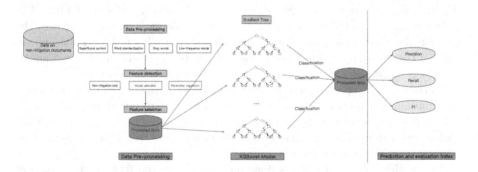

Fig. 1. XGBoost model framework diagram

As shown in Fig. 2, XGBoost has a feature importance scoring function that identifies the most informative features for the prediction of the target variable from 18 features such as applicant reason, trademark name, cited trademark name, applicant name and main reason. By calculating the split contribution of features in the decision tree, the importance of features in the model can be assessed, helping to understand what factors play a key role in refusal review prediction [30], while XGBoost employs a regularization strategy that includes the introduction of L1 and L2 regular terms during the training process to prevent the model from overfitting. This helps to maintain the generalization ability of the model when receiving new data from non-prosecution trademark instruments, thus improving the accuracy of prediction. While there may be category imbalance in trademark instruments, such as a large difference in the number of rejected and passed cases, XGBoost is able to improve the model's prediction accuracy for a few categories by adjusting the sample weights. Finally, XGBoost is able to automatically capture the interaction and non-linear relationship between the features of trademark instruments, so as to more accurately describe the complexity of the data. The complex interactions between various factors in trademark instruments for refusal review need to be fully considered, and XGBoost is able to better fit these relationships.

In summary, XGBoost acts as a powerful prediction tool in the trademark instrument prediction model for refusal reexamination, and through its features such as feature importance scoring, regularization, category imbalance processing, and feature interaction, it is able to improve the performance of the model and predict the refusal or refusal of the case more accurately.

4   Experiments and Analysis of Results

4.1   Experimental Procedure

This experiment consists of 4 steps, as shown in Fig. 3:

The first step is to collect Trademark Refusal Appeal Decision Data from the Trademark Office of the State Intellectual Property Office of China. The collected and extracted data fields include the trademark name, content of the decision document, name of the cited trademark, result of the refusal appeal, basis of the decision, rejected goods, passed goods, applicant, authorized agent, main reason for the applicant's appeal, and other relevant information.

In the second step, data processing is performed on the collected instrumental information, which is manifested in fields with missing values above 90% threshold are deleted, and the remaining missing values are imputed with appropriate strings, and categorization labeling to change the appeal results to categorization labels 0, 1, and 2, and so on.

In the third step, TF-IDF is done on several selected features to construct the word vector space. Use cosine similarity in the constructed word vector space to calculate the vector inner product between the trademark name and the cited trademark name features as similarity features to be added to the dataset.

In the fourth step, the training set, validation set, and test set are divided according to 70%, 15%, and 15% of the data proportion. Also use GridsearchCV hyperparameter optimization to seek the best hyperparameters.

4.2   Parameter Tuning

In this paper, we utilize GridSearchCV to determine the optimal hyperparameters for the prediction model. We train and test the model with different hyperparameter combinations by a k-fold cross-validation (k = 5) method to ensure

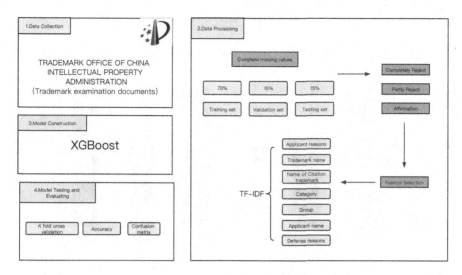

Fig. 2. Experimental Flowchart

the robustness and generalization ability of the model. The model performance is evaluated based on the confusion matrix (Fig. 6) and its derived metrics (Fig. 3) such as accuracy (P), recall (R), and F1 value. In addition, the prediction model constructed in conjunction with the XGBoost algorithm will be analyzed in comparison with the previous results obtained through mutual information scoring, as shown in Fig. 5. The final experimental results will be shown in detail in the next subsection.

4.3   Feature Analysis

In the feature selection process, mutual information scores reveal to us the interdependence between features and target variables, providing us with insights about which features are likely to have a significant impact on the prediction results. However, it is important to note that a high mutual information score does not equate to predictive capability and does not directly translate to better or more effective model performance. For example, certain features that have high mutual information scores with the target variable may not function as expected in model training due to interactions with other features or model-specific processing.

In this study, we utilized key features such as trademark name, applicant identity, etc. to train the XGBoost model and observed that there is a difference in feature ranking between feature importance and mutual information score after model fitting. As shown in Fig. 5, this divergence underscores the XGBoost model's sophisticated handling of features, which may involve intricate interactions within the model's structure, that are not necessarily captured by mutual information scores alone.

A comparative analysis of the features shows:

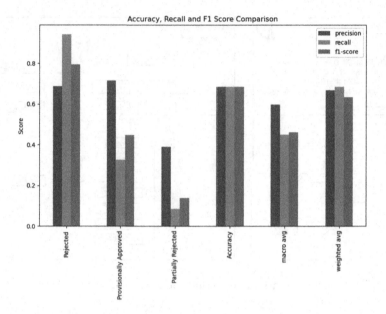

Fig. 3. Comparative Analysis of Model Evaluation Metrics. This bar chart represents the precision, recall, and F1-score for each decision category within the classification model. The three categories assessed are 'Rejected', 'Provisionally Approved', and 'Partially Rejected'.

- **Principal reasons for applicant's review**: unlike in the Mutual Information Score where this feature is ranked fourth, in the XGBoost model this feature is ranked second in importance and is not far from the importance of the feature ranked first. This emphasizes the importance of the details of the reasons for review and their interaction with other features in predicting the outcome, in line with the general perception of legal practice.
- **delegated agent**: ranks first in XGBoost's feature importance ranking, despite being ranked lower in the mutual information score. This contrast implies that although the statistical correlation between the delegated agent and the target category is not significant, it plays a key role in model prediction. This may be related to the significant influence of the agent's expertise and experience on the review results.
- **similarity score**: the mutual information score is close to zero, while it shows some significance in XGBoost. This reflects the strength of the XGBoost model in capturing data details and making predictions.

In summation, the disparity between the rankings of feature importance and mutual information scores underscores their distinct methodologies in evaluating feature relevance and underscores the adeptness of XGBoost in harnessing features for predictive endeavors. The prioritization of features is likely influenced by the multifaceted nature of the dataset and the sophisticated design of the model, particularly evidenced by the marked discrepancies in the significance

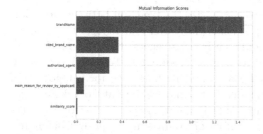

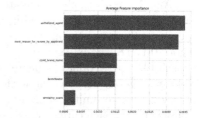

Fig. 4. mutual information score

Fig. 5. Ranking of XGBoost feature importance.

accorded to the primary reasons for review presented by the commissioning agent and the applicant. The noted relevance of similarity scores further corroborates the prognostic efficacy of the XGBoost algorithm (Fig. 4).

In the experiments, the evaluation metrics for a classification task are analyzed and presented in detail in this study. The evaluation metrics involved include precision, (precision), recall, and F1-score which are crucial for evaluating the performance of a classification model. We conducted experiments based on a dataset containing three categories, namely category 0, category 1 and category 2:

mapping =
'Rejected': 0,
'Provisionally Approved': 1,
'Partially Rejected': 2.

Table 2. Model Performance Evaluation Metrics

result_classification	precision	recall	f1-score	support
Category 0 : Rejected	0.7	0.93	0.8	1713
Category 1 : Provisionally Approved	0.71	0.33	0.45	803
Category 2 : Partially Rejected	0.39	0.08	0.14	253
accuracy	0.68			
macro avg	0.6	0.45	0.46	2769
weighted avg	0.67	0.68	0.63	2769

According to Table 2, we observe that category 0 ('Rejected') reaches 0.7 in terms of precision, indicating that the model has a high accuracy in predicting this category. The recall is 0.93, pointing out that the model is able to detect the majority of samples in this category relatively well. In contrast, the recalls for categories 1 ('Provisionally Approved') and 2 ('partially rejected') were 0.33 and 0.08, respectively, implying that the model was far more effective in detecting the rejected category than the other two.

An F1-score was also calculated to synthesize the trade-off between precision and recall. The F1-score combines precision and recall and reflects the model's performance in balancing accuracy and comprehensiveness. The F1-score for category 0 is 0.8, the F1-score for category 1 is 0.45, and the F1-score for category 2 is 0.14. These values indicate that the model performance for category 0 is relatively good, whereas the performance of the models for category 1 and category 2 is more limited. Further, this study focused on the number of supports (supports) for each category, i.e., the number of samples in the dataset belonging to each category. The number of supports for category 0 is 1713, the number of supports for category 1 is 803, and the number of supports for category 2 is 253. These differences in the number of supports indicate a possible category imbalance in the dataset, which may affect the performance of the model on different categories.

The confusion matrix (Fig. 6) and the comparison of the metrics (Fig. 3) further validate these findings.

4.4   Experimental Results

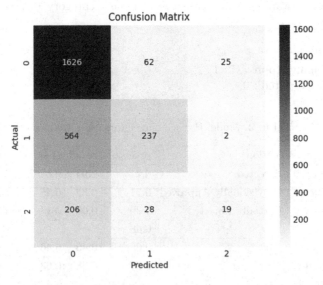

Fig. 6. Confusion Matrix Visualizing Classification Performance. The matrix delineates the number of true positives, false positives, false negatives, and true negatives for each category predicted by the model. The shades of blue represent the magnitude of instances in each category, with darker shades indicating higher frequencies. Category '0' denotes Rejected, '1' corresponds to Provisionally Approved, and '2' signifies Partially Rejected cases. The vertical axis represents the actual classes, while the horizontal axis indicates the predicted classes by the model. (Color figure online)

Taken together, the overall accuracy (ACCURACY), which serves as the overall performance metric of the model on all categories, has an overall accuracy of 0.68, which indicates that the model performs well overall. Also, two evaluation metrics, macro avg and weighted avg, were calculated to better understand the overall performance of the model on each category. The precision, recall, and F1-score of the macro avg are 0.6, 0.45, and 0.46, respectively, while those of the weighted avg are 0.67, 0.68, and 0.63, respectively. These avg metrics provide a comprehensive view of the model's performance on different categories, revealing differences in the model's performance when dealing with samples from different classes

In conclusion, through the above analysis of the experimental results, it can be concluded that the model performs relatively well in detecting the rejected review category. However, enhancements are warranted to bolster predictions for 'Provisionally Approved' and 'Partially Rejected' categories. Potential optimizations include augmenting the dataset, addressing class imbalances, or implementing more sophisticated model architectures to enhance the model's efficacy with challenging samples.

5   Limitations and Future Work

This study develops the first trademark refusal review prediction model and empirically demonstrates its accuracy and effectiveness. The model provides support for trademark applicants' decision-making and optimizes the trademark representation and review process. However, due to the limitation of the dataset, the model may not be comprehensive enough to predict new cases. Future research needs to expand the dataset, enhance the adaptability of regulatory changes, and improve the generalizability of the model. Currently, the model mainly uses TF-IDF for text vectorization, and the representation of semantic features can be optimized in the future by using a larger corpus and word embedding methods. In addition, the model has not yet integrated the visual features of graphical trademarks, which can be extended in the future through complex similarity analysis [17] and multimodal feature fusion algorithms [18,19]. The ultimate goal is to create a predictive model that is more efficient, interpretable and applicable to a wider range of trademark types.

Acknowledgements. The authors would like to express their sincere gratitude to Allen Xiao for his invaluable assistance in data collection and dataset construction. Special thanks are also extended to HuaiCheng Hui for his contributions to the documentation of our open-source project and the subsequent organization of the open-source dataset.

References

1. The Central People's Government of the People's Republic of China. (2020). Trademark law. Accessed 08 Nov 2023. https://www.gov.cn/guoqing/2020-12/24/content_5572941.htm

2. People's Information. (2021). China National Intellectual Property Administration answers poster news: Generally, the trademark registration cycle will be shortened from 8 months to 7 months at the end of the year. Accessed 8 Nov 2023. https://baijiahao.baidu.com/s?id=16991956680266697884

3. Ban, W.J., Jiang, Q., Zhao, W. Research on precise prediction of online learning performance based on multi-algorithm fusion. Mod. Distance Educ. (03), 37–45 (2022). https://doi.org/10.13927/j.cnki.yuan.20220414.003

4. Trademark Office of the State Administration for Industry and Commerce, Trademark Review and Adjudication Board. Annual Report on China's Trademark Strategy (2017). China Industry and Commerce Press, 1 (2018)

5. Wen, S.H., Huang, X.J., Zhou, L., et al.: Text mining of reasons for criminal case appeals: based on second instance document data. Inf. Eng. **7**(01), 113–121 (2021)

6. Yang, S.L.: The application and regulation of predictive algorithms in china's criminal judicial procedures. Chongqing Soc. Sci. (08), 101–111 (2022). https://doi.org/10.19631/j.cnki.css.2022.008.008

7. 'XGBoost', Wikipedia. 30 Oct 2023. Accessed 8 Nov 2023. https://en.wikipedia.org/w/index.php?title=XGBoost&oldid=1182656931

8. Sagi, O., Rokach, L.: Approximating XGBoost with an interpretable decision tree. Inf. Sci. **572**, 522–542 (2021). https://doi.org/10.1016/j.ins.2021.05.055

9. Goyal, K., Dumancic, S., Blockeel, H.: Feature Interactions in XGBoost. arXiv, 11 Jul 2020. Accessed 08 Nov 2023. http://arxiv.org/abs/2007.05758

10. Chen, T., Guestrin, C.: XGBoost: a scalable tree boosting system. In: Proceedings of the 22nd ACM SIGKDD International Conference on Knowledge Discovery and Data Mining, pp. 785–794, August 2016. https://doi.org/10.1145/2939672.2939785

11. Shi, Z., Hu, Y., Mo, G., Wu, J.: Attention-based CNN-LSTM and XGBoost hybrid model for stock prediction. arXiv preprint arXiv:2204.02623 (2023). http://arxiv.org/abs/2204.02623

12. Yang, G., et al.: Language model classifier aligns better with physician word sensitivity than XGBoost on readmission prediction. arXiv preprint arXiv:2211.07047, http://arxiv.org/abs/2211.07047 (2022)

13. Roustazadeh, A., et al.: Estimating oil recovery factor using machine learning: applications of XGBoost classification. arXiv preprint arXiv:2210.16345, http://arxiv.org/abs/2210.16345 (2022)

14. Zhang, H., Wang, X.: Dynamic application of trademark preliminary publication notification data in tobacco enterprises based on semantic similarity and machine learning method. In: Proceedings of the 12th International Conference on Logistics and Systems Engineering, pp. 475–488 (2023)

15. Li, A.S., Trappey, A.J.C., Trappey, C.V.: Intelligent identification of trademark case precedents using semantic ontology. In: Transdisciplinary Engineering for Complex Socio-technical Systems, IOS Press, pp. 534–543 (2020)

16. Chen, H.: Trademark detection algorithm based on artificial intelligence. In: 2023 4th International Conference for Emerging Technology (INCET), pp. 1–6 (2023)

17. Cao, J., Huang, Y., Dai, Q., Ling, W.-K.: Unsupervised trademark retrieval method based on attention mechanism. Sensors **21**(5), 1894 (2021)

18. Trappey, C.V., Trappey, A.J.C., Lin, S.C.-C.: Intelligent trademark similarity analysis of image, spelling, and phonetic features using machine learning methodologies. Adv. Eng. Inform. **45**, 101120 (2020)

19. Lourenço, V.N., Silva, G G., Fernandes, L.A.F.: Hierarchy-of-visual-words: a learning-based approach for trademark image retrieval. arXiv preprint arXiv:1908.02786, http://arxiv.org/abs/1908.02786 (2019)

20. Leng, Y., Fan, Q.: A real-time trademark detection method. J. Phys. Conf. Ser. **1693**(1), 012106 (2020)
21. Qu, H., Zhao, W., Zhang, S., Han, J.: CBIA image recognition trademark implementation on mobile. In: 2022 International Seminar on Computer Science and Engineering Technology (SCSET), pp. 129–132 (2022)
22. Trappey, A.J.C., Trappey, C.V., Lin, E.: Intelligent trademark recognition and similarity analysis using a two-stage transfer learning approach. In: Advanced Engineering Informatics, vol. 52 (2022)
23. Zhang, W.: Application research of trademark recognition technology based on SIFT feature recognition algorithm in advertising design. Int. J. Adv. Comput. Sci. Appl. **13**(12), 821–829 (2022)
24. Vesnin, D., Levshun, D., Chechulin, A.: Trademark similarity evaluation using a combination of ViT and local features. Information **14**(7), 398 (2023)
25. Agarwal, A., Agrawal, D., Sharma, D.K.: Trademark image retrieval using color and shape features and similarity measurement. In: Proceedings of the 2021 8th International Conference on Computing for Sustainable Global Development, INDIACom 2021, pp. 486–490 (2021)
26. Ronkainen, A.: Intelligent trademark analysis: experiments in large-scale evaluation of real-world legal AI. In: Proceedings of the Fourteenth International Conference on Artificial Intelligence and Law, pp. 227–231. ACM (2013)
27. Gao, J., Huang, H.: Text emotion analysis based on TF-IDF and multihead attention transformer model. J. East China Univ. Sci. Technol. https://doi.org/10.14135/j.cnki.1006-3080.20221218002
28. Rahman, W., et al.: Integrating multimodal information in large pretrained transformers. In: Proceedings of the Conference. Association for Computational Linguistics. Meeting (2020)
29. Li, H., Cao, Y., Li, S., et al.: XGBoost model and its application to personal credit evaluation. IEEE Intell. Syst. **35**(3), 52–61 (2020)
30. Meng, Y., Yang, N., Qian, Z., et al.: What makes an online review more helpful: an interpretation framework using XGBoost and SHAP values. J. Theor. Appl. Electron. Commer. Res. **16**(3), 466–490 (2020)

Review of Big Data Evidence in Criminal Proceedings: Basis of Academic Theory, Practical Pattern and Mode Selection

Yicheng Liao[✉]

Zhejiang University, Zhejiang Province, Hangzhou, China
Lyc19960809@163.com

Abstract. Big data materials from emerging technologies such as machine algorithms, robots and advanced artificial intelligence are rushing to the courts. Big data evidence as the product of electronic evidence iteration, mainly in the form of analysis results or reports, the use of big data evidence presents more collaborative advance and the characteristics of stage circulation, but big data evidence in technical dimension present data source and algorithm model level defects, in the legal dimension is beyond a reasonable doubt of proof standard, proof pattern and criminal presumption contradiction. The litigation procedure regulation should be carried out by the path correction of big data evidence transmission, the proof mode should be optimized by the probabilistic analysis of big data evidence and reinforcing evidence, and the illegal exclusion rules of big data evidence should be constructed by the illegal infringement of the digital basic rights as the judgment standard.

Keywords: big data evidence · judicial certificate · algorithm · exclusion of illegal evidence

1 Introduction

With the infiltration of big data, blockchain, artificial intelligence and other emerging technologies into the field of judicial certification, the traditional judicial certification model is transforming to the emerging certification mode represented by big data certification. British scholar victor 2013 had predicted: "big data opened a major era transformation, like a telescope allows us to feel the universe, the microscope allows us to observe microorganisms, big data is changing our life and understand the way we have the world, become the source of new inventions and new services, and more change is gaining momentum". In the judicial world, Victor's predictions are gradually becoming a reality, and in recent years various big data materials from emerging technologies such as machine algorithms, robots and advanced artificial intelligence are rushing to the courts. As the product of electronic evidence iteration, big data evidence is mainly presented in the form of analysis results or reports, and plays a supporting role in the relevant understanding of massive electronic data.

C. Cruz et al. (Eds.): IC 2023, CCIS 2036, pp. 344–356, 2024.
https://doi.org/10.1007/978-981-97-0065-3_26

So, should the traditional evidence review model still adhere to the evidence review and exclusion under the background of big data? What kind of review and exclusion model should be adopted? On the basis of systematically combing the principles and practices of big data evidence review and exclusion, the author analyzes the difficulties encountered by the current big data evidence review and exclusion mode. On the other hand, it puts forward the new paradigm of the review and exclusion of evidence under the background of big data, and analyzes its operability and feasibility, in order to provide ideas and enlightenment for the review of big data evidence.

2   Practice Pattern of Big Data Evidence Review

Research and application of big data evidence mainly focus on the following aspects: the first is relying on big data analysis technology to evidence with the facts, for example in the United States, represented by TrueAllele technology companies through professional algorithm model of mixed DNA mass data analysis, for the same person, and thus determine the criminal cases. TrueAllele The company aims to solve the complex interpretation puzzle of testing low-level or complex DNA mixtures.

This computer software uses and runs an clever mathematical model to estimate the statistical probability of a given individual's DNA consistent with data from a given sample by alignments to the genetic material of another unrelated individual from a wider range of relevant populations. Because of this, TrueAllele and other probabilistic genotyping software marked a profound shift in DNA forensics.

Secondly, big data evidence is used to prove the correlation between the case facts. Big data technology has been applied in many fields. On the whole, the tasks of big data analysis can be basically classified as follows: classification, clustering, regression, association rule mining, and so on. So through big data analysis method of evidence and the correlation between the facts and traditional judicial certification activities rely on logical reasoning that the causal relationship between the evidence and the facts is not the same, but in big data proof, causality is not exist, is not important, correlation is actually a causal relationship. In the process of big data proof, it is essentially reflected in the correlation of data. In some cases, judicial personnel have begun to prove causality by seeking for strong associations between variables. In Miami v. Bank of America (Miami v. Bank of America), the plaintiff proved through the algorithm that the loan policy of the defendant Bank of America led to differentiated treatment, and that race occupies an important proportion in its loan issuance. The case was appealed to the Supreme Court, and the case was returned to the Eleventh Circuit for a determination of whether Bank of America's policies and the plaintiffs' alleged racial discrimination; in May 2019, the Eleventh Circuit confirmed the existence of a "direct relationship" and thus demonstrated the causal relationship between the defendant's loan policy and discriminatory, differentiated treatment.

Finally, to predict the future through the analytical means proved by big data. Prediction is the most valuable application of big data. In the era of big data, human activities turn to creativity, that is, pioneering activities according to human needs and development, and the effectiveness of use as the measure. Creative structure activities are mainly based on big data prediction, and they can achieve the expected results through the grasp and intervention of related factors. For example, the Public Safety Assessment System (Public Safety Assessment, PSA) is the most widely used in various states of the United States. On the basis of collecting 75 0,000 cases in the United States, it calculates whether the bail of the suspect can be released based on nine indicators such as the age of the suspect, pending charges and not appearing in court. Another example is the Federal Pre-trial Risk Assessment Tool (PTRA), which determines the risk level of a criminal suspect and evaluates the possibility of arearrest for any crime type.

3   Multidimensional Flaws of Big Data Evidence Review

Big data evidence as driven by practical tool values, with machine learning, algorithm as a technical basis, and under the concept of the birth of a new product, although it has the characteristics of "technical implementation", to use emerging technology to analyze massive evidence, summarizes the regularity of cognition. But this is only a technical element. Big data evidence proof is reliable should not only can be implemented in the technical level, but also should be confirmed in law, big data evidence is not directly generated in the case facts in the process of evidence, but with the help of big data algorithm on the original born in the case of massive electronic data, data collection, mining, cleaning, sorting and calculation generated after a series of analysis. In other words, the content presented by big data evidence is not the massive information contained in the massive electronic data content, nor is it big data, artificial intelligence and algorithm, but the analysis result of cleaning, collision and standardization of massive electronic data with big data technology. Some scholars pointed out: "the big data set is not read directly by human, not because of its difficult obscure, only by filling the professional knowledge, but only because of its large size, such as in the judicial practice of disposable human to read, will not be able to complete in the tolerable time range, therefore must be accelerated by computer force". Therefore, there will be defects in the technical dimension and the legal dimension in the process of identifying the case facts based on the analysis results of big data evidence.

3.1   Defects in the Technical Dimension of Big Data Evidence Review

First is the data source defects, specifically, one is on the data information resource allocation, there is a "digital divide" between debate, that one is the ability to collect, store, mining massive judicial data of government agencies and network information practitioners, the other party is the object of the data collection, only enjoy strictly restricted data access rights. Due to the consideration of national security and personal information protection, the legislation will prosecute the relevant information and data in criminal activities, resulting in the data access barriers to the relevant departments to obtain relevant data. On the contrary, the investigation and control organs can obtain massive judicial data through the exchange of big data comprehensive integration platform jointly built by government departments, and can also require private institutions such as network information operators to assist in law enforcement to greatly improve the ability to obtain information. Second, even if the prosecution and defense obtain the same original data, there is still a "data analysis gap" in terms of data application. That is, in the face of massive judicial data, only the controller with relevant infrastructure and big data analysis skills can obtain valuable information resources from data fragmentation. In Skilling 2009, the U. S. federal court held that the government could fulfill its discovery obligation by opening file access to the original database to the defense, without considering the large number of documents, because the defendant was able to search documents like the government.

This "document dumping" (document dump) behavior, which does not consider the difference in data analysis ability between the two parties, further aggravates the substantive inequality of both parties. Third, where is the source of the massive electronic evidence before cleaning the big data set? There is the possibility of being tampered with? In addition, in the process of collecting massive evidence, it can not exclude the party providing "false" evidence and concealing key evidence, nor exclude the situation that the appraisal institution or data analysis company hired by the party unilaterally takes risks to serve the interests of one party.

The second is the defects of the algorithm model in the process of cleaning and induction. The first is the potential bias of the algorithm model itself. The scientificity and accuracy of the algorithm are the core factors for the reliability of big data. The reliability of the algorithm depends not only on the design accuracy of the algorithm model itself, but also depends on whether it is fair and just in the operation process of the judicial scene. Whether the source code and design of the algorithm model are accurate, whether the values upheld are fair, and whether the position is neutral will have an impact on the analysis results of big data. As a technology, the algorithm itself is neutral and non-biased, and the designer behind the algorithm is still human.

Human designers can integrate their own values, positions and ideas into the algorithm code, resulting in race, region, gender and other discrimination and prejudice. To some extent, algorithmic bias is an unavoidable problem. From the human cultural prejudice, the characteristics of the algorithm itself to analyze the connotation of big data algorithm discrimination, can be found even in the technical level algorithm designers will not bias into the subjective will of big data algorithm, but because of the "GIGO law" (Garbage In, Garbage Out), data samples of natural weight difference and the extension of large data attributes, big data algorithm also has the inevitability of prejudice or discrimination. When the artificial intelligence technology introduced the judicial referee, limited by the wrong case responsibility pressure and nature, the judge in the use of criminal cases intelligent auxiliary decision-making system to strengthen the results of the referee, there is no doubt that pollution by algorithm, prejudice, the case parties prejudice or broken, natural against the referee neutral principle. Big data analysis under the wrong algorithm model is likely to cause unjust and erroneous cases. In Lomis, Wisconsin, for example, the black defendant argued that a sentencing sentence rendered using an algorithm was unconstitutional. Although the Wisconsin Supreme Court concluded that the algorithm was not unconstitutional, the judge issued a written "warning" that it should not fully accept the AI assessment, but only as one of the basis, and to be aware that the results could contain errors and bias. In 2015, authorities in Queensland, Australia, confirmed errors in at least 60 cases of error codes found in the mixed DNA analysis software STRmix. The second is the contradiction between the algorithm black box and the transparency and openness of judicial decision and procedure. "Between the data input by the AI system and its output results, there is a 'hidden layer' that people cannot understand, which is the algorithm blackbox". The algorithm blackbox is obviously inconsistent with the theory of procedural justice for the transparency of judicial decision-making, resulting in people may be unable to understand how the AI machine gets the results, unable to determine whether the arguments of the parties are fairly considered, affecting people's sense of identity to criminal justice AI. Even with fairness aside, the lack of transparency in algorithmic decision-making procedures could lead to excessive trust in the reliability and scientific nature of automated judicial conclusions, resulting in an "automation bias". There have a cognitive tendency to believe too much in the automatic judgments made by the computer and accept them completely [6]. Carnieman (Daniel Kahneman) to human cognitive system as "dual processing system" (Dual-Processing), "system 1" as a perceptual cognitive system, performance by fast, automatic, unconscious, parallel, no effort, lenovo, slow acquisition, "system 2" as a rational cognitive system, characterized as slow, controlled, conscious, serial, effort, rule, relatively flexible and neutral. The former is a heuristic (heuristic) judgment, while the latter is a deliberate judgment, which monitors both and corrects psychological activities and external behaviors [7]. "The automated decision-making procedure of the algorithm eliminates the statement and defense of the parties in the litigation procedure, resulting in the lack of the principle of procedural participation". The problem of lack of transparency of the algorithm can be summarized as subjective reluctance and objective undisclosure. In particular, companies that design algorithms try to hide the internal logic of their algorithms from the perspective of the right to maintain their business secrets. For example, in Loomis, Wisconsin, the judge regarded the algorithm of the C OMPAS crime risk assessment tool as a trade secret, and exempted the algorithm owner from the obligation to disclose the algorithm code and explain how the algorithm works from the perspective of protecting intellectual property rights [8].

3.2 Defects in the Legal Dimension of Big Data Evidence Review

First of all, big data, artificial intelligence and other technologies are used to prove the facts of the case. Is essentially using algorithm and other technical means of subjective standard of objective packaging, its purpose is to enhance meet the proof standard operability, the use of large data evidence in the referee has the purpose of subjective judgment to prove the case of case, and does not help to make the facts that the unity of subjective and objective of criminal standard. Algorithmic technology is used for fact determination, and may interfere or even "usurp" the judge's fact determination power. Therefore, although the use of big data evidence can rely on intelligent algorithms to prove that activities can surpass human cognitive mode, break through the new field that human cognition cannot get in, and solve the proof dilemma that human experience is difficult to complete in some cases. However, the big data proof is still an objective mode of proof in essence, which still does not help to realize the subjective and objective unity of the criminal proof standard. Therefore, even the repeated emphasis on the subsidiarity of the corresponding software system will not help to change the reformers' tendency to strengthen the objectivity of criminal certification standards. Second, with big data evidence to confirm the contradiction between patterns, using big data, emerging technologies such as artificial intelligence to determine the case of the facts in the era of big data, the trend of quantitative proved object, especially in the process of some network crime proof, extraction of electronic data at hundreds of millions of article. So it's impossible to browse data sets based on human experience, let alone evidence reasoning. In this case, the big data proof model provides an intelligent proof method beyond the human experience. The algorithm can model the proof problem, extract the general proof rules and characteristics in the class cases, and replace the subjective reasoning with the mathematical model. For example, the Internet financial case algorithm proof model developed by China's public security organs, and the big data legal supervision model developed by the procuratorial organs. For example, in cases involving online pyramid schemes, the digital sources of big data mainly include the identity and communication data of online MLM organization leaders, participants and communication data, vehicle operation trajectory data, bank transaction data, and capital flow data. Its number of types and network pyramid selling crime "four elements" —— personnel flow, logistics, information flow, capital flow, and through the investigation in the production, circulation, distribution, consumption of funds, from the data in the management system, financial institutions and other circulation channels within the phenomenon characteristics, can find network pyramid selling clues. These are potential connections between the inability to apply human rules of thumb and the facts of the case that can be found. However, in cases requiring fact identification of massive evidence, it is difficult to form the evidence chain, and the rules of verification can not always be applied in such cases. For example, in the case of network crime need to extract number at tens of millions of evidence, the case facts in addition to need to use large data analysis means of massive electronic evidence analysis, also need to the defendant confession and excuse verbal evidence to collect, and in judicial practice, network crime cases have excessive reliance on the defendant confession and defense, the testimony of witnesses and other verbal evidence.withal, Some legal norms reduce the conditions for identifying the amount of funds in information network crime cases, In August, 2022, Article 21 of the Opinions on Several Issues concerning the Application of Criminal Procedure for Handling Information Criminal Crime Cases (hereinafter referred to as the Opinions on Criminal Procedure of Internet Crimes): "For information online crime cases involving a particularly large number of cases, It is impossible to collect evidence to prove one by one and verify the source of funds of the account involved, However, based on transaction records such as bank accounts, non-bank payment accounts and other evidential materials, Is sufficient to determine that the relevant account is mainly used for receiving and transferring the funds involved, The amount of the crime may be determined by the amount of funds received in the account, Except where the criminal suspect or defendant can make reasonable explanations. If an outsider raises an objection, it shall be examined according to law". Academic this method is called comprehensive model, comprehensive recognition law can become the fact of information network crime pattern, the reason is that the first is the evidence is difficult to identify, information network crime as under the background of large data of typical criminal means, the criminal suspect in the process of the mass of evidence if all to verify is difficult.

Secondly, it is difficult to determine whether the massive electronic evidence carried by information network crime cases is substantially related to the facts to be proved.

Some false correlations are often hidden in the results of big data analysis. Although it has statistical correlation characteristics, it cannot stand the test of causal logic.

Finally, in the information network crime, the analysis results obtained through the big data cannot be mutually corroborated with the confession, defense and witness testimony of the criminal suspects. The reason is that the big data evidence obtained through the means of big data analysis is more the actor and the same role as the evidence of the words of the facts of the case, while it is rare to judge whether the big data evidence can be mutually verified with other evidence. Therefore, some scholars pointed out that "the comprehensive identification method, as a quantitative method of identification factors of cyber crimes, does not require strict evidence, does not force" clear facts of the case, "nor" true and sufficient evidence". Reason is represented by information network crime evidence of big data evidence review is not and prove rules between contradiction and conflict, but if deliberately pursue big data evidence and facts can confirm each other in the present rely on human experience and cognition, comprehensive method is more in the face of information network crime mass evidence to pursue facts and big data analysis method using a balance.

Finally, there is a contradiction between the examination of big data evidence and the criminal presumption. The presumption refers to the "creation of a specific legal relationship between one proven fact A——causing the presumption and another constructive fact B. In judicial practice, the use of big data investigation and big data legal supervision model screening clues after transferred to the public security judicial organs for its directly for the facts of the case, obviously, in the big data investigation and big data under the tide of digital procuratorial legal supervision model of the background, the public security judicial organs may have thought that big data can reveal all problems. This is clearly a "myth" about data science. In addition, the use of big data evidence to prove the facts of the case needs to be comprehensively determined from the extraction of massive electronic evidence, the integration, cleaning, collision and comparison of the big data sets in the event, the reliability of the algorithm model, and the objectivity and the authenticity of the big data analysis results. Due to the knowledge gap and the trial efficiency, the judges take the results of big data analysis as the main basis for fact determination, and the technical facts such as the extraction of massive evidence and the algorithm model are true. In other words, if the referee only review of big data analysis results, its practice essentially "fitting" to generate large data analysis report based on the fact of massive electronic evidence and the reliability of the algorithm model, and the obligation and the burden to the defendant, if the defendant party not objection and front program of massive evidence source and algorithm model defects, is likely to rely on big data analysis results of the fact that there is the risk of error. The dominant bias of the pretrial procedure and the institutional path of bias transmission may lead to the cognitive bias in the pre-trial procedure to the subsequent criminal procedure and then have a systematic negative impact.

4 Mode Selection for Big Data Evidence Review

4.1 Regulation of Litigation Procedure: The Path Correction of Big Data Evidence Transmission

In 2007, Professor Hitren proposed the concept of "" (Technological Due Process), which aims to adhere to the judicial concept of neutrality, openness, equivalence and participation in due process, but also advocates the development of technological rationality and technological innovation, and emphasizes the fairness, transparency and accountability of automated decision-making procedures through optimization design. Although technical due process is mostly used in the automatic decision-making process of administrative procedures, it also has a certain adaptability in criminal proceedings, because both actually realize procedural justice by means of strict protection of individual rights.

4.1.1 Pre-trial Procedures: The "Cognitive Bias" to The "Cognitive Interaction" of The Big Data Evidence

Big data evidence in the criminal procedure of the transmission path for the public security organs and procuratorial organs in the pre-trial investigation and review stage using large data analysis of data comparison, and to investigation, form a big data analysis report to the trial field, and by the judge in the process of trial of big data analysis of the facts. And the source of massive electronic evidence, quality and cleaning of large data set comparison process completed before issuing big data analysis report, so cannot present big data evidence in the trial process in the process of the transfer process of "blackbox" and the lack of cognitive interaction program, the characteristics of the transmission path determines the if the analysis of large data set process cognitive bias is likely to spread to the subsequent trial procedure, namely throughout the criminal procedure, so the pre-trial phase of large data evidence of cognitive bias is transmission, has a negative impact on the criminal procedure process.

Therefore, one is to carry out litigation transformation in pre-trial procedures to promote cognitive interaction to eliminate cognitive bias. In judicial practice, explore more is to review the arrest of hearing procedure and "meeting before litigation", presided over by the prosecutor, investigators, criminal suspects and their defense lawyers respectively, prosecutors on the basis of the necessity of approval of arrest or prosecution facts and evidence of judgment or review, the two concrete measures is with the aid of "quasi judicial certificate", the implementation of the "quasi judicial review" practice exploration. In pre-trial procedures on the review of big data evidence, on the one hand, can explore to set up large data evidence review hearing procedure, give the criminal suspect and lawyer to the source of large data set, quality and algorithm method and related data information right to know and marking, on its reliable, and the parties in the hearing process to record, with the files transferred to the court. In the pre-trial procedure, opinions on the formation process and analysis of big data evidence results, which can effectively obtain cognitive hints from investigators, criminal suspects and their defense lawyers in the process of cognitive decision-making, and form cognitive interaction in a substantial sense. On the other hand, the data sharing platform should be built based on the existing big data investigation data and the data collected by the legal supervision of big data. At present, there are political and legal data centers in many places across the country, and the data stored in the center includes political and legal management data and the data of specific cases handled by the case handling organs. However, there is a lack of a big data and evidence sharing platform built for the purpose of data comparison and analysis in the past crime investigation and legal supervision process. The big data evidence sharing platform collects the evidence and clues collected by the past big data analysis and comparison, and builds relevant models, which can realize the cognitive interaction of pre-trial procedures at the data level.

4.1.2 Trial Procedure: To Construct the Algorithm Limited Disclosure System in the Trial Field

In the trial field, the technical due process needs to be guaranteed by the legitimacy and credibility of the technology itself. Specifically, the parties can disclose the algorithm and data information by forming a court hearing and signing a confidentiality agreement. But be aware that the confidentiality measures should not be excessive. For some general big data proof models, once the risk of code, data errors and bias are found in the case, the error information of the algorithm should no longer be kept confidential, but should be disclosed in due time to prevent the continued use of the model in the judicial field. At the same time, in other cases using the same algorithm model, the parties can take the error information of the algorithm as the basis for cross-examination, or as a reason to request the presentation of the algorithm in this case. The behind-the-scenes researchers of the algorithm model shall appear in court as expert witnesses to explain the algorithm principle, code and training data; the other party may also apply for experts in relevant fields as expert assistants, and both parties may conduct cross-examination on the reliability and data accuracy of the algorithm. In terms of the scope of algorithm disclosure, some scholars argue that disclosing the underlying data and code of the algorithm is neither feasible nor necessary. On the one hand, from a technical perspective, algorithms and data seem to most people like "mysterious" scrambled codes, and the role of public supervision that requires mandatory disclosure is very limited. On the other hand, program coding is the most valuable and important information of any algorithm. Developers almost never share coding information for the sake of competitive interests, and adopt the trade secret legal system for strict protection. Open program coding will lose the confidentiality of the code, and the competitors who discover and master the code will complete the code replication and use, so that the code is not possible to be protected by trade secrets. Therefore, if the parties apply for the disclosure of the algorithm source code and procedures that form big data evidence, it is only necessary to disclose the proportion of the algorithm used, the weight allocation and the historical accuracy of the algorithm results in visual and understandable issuing methods, rather than to disclose the underlying data. This helps to improve the interpretability and execuability of the algorithm opening system, the effective guarantee of the algorithm coding trade secrets and promote the realization of technical due process.

4.2 Optimization of Proof Mode: Probabilistic Analysis of Evidence From Big Data

Big data evidence can be divided into two types: big data generated based on massive database comparison and big data evidence based on algorithm model. The evidence needs to be matched based on the massive database comparison, and the massive electronic evidence is digithrough algorithm analysis, and compared with the samples in the database to output the similarity conclusion. Big data evidence based on massive database alignment includes face recognition, gait alignment, DNA alignment, (vehicle or human) trajectory information, etc. Among them, face recognition, gait comparison and other proof objects are the same identification, while (vehicle or person) track information and other proof object is the connection between the suspect and the scene of the crime. Therefore, the big data evidence based on the massive database comparison can be further divided into the same identification category and the trajectory category. Traditional proof power judgment is empirical judgment, but in the big data evidence based on massive database comparison, it is necessary to judge the probabilistic proof power of the evidence. The improvement of storage capacity brought by the era of big data and the perfect construction of relevant databases make the calculation of likelihood more accurate, thus providing a tool for the correlation judgment of big data evidence.

The probabilistic method (probabilistic approach) is applied to criminal evidence reasoning dating back to the second half of the 20th century, and has now become an important tool for court scientists to analyze scientific evidence such as DNA, speech, and fingerprints. Bayesian Network (Bayesian networks) is a theory of illustrated normative framework based on probabilistic methods. The work of Terrorone (Taroni) and others details the theoretical system of Bayesian network and its application in the field of court science [10]. The probabilistic approach is based on Bayesian subjective probability, in which the famous Bayesian formula plays a key role. In the probabilistic method, both the evidence of the case and the assumed explanatory facts are expressed as logical propositions. However, the application of Bayesian network in evidence reasoning requires massive evidence as its statistical basis, while in real cases, some probability values are difficult to obtain. Where does the probability value come from? How to accurately assign the probability value? This "numerical puzzle" has long plagued proponents of probabilistic methods. The data platform and computing power basis on which the big data evidence relies provide the basis for the refinement and objectification of the probability value. Judicial organs have the ability to collect, store and mine massive judicial data, and have a strong ability of information collection and analysis. But it should be noted that the probability method requires fact to have certain knowledge of mathematical statistics, but most does not have the relevant professional and knowledge background, even if can rely on large data analysis means about the probability value, may also be because of knowledge and analysis probability ability defects lead to misunderstanding about the probability value. Therefore, the key to the probabilistic analysis path of big data evidence is to eliminate the knowledge gap between fact cognition about the probability numerical cognition based on Bayesian network to accurately identify the facts of the case.

Wilek et al. argue that combining the requirements of the indictment, they extract a consistent story from the Bayesian network and construct the story to explain the case

[3]. Specifically, in the process of construction and optimization of judicial big data platform, the algorithm designer and judges, prosecutors can construct story algorithm model build consensus, through the Bayesian network extract consistent story algorithm, makes the probability of numerical in the form of demonstration report, improve big data probability analysis path interpretability.

4.3 Digital Rights Protection: The Construction of Illegal Exclusion Rules for Big Data Evidence

Big data evidence with huge amounts of electronic evidence for data sources, through the construction of analysis model or machine algorithm to form the analysis results, in other words, big data evidence with electronic evidence as the smallest unit to case facts electronic data as the network information age of "king of evidence", it may carry the property rights, privacy, communication freedom, freedom of speech and other basic rights. However, in judicial practice, many electronic data collected that do not conform to legal procedures are technical violations, and will not infringe on the basic rights of criminal suspects and defendants. However, some scholars point out that many collections of electronic data that do not conform to the legal procedures are technical violations, and will not infringe on the basic rights of criminal suspects and defendants. "For procedural defects with technical violations, if evidence is often excluded, it will violate the principle of proportion, so that the procedural sanctions do not adapt to the severity of illegal evidence collection, and easy to make other important interests suffer undue damage."

So should be digital rights are illegal violation as the standard to construct large data evidence illegal exclusion rules, and based on data sources, algorithm model technical considerations to review large data evidence is based on the authenticity of the review, in other words, big data evidence review should be in technical dimension to construct its adopt rules, and to right dimension to construct large data evidence of illegal exclusion rules. To judge whether the digital rights of citizens are violated, it should be included in the exclusion of illegal evidence during the period from data collection to the generation of analysis conclusions. If there is big data evidence that seriously infringes on citizens' digital rights and affects judicial justice, it should be excluded and cannot be corrected and reasonably explained. This is the absolute exclusion rule for big data evidence. However, if the collection process of big data evidence is in violation of legal procedures and minor infringement of citizens' digital rights, it should be allowed to be correction and reasonable explanation. If it cannot be corrected and reasonable explanation, it should be excluded. This is a relative exclusion rule for big data evidence.

5   Conclusion

Michele Panzavolta, a professor at the University of Leuven in Belgium, started from two judgments made by the European Court of Justice, In the era of big data, The evidence of criminal suspects collected by the national judicial organs includes yuan-evidence, subject evidence and derivative evidence, Whether and beyond all the three types of evidence should be excluded, The principle of reliability should be followed first,

Determine whether the evidence obtained under mass surveillance is reliable, Secondly, from the perspective of limited protection principle, the evidence obtained through large-scale monitoring is judged to be the infringement of the rights and interests of individual individuals or of the rights and interests of the public.

Finally, we should follow the perspective of the overall subjectivity to judge whether the acceptance of illegal evidence will endanger the judicial justice? Especially when the access to others' information and data is not obvious, and the infringement of everyone's rights is minimal, will there be a thousand-li dike destroyed in a nest? [12].

Michele Panzavolta Professor considers principles from the perspective of concentric circles (concentric circles), each corresponding to varying degrees of judicial discretion——may be preferable to constructing exclusionary discretion and making the judge's reasoning transparent. This will lead to the adoption of a cascading principle system in which each layer corresponds to a specific principle.

Michele Panzavolta Big data era illegal evidence exclude concentric rules with large data evidence review of three modes of selection have the same, also for us in conflict between three patterns should choose how to provide enlightenment, the optimization of big data evidence model helps to improve the big data evidence and reliability, and from the perspective of digital rights can judge in the big data evidence at the same time, the criminal suspect's basic rights are violated, and as big data evidence of illegal exclusion criteria. Finally, from the perspective of due process, the right to know and review of the criminal suspects and their defense lawyers should be guaranteed in the trial process. In addition, the limited disclosure system of algorithms in the trial process is constructed to ensure judicial justice. The three modes all correspond to different rules but are organically unified, providing the theoretical basis and practical paradigm for the review of big data evidence.

References

1. Kwong, K.: The algorithm says you did it: the use of black box algorithms to analyze complex DNA evidence. Harvard J. Law Technol. **31**, 281–282 (2017)
2. De Miguel Beriain, I.: Does the use of risk assessments in sentences respect the right to due process: a critical analysis of the Wisconsin v. Loomis Ruling. Law Probabil. Risk **17**, 45–54 (2018)
3. Panzavolta, M., Maes, E.: Exclusion of evidence in times of mass surveillance. In search of a principled approach to exclusion of illegally obtained evidence in criminal cases in the European Union. Int. J. Evid. Proof **26**(3), 199–222 (2022)
4. Citron, D.K., Pasquale, F.: The scored society: due process for automated predictions. Washington Law Rev. **89**, 8–19 (2014)
5. Andrejevic, M.: Big data, big questions| the big data divide. Int. J. Commun. **8**(1), 17 (2014)
6. Wexler, R.: Life, liberty, and trade secrets: intellectual property in the criminal justice system. Stan. L. Rev. **70**(5), 1343 (2018)
7. Kahneman, D., Slovic, P., Tversky, A.: Judgment under Uncertainty: Heuristics and Biases, pp. 3–20. Cambridge University Press, Cambridge (1982)
8. Vlek, C.S., et al.: Extracting scenarios from a bayesian network as explanations for legal evidence. In: Hoekstra, R. (ed.) Legal Knowledge and Information Systems:JURIX: The Twenty-Seventh Annual Conference, pp. 150–159. IOS Press, Amsterdam (2014)

9. Taroni, F., et al.: Bayesian Networks and Probabilistic Inference in Forensic Science, p. 33234. John Wiley Sons, Ltd., Hoboken (2006)

10. Williamson, B.: Who owns educational theory? big data, algorithms and the expert power of education data science. E-learn. Digital Media **14**(3), 105–122 (2017)

Abstracts

Application of Machine Learning-Based Neural Networks in Positron Annihilation Spectroscopy Data Analysis

Jiayi Xu[1,2], Xiaotian Yu[2], Haiying Wang[1], Peng Kuang[2], Fuyan Liu[2], Peng Zhang[2], Xingzhong Cao[2(✉)], and Baoyi Wang[2]

[1] School of Science, China University of Geosciences, Beijing, China
[2] The Institute of High Energy Physics of the Chinese Academy of Sciences, Beijing, China
caoxzh@ihep.ac.cn

Abstract. Positron Annihilation Spectroscopy (PAS) is a well-established technique for material inspection and characterization in various fields, including material science, chemistry, and biomedical research. This area of study encompasses several fundamental experimental methods, such as measuring positron annihilation lifetime, positron annihilation angular correlation, positron annihilation Doppler broadening, coincidence Doppler broadening, and slow positron beam methods. However, conventional methods for processing spectroscopic data are still carried out by iterative parameter fitting and parametric analysis, which are time-consuming and require expertise and initial parameter estimates. In this work, a machine-learning based method is proposed to analyze positron annihilation spectroscopy data, to surmount the limitations of conventional analytic approaches. Through training, the neural network has the ability to automatically analyze the Doppler broadened spectrum, separating various components and noise, and denoising the data, and achieving more accurate prediction results than the least square method fitting. Several related factors such as backscattering, combinations of Compton effects, pileup, ballistic deficit, and pulse-shaping problems of positron annihilation were duly considered in the simulation framework. The proposed method shows promise for application in the automatic data-processing pipeline for fast analysis of the complex spectroscopic data collected by photon detectors. The neural network method based on machine learning proposed in this study is expected to be applied to automatically process positron spectroscopy data to quickly analyze the energy spectrum containing various components and noise collected by photon detectors, and achieve more accurate prediction results than the least square method, providing useful reference for spectral data methods.

Keywords: Positron annihilation spectroscopy · Neural networks · Data processing

---

Supported by The Institute of High Energy Physics of the Chinese Academy of Sciences.

C. Cruz et al. (Eds.): IC 2023, CCIS 2036, pp. 359–360, 2024.
https://doi.org/10.1007/978-981-97-0065-3

References

1. Stepanov, P.S., Stepanov, S.V., Byakov, V.M., Selim, F.A.: Developing new routine for processing two-dimensional coincidence doppler energy spectra and evaluation of electron subsystem properties in metals. Acta Physica Polonica Series a. **132**(5), 1628–1633 (2017). https://doi.org/10.12693/APhysPolA.132.1628
2. do Nascimento, E., Helene, O., Vanin, V.R., da Cruz, M.T.F., Moralles, M.: Statistical analysis of the Doppler broadening coincidence spectrum of electron-positron annihilation radiation in silicon. Nucl. Instrum. Methods Phys. Res. Sect. A: **609**(2–3), 244–249 (2009). https://doi.org/10.1016/j.nima.2009.07.051

Machine Learning Techniques for Automatic Detection of ULF Waves

Shaofeng Fang[1(✉)] and Jie Ren[2]

[1] National Space Science Center, Chinese Academy of Sciences, Beijing, China
`fangsf@nssc.ac.cn`
[2] School of Earth and Space Sciences, Peking University, Beijing, China
`jieren@pku.edu.cn`

Abstract. Ultra-low frequency waves, also known as geomagnetic pulsations, have the lowest frequency, longest wavelength and carry the most energy. They play an important role in the multi-layer coupling system of the solar and Earth system, and are one of the important processes of magnetosphere energy and material transport. With the arrival of the big data era of space science, the ground-based observation data of the Earth's magnetosphere has shown exponential growth. The large-scale geomagnetic observation data provides a good opportunity to study the interaction between solar wind and geomagnetic, but it also brings new challenges to the automatic identification of geomagnetic pulsations. Most of the existing geomagnetic pulsation recognition system relies on traditional time-frequency analysis and carefully artificial hyperparameter setting, resulting in poor generalization performance and slow recognition efficiency. We propose and compare three different kinds of machine learning techniques for automatic detection of ULF waves, and the models are trained and tested on different stations of ground-based fluxgate magnetometer observation data from the Meridian Project. According to the results of comparison, we successfully construct an end-to-end geomagnetic pulsation event recognition framework, which does not rely on traditional time-frequency analysis, and the recognition efficiency is greatly improved. Our experiments show that the accuracy is greater than 95% for different ULF wave events (Pc3, Pc4 and Pc5). The detected events show a good correlation between different stations, which is useful for ULF wave propagation analysis.

Keywords: ULF waves · Machine learning · Meridian project

1 Main Text

Ultra-low frequency waves are the plasma waves in the Earth's magnetosphere with a frequency distribution between 1mHz and 1 Hz, and the regular pulsation can be divided into Pc1 pulsation, Pc2 pulsation, Pc3 pulsation, Pc4 pulsation and Pc5 pulsation according to the different period or frequency [1]. Traditional methods for extracting ULF signals, such as bandpass filter, power spectral density, and wavelet transform relies on time-frequency analysis and carefully artificial hyperparameter setting. Therefore, machine learning technique was first

C. Cruz et al. (Eds.): IC 2023, CCIS 2036, pp. 361–362, 2024.
https://doi.org/10.1007/978-981-97-0065-3

introduced by Balasis et al. [2] in the automatic detection of Pc3 waves in the time series of the magnetic field measurements on board the low-Earth orbit CHAMP satellite.

In this article, We propose and compare three different kinds of machine learning techniques for automatic detection of ULF waves based on fluxgate magnetometer observation data from the Chinese Meridian Project. The first type is to use the traditional machine learning techniques, such as SVM, XGBoost, with the help of human designed features extracted from original observed time series data. The second type is to transform original geomagnetic time series into time-frequency matrix by wavelet analysis, thus the original problem can be easily solved by using wonderful image classification techniques. To get rid of the restrictions of human designed features or traditional time-frequency techniques, we finally design a deep learning system that automatically extract features from raw geomagnetic time series based on 1D CNN neural network, and detect different kinds of ULF signals by adding different classification head.

The methods proposed above are trained and tested on different stations of ground-based fluxgate magnetometer observation data from the Meridian Project. We take raw geomagnetic time series from three stations dis-

Table 1. The performance of the final proposed method.

Classes	Accuracy (%)	Recall (%)	Precision (%)
Pc3	97.11	88.51	92.79
Pc4	96.10	87.25	83.68
Pc5	95.07	82.54	87.11

tributed on different latitudes, which are SYS, JFT and MHT. Furthermore, a machine learning ready dataset is built based on traditional wavelet analysis [3]. The duration of the dateset is one year. The dataset is split into train dataset, valid dataset and test dataset by three months, three months and six months. The results on the test dateset is shown in Table 1. Our experiments show that the accuracy is greater than 95% for different ULF wave events (Pc3, Pc4 and Pc5).

Acknowledgements. This work was supported by the National Key Research and Development Program of China, Grant No. 2022YFF0711400. This work was also supported by the Informatization Plan of Chinese Academy of Sciences, Grant No. CAS-WX2022SF-0103.

References

1. Jacobs, J.A., Kato, Y., Matsushita, S., Troitskaya, V.A.: Classification of geomagnetic micropulsations. J. Geophys. Res. **69**(1), 180–181 (1964)
2. Balasis, G., Aminalragia-Giamini, S., Papadimitriou, C., et al.: A machine learning approach for automated ULF wave recognition. J. Space Weather Space Clim. **9**, A13 (2019)
3. Renjie, F.S.: Geomagnetic ULF wave dataset based on Chinese Meridian Project observation (Pc3-Pc5). 1.0. National Space Science Data Center (2022). https://doi.org/10.12176/01.12.013-V01

Reconstruction of Unstable Heavy Particles Using Deep Symmetry-Preserving Attention Networks

Michael James Fenton[1], Alexander Shmakov[2], Hideki Okawa[3(✉)], Yuji Li[4], Ko-Yang Hsiao[5], Shih-Chieh Hsu[6], Daniel Whiteson[1], and Pierre Baldi[2]

[1] Department of Physics and Astronomy, University of California, Irvine, Irvine, CA 92607, USA

[2] Department of Computer Science, University of California, Irvine, Irvine, CA 92607, USA

[3] Institute of High Energy Physics, Chinese Academy of Sciences, Shijingshan 100049, Beijing, China
okawa@ihep.ac.cm

[4] Institute of Modern Physics, Fudan University, Yangpu, Shanghai 200433, China

[5] Department of Physics, National Tsing Hua University, Hsingchu City 30013, Taiwan

[6] Department of Physics and Astronomy, University of Washington, Seattle 98195-4550, WA, USA

At the Large Hadron Collider (LHC), heavy, unstable particles such as top quarks, Higgs bosons and W and Z bosons decay before reaching the detector. Recovering information of these original particles requires reconstructing their four-momenta from their immediate decay products, which are the partons (gluons and quarks), charged leptons and neutrinos. Since many partons leave indistinguishable signatures in detectors, a major difficulty lies in assigning the observed detector objects to each parton.

Traditionally, χ^2 fits or kinematic likelihoods [1] have been utilized to pursue this task. A more recent approach: Permutation Deep Neural Network (PDNN) utilizes a fully connected deep neural network which takes the kinematic and tagging information of the reconstructed objects as inputs [2]. However, all these methods require to build exhaustive permutations of the physics objects in the event, and their performance is limited by the amount of kinematic information that can be incorporated. Furthermore, at high energy hadron colliders, many extra objects are often contained in the events, leading to performance degradation of the permutation-based methods.

This work presents a sophisticated machine learning approach, named SPA-NET, utilizing a symmetry preserving attention mechanism, which allows us to incorporate all of the symmetries present in the event topology. It was first proposed in the context of reconstructing all hadronic final states [3, 4]. In this work, we extend the method (Fig. 1 left) to handle arbitrary types of objects as well as adding capabilities to pursue signal and background discrimination, kinematic regression and auxiliary outputs to classify different kinds of events. To demonstrate the capability of the new technique, we present its performance

C. Cruz et al. (Eds.): IC 2023, CCIS 2036, pp. 363–365, 2024.
https://doi.org/10.1007/978-981-97-0065-3

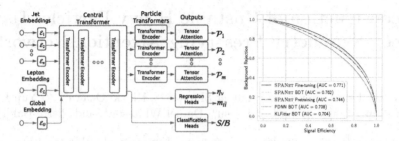

Fig. 1. Extended diagram of the new SPA-NET architecture (left) and receiver operating curves for networks trained to distinguish $t\bar{t}H$ from $t\bar{t} + b\bar{b}$ (right).

in three flagship LHC analyses: $t\bar{t}H(H \to b\bar{b})$, top-quark mass, and a search for a hypothetical Z' boson decaying to a top-quark pair.

Measuring the $t\bar{t}H(H \to b\bar{b})$ cross section faces a major challenge in separating the $t\bar{t}H$ signals from the large $t\bar{t} + b\bar{b}$ background. Exploiting SPA-NET not only allows us to assign the jets to partons with the highest efficiency ever achieved, but also provides further signal/background discrimination (Fig. 1 right) utilizing the classification head, which is a new feature added in SPA-NET. Utilizing SPA-NET with or without an additional classification head allows us to reach 5σ discovery with the Run-3 statistics, whereas with the traditional methods, they will reach only about 4σ significance at best.

The top quark mass is a fundamental parameter of the Standard Model. In this work, we perform a two-dimensional fit to the invariant mass distributions of the hadronically-decaying top quark and W boson as reconstructed by each method. The top quark mass and jet scale factor (JSF) are extracted by a template fit from Monte Carlo samples generated for this study. SPA-NET provides a 15% improvement in the top quark mass uncertainty.

Various theories beyond the Standard Model hypothesize additional heavy particles which would decay to $t\bar{t}$ pairs, such as heavy Higgs bosons or new gauge bosons (Z'). Signal sensitivity is evaluated for three benchmark signals with a mass of 500, 700, and 900 GeV, respectively. With the Run 3 condition, for all the three benchmarks, the discovery significance exceeds 5σ with SPA-NET, whereas with the baseline methods, only the high mass point reaches this threshold using the PDNN.

In addition, we also present ablation studies to provide insight on what SPA-NET has learned. To conclude, SPA-NET is the most efficient and highest performing method for event reconstruction to date. Ref. [5] presents the full description of our work and our code is available at Ref. [6].

References

1. Erdmann, J., et al.: Nucl. Instrum. Meth. A **748**, 18–25 (2014)
2. Erdmann, J., et al.: JINST **14**(11), P11015 (2019)

3. Shmakov, A., et al.: SciPost Phys. **12**, 178 (2022)
4. Fenton, M., et al.: Phys. Rev. D **105**(11), 112008 (2022)
5. Fenton, M., et al.: arXiv:2309.01886 (2023)
6. https://github.com/Alexanders101/SPANet

Interpretable Prediction of Commercial Flight Delay Duration Based on Machine Learning Methods

Lin Zou$^{(\boxtimes)}$, Jingtao Wang , Weiping Li , and Jianxiong Chen

Civil Aviation Flight University of China, Guanghan 618307, China
8712879360qq.com

Abstract. In recent years, the civil aviation industry has developed rapidly and faced increasingly complex and diverse problems. Optimizing the timely arrival of flights or making more accurate predictions of flight delays has always been a focus of research. This study takes the route from Kunming to Chengdu Tianfu Airport or Shuangliu Airport as an example, uses high-dimensional variables to predict flight delay duration, and delves into the impact of relevant variables on flight operation.

This study used relevant data from January 2023 to October 2023 to study the delay situation of flights between Kunming and Chengdu. ADS-B data from flights departing and landing from Kunming Changshui Airport in August 2023 was used to assist in analyzing the actual operation of flights. The data from January to September was randomly divided into training and testing sets for model training, and the data from October was used to verify the effectiveness of the model.

Different routes have different characteristics, and the dataset used in this study has a higher proportion of early arrivals. After removing outliers from the data used for model training, out of the 5504 flights from January to September, 5041 flights arrived early and 3967 flights arrived more than 15 min early. The International Civil Aviation Organization considers flight delays exceeding 15 min as flight delays, and the dataset considers 244 flights as delayed. Among the 439 delayed flights, Kunming Airport has counted the reasons for 401 delayed flights. Weather and airline reasons account for over 0.972, while the rest also include military activities and passenger reasons. Statistics have found that delayed flights mainly occur in July and August.

This study uses flight number, airline, destination, aircraft seat, whether it is covered by a bridge, aircraft type, number of people, children, number of pieces of luggage, luggage weight, cargo weight, mail weight, total seating capacity, seat rate, departure runway number, departure delay time, estimated stay time, planned flight time, impact of previous flight delays on this flight, actual departure time of the flight, number of planned departure flights during this time period Regression prediction is conducted on variables such as weather conditions along the route. Focus on exploring the impact of pre flight delays and the impact of weather on delays. The delay in the arrival of the preceding flights will bring pressure to the ground support after arrival, directly causing the flight to take off later than the planned departure time; The impact

C. Cruz et al. (Eds.): IC 2023, CCIS 2036, pp. 366–367, 2024.
https://doi.org/10.1007/978-981-97-0065-3

of weather on flight delay has obvious seasonal correlation. In summer, there are many rainstorm, and flights affected by weather will not only delay takeoff, but also increase flight duration to avoid thunderstorms. Long periods of thunderstorm weather may lead to airspace congestion in the terminal area, resulting in prolonged waiting times for flights in the air.

This study used machine learning algorithms for regression prediction and compared the MSE, AE, and R2 predicted by different algorithms. Secondly, this article provides a meaningful analysis of the important features that affect flight delays and a reasonable explanation of the model's prediction results. Analyzing the important features can provide reasonable suggestions for flights to arrive on time, and understanding the decision-making process of the model can help establish trust in the flight delay prediction model, and reasonably explain the reasons for a specific flight delay while ensuring the reliability of the model.

Consider carefully optimizing the final prediction results based on different airlines. Research has found that there is a certain correlation between the punctual arrival of flights on the Kunming Chengdu segment and airlines. There are differences in the flight schedules and delay times set by different airlines.

The experiment has confirmed that machine learning models can reasonably predict flight delays and provide the reasons for flight delays. Research has found that the variables that have a significant impact on flight delays are takeoff delay time, planned flight time, pre order flight delay, and meteorological factors. This study will also provide rationalization suggestions for flights arriving on time based on these factors.

Keywords: Machine learning · Flight delay · Interpretability

Semantic Retrieval of Mars Data Using Contrastive Learning and Convolutional Neural Network

Yunlong Li[✉][ID], Ci-Feng Wang[ID], Jia Zhong, and Yang Lu[ID]

National Space Science Center, Chinese Academy of Sciences, 100190 Beijing, China
{liyunlong,wcf,zhongjia,yanglu}@nssc.ac.cn

Abstract. Mars data, such as remote sensing images, are valuable for scientific research, but they are often difficult to search and filter due to their large volume and diversity. Deep learning models can learn semantic representations of data that facilitate efficient and accurate retrieval of relevant data. In this work, we propose to learn semantic features of Mars remote sensing images using contrastive learning and convolutional neural network. We use a large collection of Mars images from the MRO HiRISE 2006–2023 and the Tianwen-1 HiRIC 2021–2022 missions as the training data, which cover various aspects of the Martian surface under different viewing angles, lighting conditions, and altitudes. We fine-tune the model using several open-sourced Mars landmark datasets to align the learned features with human knowledge and concepts. We evaluate the performance of our model and the extracted features on semantic retrieval tasks, and demonstrate that they can enable high-precision and fast search of Mars data at the file level.

Keywords: Mars · Data retrieval · Machine learning

1 Introduction

The analysis of the distribution of landforms plays a crucial role in studying the past and current activity of Mars. With the rapid growth in the volume of high-resolution space-borne images available, various deep learning approaches have been proposed to improve the automated method for detecting different types of landforms on Mars [2]. Content-based search for Mars has been developed to help find useful targets based on these models [3]. However, these studies use supervised learning which need labels that are usually limited to a few types of landforms, thus difficult to find landforms of new categories defined by users.

This study proposes to develop a semantic-based search capability for Mars orbital images using the contrastive learning MoCo [1] and the CSPDarkNet [4]. The model is able to learn semantic features from the entire Mars observations in an unsupervised manner, considering various aspects of the Martian surface under different shapes, floor structures, lighting conditions, and so on. The model is then able to find similar landforms according to the sample image provided by the user, and not constrained by any pre-defined labels. We collected the train data from the MRO HiRISE and Tianwen-1 HiRIC and split them into small patches using a sliding window of size 1024×1024 and step 512×512. The final dataset consisted of 6.3 million images of Mars with a time span from 2006 to 2023.

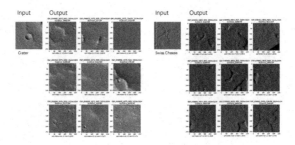

Fig. 1. Two types of Mars landforms samples are selected randomly from an open-access labeled dataset [5]. The outputs are from the unsupervised training dataset. The retrieved data have visually similar semantic features to the inputs.

The generalized pretrained model, which is not task-specific, allows for more flexible searching of the dataset compared to the supervised models. Figure 1 shows part of the Mars data search results. The pretrained model was also tested on the open-access labeled dataset [5]. The linear classification accuracy achieved 83.2% using the pretrained model, while fine-tuning resulted in an accuracy of 90.1%. The results are comparable to the classification model trained from scratch, which achieved an accuracy of 91.6%.

Acknowledgements. This work is supported by the National Key Research and Development Program of China, Grant No. 2022YFF0711400.

References

1. He, K., Fan, H., Wu, Y., Xie, S., Girshick, R.: Momentum contrast for unsupervised visual representation learning. In: Proceedings of the IEEE/CVF Conference on Computer Vision and Pattern Recognition, pp. 9729–9738 (2020)
2. Jiang, S., Lian, Z., Yung, K.L., Ip, W.H., Gao, M.: Automated detection of multitype landforms on mars using a light-weight deep learning-based detector. IEEE Trans. Aerosp. Electron. Syst. **58**, 5015–5029 (2022)
3. Wagstaff, K., Lu, Y., Stanboli, A., Grimes, K., Gowda, T., Padams, J.: Deep mars: CNN classification of mars imagery for the PDS imaging atlas. In: Proceedings of the AAAI Conference on Artificial Intelligence (2018)
4. Redmon, J., Farhadi, A.: YOLOv3: an incremental improvement. arXiv preprint arxiv:1804.02767 (2018)
5. Doran, G., Dunkel, E., Lu, S., Wagstaff, K.: Mars orbital image (HiRISE) labeled data set version 3.2 (3.2.0) (2020). https://doi.org/10.5281/zenodo.4002935

Author Index

C. Cruz et al. (Eds.): IC 2023, CCIS 2036, pp. 371–372, 2024.
https://doi.org/10.1007/978-981-97-0065-3

Printed in the United States
by Baker & Taylor Publisher Services